Encyclopedia of Fire

Encyclopedia of Fire

David E. Newton

Oryx Press
Westport, Connecticut • London

The rare Arabian Oryx is believed to have inspired the myth of the unicorn. This desert antelope became virtually extinct in the early 1960s. At that time several groups of international conservationists arranged to have 9 animals sent to the Phoenix Zoo to be the nucleus of a captive breeding herd. Today the Oryx population is over 1,000 and nearly 500 have been returned to reserves in the Middle East.

Library of Congress Cataloging-in-Publication Data

Newton, David E.
    Encyclopedia of fire / David E. Newton.
      p. cm.
    Includes bibliographical references and index.
    ISBN 1–57356–302–1 (alk. paper)
    1. Fire—Encyclopedias. 2. Fires—Encyclopedias. 3. Fire prevention—Encyclopedias. I. Title.
TP265.N48 2002
363.37'03—dc21 2001036180

British Library Cataloguing in Publication Data is available.

Library of Congress Catalog Card Number: 2001036180
ISBN: 1–57356–302–1

First published in 2002

Oryx Press, 88 Post Road West, Westport, CT 06881
An imprint of Greenwood Publishing Group, Inc.
www.oryxpress.com

Printed in the United States of America

The paper used in this book complies with the Permanent Paper Standard issued by the National Information Standards Organization (Z39.48–1984).

10 9 8 7 6 5 4 3 2 1

Publisher's Note on Safety: Fire and heat, in all their manifestations and applications, can be dangerous. This book is meant to provide background material on a variety of subjects, not provide directions on how to work with, prevent, or extinguish fire or treat burns. Readers are encouraged to respect the manifestations and applications of fire and should never experiment with any of them. Techniques for, and theories of, fire prevention, protection, and safety, as well as burn treatments, change over time as a result of new research, discoveries, and applications. Readers should use the information as a general guide; for current and definitive information on fire safety, prevention, and protection and burn treatments, they should consult experts in the field such as medical doctors, emergency medical technicians, firefighters, police officers, and disaster-relief specialists.

To Dr. Howard Miller
and the staff at the Bear Creek Animal Clinic
on behalf of all those animals
who cannot say "Please" or "Thank You"

In Appreciation

Although written prior to September 11, 2001
this book is dedicated to all courageous firefighters,
many of whom gave their lives to save the lives of others
following the attacks on the World Trade Center.

CONTENTS

GUIDE TO SELECTED TOPICS

Cultural Issues And Allusions
"The Firebird"
Hall of Flame
Hearth Fire
Hunting with Fire
Intimidation Fires
Motion Pictures and Videos
Musical Allusions to Fire
Nazi Death Camps
Need-Fire
New Fire
Ordeal by Fire
Origins of Fire
Protest and Defiance Burnings
Shifting Cultivation
Smoke Signals
Smoking (Tobacco)
Visual Arts
Yule Log

Environmental Issues
Air Pollution
Climate Change
Fire Damp
Fire Ecology
Fire Regime
Forest Fires
Global Fire Monitoring Systems
Ignis Fatuus
Intermix Fire
Mine Fires and Explosions
Nuclear Winter
Prairie Fire
St. Elmo's Fire
Santa Ana Winds

Famous Fires
Assiut (Egypt) Fire of 1994
Berkeley Hills (California) Fire of 1991
Burning of Rome
Chernobyl Nuclear Reactor Fire
Coconut Grove Nightclub Fire of 1942
Cuyahoga River Fires
Eurotunnel Fire
The Great Chicago Fire of 1871
The Great London Fire of 1666
The Great San Francisco Earthquake and
    Fire Of 1906
Iroquois Theater Fire of 1903
Peshtigo Fire
Space Exploration Fires
Triangle Shirtwaist Fire
Waco (Texas) Fire of 1993
Yellowstone Fires of 1988

Festivals and Special Events
Bonfires
Burning Man
Campfire
Eternal Flames
Fire Resistance
Firewalking
Fireworks
Olympic Torch

Firefighting Devices and Technology
Arson
Arson Dogs
Fire Alarm Systems
Fireboats
Fire Code

Fire Drill
Fire Engine
Fire Extinguisher
Fire Hydrant
Fire Insurance
Fire Management
Fire Modeling
Halons
Lookout Towers
Prescribed Burn
Smoke Detector
Sprinkler Systems
Wildfire Landscaping

Firefighting Profession
Firefighters
Firefighting
Fire Protection Handbook
Maltese Cross
St. Florian (ca. 250 A.D.–304)
Smoke Jumpers

Fire in the Home
Candle
Cooking
Fireplace
Lamps and Lighting
Matches
Stoves

Health and Safety
Burns and Scalds
Davy Safety Lamp
Death by Fire
Lighthouses
Pyromania

Historical Devices and Concepts
Fire Clocks
Phlogiston Theory
Roman Baths

Industrial Applications
Ceramics
Fire Assaying
Fire Brick
Fireclay
Firing of Pottery
Flares
Forges and Blacksmithing
Glassmaking
Internal Combustion Engine
Jet Engines

Metallurgy
Oxyacetylene Torch
Smoke Generation
Steam Engine
Thermite
Welding

Language and Literature
Aphorisms and Sayings
The Chemical History of a Candle
Dante's *Inferno*
Literature
Metaphors
Psychoanalysis of Fire
World Fire

Military Applications
Flame Throwers
Greek Fire
Incendiary Bombs
Napalm
Scorched Earth Policy
Tracer Bullets

Organizations and Programs
Building and Fire Research Laboratory
Camp Fire Boys and Girls
Chimney Safety Institute of America
Congressional Fire Services Institute
Firenet: The International Fire Information
 Network
Fire Safety Institute
International Association of Fire Chiefs
International Association of Wildland Fire
Learn Not To Burn®
Missoula - Intermountain Fire Sciences
 Laboratory
National Fire Academy
National Fire Prevention Week
National Fire Protection Association
National Interagency Fire Center
National Park Service
Riverside Fire Laboratory
Smokey Bear
Sparky™ The Fire Dog
Tall Timbers Research Station
U.S. Fire Administration
U.S. Forest Service
Wildland Fire Assessment System

People
Becher, Johann Joachim (1635–1682)
Bessemer, Sir Henry (1813–1898)

INTRODUCTION

Humans have always been intrigued by the natural world in which they live. Part of that fascination seems to involve nothing more than simple curiosity. We want to know why things are as they are. But a knowledge of nature also has its practical aspects. Understanding phenomena such as day and night, lightning and thunder, floods, and fire can help us prepare for both the everyday events of life as well as its disasters and catastrophes.

Certain natural events, however, have always been held in special awe. Among the most common of these events is fire. Most human cultures throughout history have accorded fire a special place in stories of the origin of the Earth, the origin of the human race, and the origin of particular tribes. These cultures have often invented supernatural beings who are assigned control over the creation and distribution of fire. Many of the great gods and goddesses of fire—such as Prometheus and Agni—are still part of our art, culture, and imagery.

Today, few cultures rely on beliefs in gods and goddesses of fire. They recognize that fire is a natural phenomenon that can be explained scientifically. While natural and human-set fires are still abundant throughout the world, it is now possible to understand quite clearly how those fires began and how they can be brought under control.

Our scientific understanding of fire has not diminished our fascination with the phenomenon, however. The siren of a fire engine or the announcement of fireworks still stirs a passion in the hearts of many men, women, boys, and girls. Fire engines, fireboats, firewalking, lookout towers, and the other aspects of firefighting are still subjects of interest for many people, not just because of their role in firefighting itself, but also because of the drama surrounding fire itself.

The purpose of the *Encyclopedia of Fire* is to review some of the most important subjects related to the topic of fire. A primary focus, of course, is on combustion, the scientific explanation for fire. A variety of entries discuss the factors required for combustion and the methods used to promote, prevent, or suppress fire. The role of fire in everyday life also receives extensive attention, with entries covering forest fires, prairie fires, and fires in other natural settings, as well as those dealing with urban, suburban, and rural fires that affect human habitations. The role of fire in the development of modern ecosystems and modern societies is a subject of ongoing interest not only for experts, but also for many ordinary citizens.

But this encyclopedia is more than a scientific and technical treatise on combustion and its role in human life. It is a multidisciplinary attempt to show how human societies in all parts of the world throughout history have incorporated images of fire into their philosophies, mythologies, religions, music, art, and other areas of human thought and action. The reader will find, therefore, essays on pollutants produced by combustion and their effect on the human and natural environment, the connection humans have made between natural fire and biological phenomena with fire-like properties, the use of fire-related language in aphorisms and sayings, the effects of natural fires on human societies and the types of governmental agencies designed to deal with fire, the use that humans have made of fire to shape the environment in which they live, and the ways in which religions have incorporated the properties of fire into their traditions.

Views of fire as a supernatural force that controls human life have largely disappeared from modern society. But our respect for fire as a natural force that brings both benefits and tragedies to our lives remains. And that respect still stirs human emotions and creativity as do few other natural forces.

A Natural Phenomena

From a technical standpoint, fire is combustion—a process of rapid oxidation in which noticeable heat and light are produced. Fire is one of the most common natural phenomena on Earth. Satellites that orbit the Earth constantly detect hundreds of forest fires, grass fires, brush fires, and other types of fire. These fires may be produced by natural forces, such as lightning or volcanic eruptions, or by human activities. In some cases, anthropogenic fires begin accidentally, with a dropped cigarette or a campfire not completely extinguished. In other cases, they are caused intentionally, to clear a section of land or to assist in the regeneration of a forest or other wildland.

Fires have existed on Earth as long as our atmosphere has contained oxygen. Thus, fires are a normal and natural part of the ecosystem. Indeed, some plants and animals depend for their survival on the regular recurrence of fire. When humans attempt to interrupt the natural cycle of fire, they are often thwarted because of the accumulation of dead and dying—but not burned—plant material.

Indeed, the question as to how humans should attempt to control natural fires, if at all, is one of the basic issues facing those responsible for the management of public and private lands. The debate over prevention and suppression versus controlled burns versus some other policy about natural fire has been going on for hundreds of years in nearly every part of the world.

No one knows when humans first learned to use natural fire or when they learned to make their own fire. Records suggest that early humans were using fire to cook their food, warm their caves, light up the nights, and frighten away wild animals more than a million years ago. Slowly, humans learned to build more sophisticated cooking devices, such as stoves and ovens; more efficient lighting devices, such as lamps and torches; and more practical heating devices, such as furnaces and fireplaces.

Late in human history humans also learned how to use fire to do work for them. Ancient Greek philosophers had discovered that heating water produced steam, which could be used to operate simple machines and toys. But it was not until the seventeenth century that steam-operated devices were invented that could be used for practical purposes, such as pumping water out of mines and running railroad engines and boats. The invention of the internal combustion engine shortly thereafter revolutionized some of the most fundamental aspects of human society, including production and transportation.

There can be little doubt, therefore, that the development of machines that harness and make use of the power of fire has been one of the most important ongoing threads in the history of human civilization.

But fire has had other equally important impacts on other aspects of human society. Early humans viewed fire not only as a practical tool, but also as a fundamental force whose influence extended beyond the visible and tangible world. As long as there have been records of humans using fire for heating, lighting, and cooking, there have also been records of humans honoring and worshipping fire. Indeed, most cultures include in their pantheon of gods and goddesses one or more deities specially related to fire. In some religions, in fact, the fire god or fire goddess is the primary deity to whom worship is made.

Modern religions honor fire just as did older mythological systems. In Judaism and Christianity, for example, fire is a representation of the power and glory of God, a method of purification, and a system for conveying prayers and respect to God, as well as having other functions.

Given its fundamental role in nature and in human civilizations, it is not surprising that fire has long been a common and powerful theme in the arts. Musical compositions, paintings, and literary works have all taken one aspect or another of fire as their focus. In some cases, it is the physical phenomenon of fire that becomes the theme of artistic work. In other cases, it is some idea or emotion represented by fire—passion, punishment, or revenge, for example—that drives the work.

When the first natural philosophers began to think about the forces of nature that can really be said to be elemental or fundamental, they often chose earth, air, water, and fire. Even in today's world, it would be difficult to argue against the dominating importance of these four factors.

AIR POLLUTION

The term *air pollution* refers to the contamination of air by waste gases produced by industrial operations, motor vehicles, community or private incineration, and similar processes. By far the greatest amount of air pollution is produced as a result of the combustion of fossil fuels.

Air pollution is certainly not a strictly modern phenomenon. Indeed, almost any description of large cities in the past includes some reference to dirty, smoke-filled skies, especially in the centuries following the Industrial Revolution. Air pollution has become a social issue of major concern, however, primarily since the 1960s, when citizens in many nations of the world began to have a more profound understanding of the nature of the natural environment and of anthropogenic effects on that environment. Today, many nations have detailed laws and regulations dealing with the problems posed by air pollution.

Some of the major air pollutants are carbon monoxide, particulates, sulfur dioxide, oxides of nitrogen, and smog. These pollutants are formed when fossil fuels—coal, oil, and natural gas—are burned in power plants, motor vehicles, factories, and other sources.

Primary Pollutants

In general, the combustion of fossil fuels can be represented by a simple chemical equation:

$$C_xH_yO_z + O_2 + heat = CO_2 + H_2O$$

where $C_xH_yO_z$ represents the fossil fuel. The problem with this equation is that it represents an ideal condition that is seldom, if ever, encountered in the real world. For example, there is probably never an occasion when oxygen can be supplied to the burning fuel rapidly enough to permit complete combustion of the fuel; that is, to convert the carbon in the fuel completely to carbon dioxide (CO_2). Instead, the more likely case is that some of the carbon is oxidized only partially, to carbon monoxide (CO), and some is not oxidized at all. It is released to the air in a finely divided form known as *soot* or *particulates*. A better representation of fuel combustion in the real world, then, would be:

$$C_xH_yO_z + O_2 + heat = CO_2 + H_2O \\ + CO + C$$

Even this statement is incomplete and inadequate, however, because of impurities present in all fossil fuels. The two most common of these impurities are usually compounds of sulfur and nitrogen. When a fossil fuel is burned, then, so are the sulfur and nitrogen compounds present in them. The products of these reactions are the oxides of sulfur and nitrogen:

$$S + O_2 = SO_2$$

$$N + O_2 = NO$$

The products of fuel combustion—carbon monoxide, particulates, and oxides of sulfur and nitrogen—can be regarded as *primary pollutants* because they are released directly from the combustion process into the atmosphere. But these substances create additional problems when they react with other substances in the atmosphere, producing *secondary pollutants*.

Secondary Pollutants

One of the best known secondary pollutants is *acid rain*. Acid rain is formed when oxides of sulfur and nitrogen react further with oxygen in the air to form higher oxides:

$$2SO_2 + O_2 = 2SO_3$$
$$2NO + O_2 = 2NO_2$$

(The mechanisms by which these reactions occur is actually more complex than represented by these equations, although the net results are essentially the same.)

Both sulfur trioxide (SO_3) and nitrogen dioxide (NO_2) dissolve readily in water droplets present in air to form acids:

$$SO_3 + H_2O = H_2SO_4$$
$$4NO_2 + O_2 + 2H_2O = 4HNO_3$$

These acids are carried to earth by rain, snow, fog, and other forms of precipitation. When they reach the Earth's surface, they may damage plant material, certain forms of aquatic life, and some kinds of rock and mineral. For example, they may attack and dissolve the limestone (primarily calcium carbonate; $CaCO_3$) out of which buildings and statuary are made:

$$H_2SO_4 + CaCO_3 = CaSO_4 + H_2O + CO_2$$

Acid rain is also thought to be responsible for raising the acidity (pH) of lakes and ponds in certain parts of the world, making them unsuitable for the survival of microorganisms and fish.

A second familiar type of secondary pollutant is *smog*, of which there are two general types. The first type of smog is known technically as a *reducing smog* or, more familiarly, as *London smog*, after one of the cities where it has traditionally occurred most commonly. The second type of smog is called an *oxidizing smog*, or a Los Angeles smog. Many locations experience both types of smog, although one or the other form tends to predominate in other places.

London fog tends to occur in cool, humid climates. It is caused primarily by the combustion of coal. When coal burns, it releases both smoke (soot, particulates) and sulfur dioxide. The two combine to produce a combination of *smoke* and *fog*. The sulfur dioxide combines with water droplets in the air to produce sulfuric acid, which accounts for sore throats and a burning sensation often typical of a London smog.

Los Angeles smog is also called *photochemical fog* because it is formed partly as the result of light ("photo-") acting on certain chemicals produced during the combustion of petroleum products. It is most likely to be found in warm, dry, and sunny environments, especially those with a large volume of motor vehicle traffic. Compounds produced during the combustion of motor fuels escape into the air and react with substances present in the atmosphere to form the components of photochemical smog.

The detailed steps by which all components of photochemical smog are formed are not completely understood. However, the general outline of the process has been worked out. The following are some steps in that process.

Nitrogen dioxide (NO_2) in the air breaks down to form nitric oxide (NO) and free atomic oxygen (O):

$$NO_2 = NO + O$$

The atomic oxygen combines with normal, diatomic oxygen to form ozone (O_3):

$$O + O_2 = O_3$$

Ozone may then react with unburned hydrocarbons that escape from the exhaust system of a motor vehicle. As an example, ozone, propene ($CH_3CH = CH_2$; a compound present in exhaust gases), and nitrogen dioxide may react to form a product known as peroxyacetylnitrate (PAN), a compound that irritates the eyes and the mucous lining of the nose:

$$CH_3CH = CH_2 + O_3 + NO_2 =$$
$$CH_3COOONO_2 \text{ (PAN)}$$

Photochemical smog contains dozens, if not hundreds, of other compounds formed by similar mechanisms. Some of these compounds have temporary, irritating effects, like those of PAN, while others may have long-term, more serious effects.

Pollution Prevention

Over the past fifty years, a number of systems have been developed for the purpose of reducing air pollution and its effects on humans, other animals, plants, and physical structures. For example, one possibility is to reduce or eliminate substances in fuels that lead to pollution in the first place. Where possible, industries have started to use "sweet" fuels, with relatively low levels of sulfur, rather than "sour" fuels, with higher levels of the element.

Many methods have been invented for preventing or reducing the release of pollutants once they have been produced by combustion. For example, many factories have installed devices in their smoke stacks to capture potentially polluting gases. Some types of "scrubbers," for example, contain basic substances that react chemically with sulfur and nitrogen oxides:

$$2SO_2 + O_2 + 2Ca(OH)_2 = 2CaSO_4$$
$$+ 2H_2O$$

Today most automotive vehicles are also outfitted with catalytic converters, devices that increase the efficiency of combustion of motor fuels. A catalytic converter contains a catalyst of finely divided rhodium or ruthenium metal. As exhaust gases pass through the catalytic converter, they more completely react with oxygen in the air, reducing the amount of unburned pollutants in the final exhaust stream. *See also* CARBON DIOXIDE; CARBON MONOXIDE; CLIMATE CHANGE.

Further Reading: Bunce, Nigel. *Environmental Chemistry*, 2nd edition. Winnipeg: Wuerz Publishing, 1994, Chapters 3 and 6; Liu, David H.F., and Louise A. Berry, eds. *Air Pollution*. Boca Raton, FL: Lewis Publishers, 1999; Manahan, Stanley E. *Environmental Chemistry*, 6th edition. Boca Raton, FL: Lewis Publishers, 1994, Chap-

ters 9–13; Miller, Christina G., and Louise A. Berry. *Air Alert: Rescuing the Earth's Atmosphere*. New York: Atheneum, 1996; Newton, David E. *Environmental Chemistry*. Portland, ME: J. Weston Walch,, 1991, Chapters 3 and 4; Rossotti, Hazel. *Fire*. Oxford: Oxford University Press, 1993, Chapter 17; Turco, Richard P. *Earth under Siege*. Oxford: Oxford University Press, 1997, Chapters 5–7.

APHORISMS AND SAYINGS

An aphorism is a short statement designed to convey some fundamental truth. Many aphorisms include some mention of fire or a related topic. In many instances, the origin of these aphorisms is not known. The following are some examples of fire-related aphorisms.

"Burn Your Bridges Behind You"

One authority on the origin of aphorisms (Robert Hendrickson 1997) surmises that this aphorism may have begun as "burn your boats behind you." He points out that Roman generals often set fire to the boats in which they had crossed a river in order to impress on their troops that there was no going back, that is, no hope of retreat. The phrase has, of course, the same meaning today in that it suggests that a decision once made cannot be reversed.

"Burning the Candle at Both Ends"

The phrase "burning the candle at both ends" refers to a person who works very hard and tends to become very tired. One possible source of the quotation may be a custom used by clerks who had to work late into the evening, with only a single candle to see by. By lighting both ends of the candle, the clerks would get more light. The candle, like the clerks, may have been more productive for a short period of time, but they both also tended to "burn out" more quickly. Another interpretation of the aphorism points out that the complete statement adds the words "of the day" at the end, suggesting that a person gets up before dawn and works until after dark. By "burning the candle" both very early and very late, one is more productive but, also, more likely to become worn out quickly.

"Burning the Midnight Oil"

Before the availability of natural gas or electricity, oil lamps were a common form of lighting. As far back as the seventeenth century, the expression "burning the midnight oil" was being used to describe people who worked late into the night. The phrase was often applied particularly to writers who wrote far into the night.

"Can't Hold a Candle"

A person who "can't hold a candle" to some standard, including the accomplishments of another person, is unable to do as well as the person to whom he or she is being compared. One explanation for the origin of this phrase relates to an old custom of having young boys hold candles in theaters and other places of entertainment. Only the most qualified boys were chosen for this task. Other boys were not thought to be good enough to hold a candle for the performers. Another explanation draws on the religious custom of placing a candle before an icon, a picture, or a statue to give honor to the being represented.

"Fiddling while Rome Burns." See
Burning of Rome

"Fighting Fire with Fire"

This phrase appears to have arisen in the United States during its early history. A reasonable explanation for the term comes from the well-known practice used by early settlers to contain wildfires by burning off grass, brush, bushes, and other vegetation near the natural fire. By depriving the natural fire of additional fuel, the settlers hoped to stop the advance of the fire. The risk was, of course, that the intentionally-set fire might, itself, get out of control. "Fighting fire with fire" represents, therefore, a last-ditch effort to stop some type of action at some risk to oneself.

"Flash in the Pan"

This saying grew out of the method by which flintlock guns, first used in the seventeenth century, were fired. The first step in firing a flintlock gun was to release a hammer, which struck a piece of flint, producing a spark. The spark, in turn, would (at least in theory) ignite a small explosive charge in a small bowl (the "pan") that propelled the bullet out of the gun. In many cases, however, the charge would not ignite, but would simply burst into flame briefly, then die out. The expression "flash in the pan" became applicable, therefore, to any person who rose quickly to the public limelight then, almost as quickly, disappeared from view.

"Hauled over the Coals"

We sometimes say that a person has been "hauled over the coals" if they have been brought to account for some misdeed that they conducted. The expression is thought to have originated from an old English custom in which people who did not follow orders were literally dragged slowly over a bed of hot coals. The practice is said to have been used by English rulers as an inducement to get Jewish money lenders to loan them money.

"Out of the Frying Pan into the Fire"

One possible origin of this aphorism may be a Latin phrase, *de fumo in flammam*, or "from the smoke into the flame." The first mention of the aphorism in English may have been a line from Sir Thomas More (1478–1535): "Leapt they like flounder out of a frying pan into the fire." A nearly identical sentiment was expressed by John Heywood in his book *Proverbs*, published in 1546. As Heywood put it: "Leapt out of the frying pan into the fire and change from ill pain into worse." The point of the aphorism is that it is possible, in trying to escape from a difficult situation, to jump into one that is even more dangerous.

Further Reading: Hendrickson, Robert. *The Facts on File Encyclopedia of Word and Phrase Origins*, revised and expanded edition. New York: Facts on File, 1997; Lurie, Charles N. *Everyday Sayings: Their Meanings Explained, Their Origins Given*. New York: G.P. Putnam's Sons, 1928.

ARSON

Arson is defined as the willful or malicious burning of property, usually with criminal or fraudulent intent. For example, a business owner might have his or her building burned down in order to collect insurance money. Or one individual might burn down someone else's building because of bad feelings between the two people. Or a person might burn down a building just because he or she gets pleasure in watching dramatic fires. Arsonists of the last type are often called *firebugs*.

Arson is a crime and is often subdivided into other categories depending on the intent behind the act and the amount of damage incurred. These definitions differ from state to state. One common practice is to designate as first degree arson any act in which the defendant is aware that one or more persons is likely to be in the building he or she ignites. That is, the arsonist sets out not only to destroy property, but also to endanger human life.

Second degree arson, by contrast, usually refers to acts in which no human lives are at stake. Finer distinctions can be made depending on the amount of damage done or intended to be done by the act of arson. A state might define second degree arson as any act in which the value of the property is $500 (or some other amount) or more, and third degree arson, any act in which the value of the property is less than that amount.

Types of Arson

Arson cases can often be subdivided into seven major categories. The first is fraud—the attempt to obtain profit by destroying one's property or having one's property destroyed. For a company on the edge of bankruptcy, for example, one possible "solution" may be to burn down a physical structure and collect the insurance money.

A second motive for arson is to cover up another crime. A person who commits a murder, for example, might be tempted to burn down the structure in which the crime was committed. In that way, much of the evidence for the first crime (murder) would be destroyed by the second crime (arson).

Three other motives involve the human emotions of jealousy, revenge, and thrill. The arsonist may want to "get back" at another person by destroying that person's property

or may just derive pleasure from watching a huge fire.

Riots and vandalism are yet another type of arson. Individuals involved in a riot often perform acts of anger and revenge that they would otherwise never perform. They might attack the symbols of authority or oppression by burning down their property.

Finally, some cases of arson are acts of terrorism. Groups may set fire to the property of others with whom they have political differences. For example, Palestinian guerillas have often used arson as a weapon against Israelis in the Middle East, and the Irish Republican Army has used similar tactics against the British government in Northern Ireland.

Statistical Data

Arson is the second leading cause of fires in the United States today, exceeded only by smoking. An estimated 500,000 arson fires are set each year, about one in every four fires. These fires annually result in about 500 deaths and property losses of about $3 billion (Arson Prevention, 1).

About half of all arson arrests involve fires in structures, such as residential, industrial, commercial, and public buildings. Single-occupancy residences account for about 40 percent of these fires. Another 30 percent of arson fires occur in mobile structures, 95 percent involving motor vehicles. The remaining arson fires involve other types of property, such as crops, timber, signs, and merchandise (*Crime in the United States 1998*, 54–55).

The arrest and conviction rates for arson fires are very low. An arrest is made in only about one-sixth of all arson fires, and convictions are obtained in only about 20 percent of these cases. Young people under the age of eighteen make up 52 percent of those arrested for arson, by far the largest single age group among arrestees. Nearly 70 percent of all arson arrests involve individuals under the age of twenty-five. Males are about six times as likely to be arrested for arson as females, and Whites about three times as likely as Blacks to be arrested (*Crime in the United States 1998*, 54–56, and passim).

Arson Investigations

Instances of suspected arson are often investigated not only by local law enforcement agencies, but also by insurance investigators. Insurance companies make every effort to determine whether claims filed for fire damage are legitimate or whether they are the result of arson.

Arson investigations make use of two kinds of information primarily: fire scene and case history data. Experts know that fires that have been intentionally set have different characteristics than those that start naturally or occur as the result of an accident. Some obvious signs are the presence of fuel cans and the absence of important or emotionally valuable materials from the building, materials such as personal photographs and yearbooks. Some less obvious signs include the presence of certain distinctive burn patterns in the structure, the identification of points at which the fire began, and the lack of more obvious causes of fire.

On-site investigations of suspected arson are now a highly sophisticated procedure. Investigators not only use their knowledge about distinctive kinds of fire patterns, but also employ a variety of techniques and devices. For example, it is possible to purchase equipment that is able to detect very small quantities of vapors associated with arson fires. These vapors usually consist of hydrocarbons found in gasoline, kerosene, benzene, and other fluids used to start fires. These "mechanical sniffers" can detect the few drops of gasoline (or other liquid) left behind a few days after the fire began.

Arson dogs are also used for the same purpose. These dogs are especially trained to detect the very faint odor of accelerators (flammable liquids) left behind by the arsonists. Some experts claim that dogs are even better as "sniffers" than are mechanical devices.

Investigators can also obtain clues about possible arson cases simply by reading the claim submitted by the owner of a burned property. They look for such signs as the owner's having a foolproof alibi for the time of the fire, the recent purchase of an insur-

ance policy, a property that is insured for more than its real value, the failure to set an alarm system, recent reductions in the use of utilities, and financial problems with the business. *See also* ARSON DOGS; COMBUSTION; EXPLOSIVES.

Further Reading: "Arson and Explosives," <http://www.atf.treas.gov/core/explarson/explarson.htm>, 08/08/99; "Arson Clearinghouse," <http://www.usfa.fema.gov/usfa/>, 08/08/99; "Arson Prevention" <http://www.usfa.fema.gov/napi/stats.htm>; Bouquard, Thomas J. *Arson Investigation: The Step-by-Step Procedure*. Springfield, IL: Charles C. Thomas Publishing, 1983; *Crime in the United States 1998: Uniform Crime Reports*. Washington, DC: Federal Bureau of Investigation, October 1999; Faith, Nicholas. *Blaze: The Forensics of Fire*. New York: St. Martin's Press, 2000; Geller, J.L. "Arson in Review: From Profit to Pathology." *Psychiatric Clinics of North America*. 15, no. 3 (1992): 623–645; Infolink: "Arson," <http://www.nvc.org/infolink/info02.htm>, 08/08/99.

ARSON DOGS

Arson dogs are dogs that have been trained to detect the presence of small amounts of hydrocarbons. These hydrocarbons are the remnants, or "fingerprints," left behind when an arsonist starts a fire with gasoline, kerosene, or some other flammable liquid.

The arson dog program was begun in 1985 in a collaborative program of the Bureau of Alcohol, Tobacco, and Firearms, the New Haven (Connecticut) State's Attorney Office, the Connecticut State Police Forensic Science Laboratory, the Bureau of the State Fire Marshal, and the Emergency Services Division's K-9 Unit.

The arson dog program has now spread beyond Connecticut. Many state fire marshal offices, including those in New York, Ohio, Illinois, and Texas, now have one or more arson dogs. Arson dogs are also being used in other nations, including Great Britain and Australia. In addition, some private companies provide arson dogs for fire investigations. An example is the Blaze Fire Investigation and Consulting Service in Waukesha, Wisconsin, whose arson dog is named "Blaze."

Dogs, of course, have a powerful sense of smell. The arson dog program takes advan-

Greg Keller and Charlotte, the arson dog, search an arson-caused fire that cost over $1 million in damages. Charlotte is trained to locate flammable and combustible hydrocarbons that may have been used to set the fire. *Courtesy of Gregory Keller/Portland Fire Bureau.*

tage of that fact by teaching dogs to look specifically for the kinds of chemicals associated with arson fires. A common method is to show a dog a container holding a few drops of gasoline and to provide the dog with a food reward. Later, the target (gasoline) is hidden inside a box, behind a wall, or in a hole in the ground. Each time the dog finds the target, it is rewarded with a food treat.

An arson dog is taught two behaviors in this program. The first, called primary alert, involves the dog's coming to a sit position when it has located the target. The second, called the secondary alert, involves the dog's exhibiting some other type of behavior, such as establishing eye contact with the trainer.

Dogs selected for arson work can be either male or female, between twelve and thirty-six months of age, and at least 45 pounds in weight. Preferred breeds include members of the sporting group, such as Labrador retrievers, golden retrievers, and German shorthair pointers; working breeds, such

as German shepherds and border collies; and mixes of these breeds.

Those who have worked with arson dogs have high praise for their skills. Some trainers claim their dogs can detect a sample the size of a thousandths of a drop, a greater sensitivity than that currently available with the best mechanical device. Arson dogs also tend to give more valid responses than mechanical equipment. In one study, 90 percent of the suspected samples discovered by dogs turned out to be correct (that is, they were samples of flammable liquids used to start arson fires), compared to a rate of 50 percent for samples obtained with mechanical equipment. *See also* ARSON.

ART. *See* VISUAL ARTS

ASBESTOS

Asbestos is the general name for a large group of magnesium silicate minerals. Magnesium

7

silicates are chemical compounds that contain magnesium, silicon, oxygen, and, often, at least one more element. The chemical formula for a typical magnesium silicate is $Mg_7Si_8O_{22}(OH)_2$. All forms of asbestos have two properties in common. They occur in the form of long, fibrous crystals, and they are fire resistant.

Most forms of asbestos are classified into one of two major groups: serpentine asbestos and amphibole asbestos. Serpentine asbestos is the mineral chrysotile, $Mg_6(Si_4O_{10})(OH)_8$. Amphibole asbestos includes a variety of different minerals, most of which contain calcium, sodium, and/or iron in addition to magnesium, silicon, and oxygen. An example of an amphibole asbestos is tremolite, $Ca_2Mg_5(Si_8O_{22})(OH)_2$.

The fire-resistant properties of asbestos have been known since ancient times. In fact, the name *asbestos* comes from the Greek word *asbeston,* for "noncombustible." The mineral was used during Roman times for wicks in lamps. Wicks made of asbestos allowed the flow of oil upward without having the wick to catch fire. According to a popular ninth-century story, the Emperor Charlemagne enjoyed impressing visitors by throwing an asbestos tablecloth into the fire without its catching fire.

Asbestos occurs abundantly in the Earth's crust. According to some authorities, two-thirds of all rocks contain some form of asbestos. Primary producers of the mineral for commercial use are the Commonwealth of Independent States (Russia and some other members of the former Soviet Union), Canada, China, Brazil, and Zimbabwe.

At one time, asbestos was used in a wide variety of products, primarily because of its fire-resistant properties. Included among these products were protective clothing for firefighters, fireproof fabrics, brake and clutch linings, electrical and heat insulation, linings for chemical containers, and insulation.

Little information about the health effects of asbestos was available before World War II. One reason was that health effects develop only after long periods of time, usually twenty years or more. As the mineral became more widely used, however, these health effects began to appear.

The three most common health disorders associated with asbestos are mesothelioma, lung cancer, and asbestosis. All of these disorders occur when asbestos fibers are inhaled. They enter the respiratory system and lodge in the interstitial areas between the alveoli. Over time, they cause the development of scar tissue, which reduces the permeability of the alveoli. It gradually becomes more and more difficult for oxygen to pass into the bloodstream. Symptoms of asbestos-related disorders include coughing, shortness of breath, loss of weight, and anorexia. As the condition becomes worse, more serious respiratory problems, such as emphysema and bronchitis may result.

Workers involved in the extraction and processing of asbestos are at greatest risk for asbestos-related disorders. Without protective devices, they may be exposed to large concentrations of asbestos fibers. Today, miners and those who work with asbestos are provided with face masks that filter out almost all of the asbestos fibers with which one comes into contact.

Not all forms of asbestos are equally dangerous to human health. In fact, a rather narrow range of fiber lengths is associated with such problems, those less than two microns and between five and 100 microns in length. Of the two forms of asbestos, chrysotile is by far the safer. Studies have shown that workers who are provided with face masks and protective clothing are at virtually no risk for asbestos-related health problems.

The health problems associated with exposure to asbestos became widely known during the 1970s. Massive removal programs were initiated to remove the product from buildings in which it had been used for insulation. In retrospect, some critics have argued that the near panic leading to such cleanup programs may have caused more problems than the original asbestos insulation itself. During such programs, large amounts of asbestos dust were often released to the air, increasing the natural background level of the mineral in an area.

Today, about 90 percent of all chrysotile produced is mixed with cement or resin to make building materials. These materials are compact, do not break apart easily, and, therefore, do not release asbestos fibers to the atmosphere. The most common product, known as *chrysotile cement*, is a very popular building material because it is relatively inexpensive, durable, and less dense than pure cement. It can also be produced using less energy than is the case with comparable products. The most popular uses of chrysotile cement are in pipes, sheeting, shingles, and building blocks.

Restrictions on the use of asbestos have caused some problems for industry. No completely satisfactory substitute has yet been found for the material in some of its applications. Fiberglass is now sometimes used in place of asbestos for insulating homes and commercial buildings. While fiberglass is an effective insulator, it is not as resistant to flame as is asbestos.

For further information on asbestos, contact:

The Asbestos Institute
1200 McGill College, Suite 1640
Montreal, Quebec, Canada H3B 4G7
Tel: (514) 877-9797
Fax: (514) 844-9717
e-mail: ai@asbestos-institute.ca

Further Reading: Castleman, Barry I., and Stephen L. Berger. *Asbestos: Medical and Legal Aspects*. Gaithersburg, MD: Aspen Publishers, 1996; Corn, Jacqueline Karnell. *Environmental Public Health Policy for Asbestos in Schools: Unintended Consequences*. Boca Raton, FL: CRC Press, 1999; Tweedale, Geoffrey. *Magic Mineral to Killer Dust: Turner & Newall and the Asbestos Hazard*. New York: Oxford University Press, 2000.

ASH

The term *ash* has two slightly different meanings. In analytical chemistry, it refers to the solid residue that remains after a material has been completely burned. This form of ash usually consists of silicon dioxide, aluminum oxide, iron oxide, and oxides of other metals. Determining the amount of ash in a material is often one of the goals of analysis.

Ash is also the name given to the solid material left behind after the combustion of coal. It may occur in one of three forms: fly ash, bottom ash, and boiler ash (or boiler slag). During the combustion of coal in power-generating plants, about 90 percent of the ash formed occurs as fly ash. The remaining 10 percent of the waste material produced is bottom and boiler ash. The combination of fly ash, bottom ash, and boiler ash is sometimes known collectively as coal combustion by-products (CCBs).

Fly ash consists of tiny particles of unburned elements and compounds. On average, the most common constituent of fly ash is silicon dioxide (silica; SiO_2), which makes up about one-half of the product. The next most abundant component is aluminum oxide (alumina; Al_2O_3), which makes up about one-quarter of the fly ash. Other constituents include sulfur, iron, lime, magnesium, and alkalies.

Very large amounts of fly ash are now produced annually during the combustion of coal in power-generating plants. In 1998, an estimated 45.3 million metric tons (44.9 million tons) of the product were generated. This total makes fly ash one of the top ten most common minerals in the United States.

At one time, all forms of ash were generally regarded as a waste product for which there was no demand or use. Today that situation has changed. Fly ash is collected by means of electrostatic precipitators on the inside of smokestacks and chimneys. The ash is then used as a building material. It has many properties similar to those of Portland cement and can be used in its place in the production of concrete for roads and other structures. Concrete made with fly ash is durable, workable, and less expensive than concrete made from Portland cement.

Fly ash is also used for other purposes, such as the improvement of soils for agriculture, in water purification and waste water treatment systems, and as a source of raw materials in the chemical industry.

Ash is also formed naturally during volcanic eruptions. It consists largely of finely divided volcanic glass and pumice. Ash falls have been known to distribute these materials over hundreds of square kilometers of land surrounding a volcano. The nutrients they carry can dramatically improve the agricultural value of the soil on which they land.

For further information about fly ash, consult:

The Fly Ash Resource Center
http://www.geocities.com/CapeCanaveral/
Launchpad/2095/flyash.html

Further Reading: "Ash Development Association," <http://www.adaa.asn.au/ash.html>, 03/12/00; "What is Fly Ash?" <http://www.sefaflyash.com/SUBwhatsfa.html>, 03/12/00.

ASH WEDNESDAY

Ash Wednesday is a Roman Catholic holiday, the first day of Lent. It occurs anywhere between 4 February and 11 March, depending on the date on which Easter occurs. Its original name in Latin was *dies cinerum* (day of ashes).

The holiday appears to have originated during Roman times, probably as early as the eighth century. In its earliest form, penitents dressed in sackcloth, were sprinkled with ashes, and remained apart from the secular community until Maundy Thursday. The original purpose of the ceremony seems to have been to remind the faithful of the Biblical injunction: "Remember man that thou art dust and unto dust thou shalt return."

The ritual begins when palms used in the Palm Sunday ceremonies of the preceding year are burned. The ashes formed are then sprinkled with holy water and incense and blessed by a priest. They are used in making the sign of the cross on the forehead of church members who come to the communion rail following mass.

ASSAYING. *See* FIRE ASSAYING

ASSIUT (EGYPT) FIRE OF 1994

The fire that broke out in Assiut Province, Egypt, in November 1994 is believed to have been the worst fire in terms of loss of human life since the Coconut Grove fire in Boston in 1942. An estimated 475 people lost their lives in the Assiut fire.

The Assiut fire was the result of a strange combination of circumstances. A train carrying fuel oil derailed on 2 November 1994 during the worst rainstorm in Egypt in sixty years. Oil that poured out of damaged tank cars floated on flood waters and was ignited by electrical wires blown down by the storm. To add to the tragedy, a fuel tank near the train accident was struck by lightning and burst into flames.

Center of the fire damage was the village of Dronka, where 22,000 people lived. More than 200 homes in the village were destroyed and hundreds of residents were killed either by fire or by drowning. Nearly 300 more people were hospitalized for burns. Many residents fled the village convinced that the Day of Judgment had arrived.

AUTO-DA-FÉ. *See* CHRISTIANITY, BURNING OF HERETICS

B

"BAPTISM WITH THE HOLY SPIRIT AND WITH FIRE." *See* OLD BELIEVERS

BECHER, JOHANN JOACHIM (1635–1682)

Becher proposed a theory of combustion in the mid-1600s that was to dominate scientific thinking for more than a century. His life spanned an important period in scientific history when old ideas from alchemy were dying out and a basis for the modern science of chemistry was being born. Alchemy is a form of prescience and prechemistry that was strongly influenced by magical and mystical notions rather than by principles that could today be described as scientific.

His Life

Becher was born in Speyer in the Palatinate on 6 May 1635. He was the son of a Lutheran minister who had lost his fortune during the Thirty Years' War. Becher was an avid student, but his formal education suffered from having to help support his impoverished family.

Eventually he became a physician and found a position in 1666 as court physician in the state of Mainz. There he became embroiled in a variety of activities ranging far beyond his medical duties. For example, he argued for the construction of a canal between the Rhine and Danube Rivers as a way of increasing commerce between Austria and the Netherlands.

Becher was also engaged in some traditional forms of alchemical research, particularly the effort to find ways of transmuting simple substances into gold. He was convinced that he would be able to convert sands from the Danube River into the precious metal and when he failed in that attempt, found it wise to leave his homeland and move first to the Netherlands and then to England.

Terra Pinguis

Becher lived at a time when some of the earliest chemical research was being conducted. It could not yet be said that modern chemistry had appeared on the scene, but some of the discoveries that were to lead to the birth of the science were being announced. Becher was intelligent enough to see the need to salvage from alchemy any ideas that could be used in the development of the new science. To this end, he wrote a textbook in 1669 that described what was then known about the nature of matter.

In this book, Becher divided all known forms of matter into three general types, one of which he called *terra pinguis*, or "fatty earth." Becher believed that all objects capable of combustion contained this material. When combustion occurred, he said, *terra pinguis* escaped from the material into the atmosphere.

This theory had some validity since it explained some of the obvious observations that accompany the process of burning. For example, when a piece of wood is burned, smoke (containing *terra pinguis*?) is emitted, leaving a diminished volume of residue behind. This theory was later to be revised and embellished by one of Becher's students, Georg Ernst Stahl. In Stahl's formation, the *phlogiston theory* was to become one of the guiding (if incorrect) theories during the first century of the rise of modern chemistry. *See also* Phlogiston Theory; Stahl, Georg Ernst.

Further Reading: Asimov, Isaac. *Asimov's Biographical Encyclopedia of Science & Technology,* 2nd revised edition. Garden City, New York: Doubleday, 1982, pp. 143–144; Smith, Pamela H. *The Business of Alchemy: Science and Culture in the Holy Roman Empire.* Princeton, NJ: Princeton University Press, 1994; Weeks, Mary Elvira, and Henry M. Leicester. *Discovery of the Elements,* 7th edition. Easton, PA: Journal of Chemical Education, 1968, pp. 177–180, passim.

BEHAVE

BEHAVE is a computer program for predicting the behavior of a wildfire. It was based on an earlier and simpler program developed by Richard Rothermel of the Northern Fire Laboratory of the U.S. Forest Service. BEHAVE was developed during the early 1980s and was first put into use in 1984.

The BEHAVE program is designed to be used by individuals responsible for carrying out prescribed burns, for fire prevention planning, for projecting the future pattern of an existing fire, to suggest methods of attack when a fire has first been sighted, and for training programs.

The program is based on input data covering independent variables such as the amount of fuel in an area, the moisture contained in that fuel, wind speed and direction, topographic factors, and atmospheric temperature. From data such as these, the program is able to predict with some accuracy dependent variables, such as rate of fire spread, probable size of the fire, and intensity of the burn.

A revised version of BEHAVE, first called Beyond-BEHAVE and, later, BehavePlus, was developed in the late 1990s. The new version of the program added additional modeling factors such as soil heating, consumption of organic ground fuel, and certain characteristics of crown fires (fires that sweep across the tops of trees).

BehavePlus was tested by a group of eight selected users in February 2000, and feedback from that test was used to make further revisions in the program. The formal release of BehavePlus was scheduled for September 2000. *See also* Fire Modeling; U.S. Forest Service.

Further Reading: Agee, James K. *Fire Ecology of Pacific Northwest Forests.* Washington, DC: Island Press, 1993, pp. 47–52; "Fire Management Tools," < http://www.fire.org/perl/tools.cgi?BEHAVE= detailed>, 07/31/99; Burgan, R.E., and R.C. Rothermel. *BEHAVE: Fire Prediction and Fuel Modeling System—FUEL Subsystem.* General Technical Report INT-167. Ogden, UT: U.S. Department of Agriculture, Forest Service, Intermountain Research Station, 1984; Rothermel, R.C. *How to Predict the Behavior of Forest and Range Fires.* General Technical Report INT-143. Ogden, UT: U.S. Department of Agriculture, Forest Service, Intermountain Forest and Range Experiment Station, 1983.

BERKELEY HILLS (CALIFORNIA) FIRE OF 1991

The Berkeley Hills fire in Oakland, California took place on 20 October 1991. It occurred on the western slopes of a range of hills that separate the city of Oakland from eastern Contra Costa County. The fire began in the region of the Caldecott Tunnel, through which Interstate Highway 5 joins the two areas. Because of this fact, the fire is sometimes referred to as "The Tunnel Fire of 1991."

Background of the Fire

Conditions for the fire had been building for many years. The Berkeley Hills, as well as most of the San Francisco East Bay Area, has a classic Mediterranean climate with warm dry winters and hot dry summers. The area had been experiencing a long drought dur-

Aerial photo of Berkeley Hills, showing the aftermath of the 1991 East Bay Hills fire storm, the largest urban fire in United States history; 3,000 homes were burned. *Courtesy of Gary Keyes,* The Oakland Tribune.

the geography and climate of the area. For example, highly flammable shake roofs were widely popular, as well as were landscapes covered with highly flammable bushes and trees.

In general, the Berkeley Hills had become a perfect setting for one of the new and deadliest forms of fires, an intermix fire. An intermix fire is one in which the characteristics of both urban and extraurban (such as forest and brush) fires were present. Such fires present challenges for a number of reasons, not the least of which is that fire fighting teams are often trained to deal with one or the other type of blaze, but usually not both.

Progress of the Fire

The Berkeley Hills fire started in an area that had experienced a smaller fire on the previous day. Members of the Oakland city fire department had remained in the area overnight on 19 October, checking on hot spots from the previous day's fire. Early on Sunday morning, a glowing ember was blown from the burned-out area and into a stand of nearby trees. The trees exploded in flame and the fire reemerged in a new and more terrible form.

ing which rainfall was even less than its usual minimal level. The area was covered with dead, dry shrubs and groves of pines and eucalyptus trees that had been killed or damaged by freezing weather of the preceding winter.

In addition, the area had become a prime location for new home building over the previous decade. The Berkeley Hills provide a spectacular view of the East Bay, the San Francisco Bay, and San Francisco itself. Hundreds of spectacular homes costing many hundreds of thousands of dollars had been built in the area. Unfortunately, many had been built without consideration for the unique problems of fire safety presented by

The initial fire was rapidly swept through the hills by a combination of circumstances, including a sixty-five mile-per-hour wind from the top of the hills, to a record high temperature of nearly 100°F. Firefighters were suddenly faced with the challenge not of cleaning up after the previous day's fire, but dealing with a newer and much larger blaze. They quickly had to fall back to new positions from which to fight the fire.

Their efforts were handicapped by a number of problems. For example, roads in the area quickly became jammed with residents trying to escape the fire, firefighters arriving to do battle, and sightseers wanting to have a look at what was to be the worst fire in the area since the 1906 earthquake in San Francisco. In addition, firefighters soon lost water with which to fight the fire. Water loss occurred partly because of efforts to suppress the fire, but also because water pipes in homes in the area began to burst. Before long, reservoirs were empty, and the pumps that would refill them could not be started because of damage to the electrical system caused by the fire.

The fight to contain the Berkeley Hills fire was the largest, most complex firefighting operation in California history. Seventy-four firefighting teams of up to five engines each from all over the West Coast arrived to contain the fire. Aircraft from hundreds of miles away arrived with loads of water and fire retardants. They made numerous round-trip visits to the fire area during the height of the blaze.

The fire continued to burn through the night of 20 October and into the next day. Residents of the San Francisco peninsula had the eerie experience of finding whole burned pages of telephone directories, cookbooks, dictionaries, and other books floating down from the sky onto their own porches and yards.

Fire Losses

The statistics for the Berkeley Hills fire were truly astounding. Twenty five people were killed and 150 were injured by the fire. A total of 2,843 single family dwellings were destroyed, and another 193 were damaged. Overall, 3,469 living units were either damaged or destroyed. The estimated dollar cost of the fire was eventually set at $1.537 billion.

Lessons Learned and Future Plans

A fire the size of the Berkeley Hills fire could hardly occur without providing a host of lessons for fire management. For example, the loss of water with which to fight the fire has caused the Oakland Fire Department to explore a variety of ways in which to ensure that there is not a repeat of the Berkeley Hills

problem. The department also discovered that its system of radio communications was inadequate to deal with the demands placed on it during the fire. Not only were there too few channels for the number of messages that needed to be sent, but the geography of the Hills was such that messages could not be sent or received at very great distances.

In some ways, however, it is not entirely clear that the risk of another Berkeley Hills fire has completely disappeared. The vast majority of the homes destroyed in the fire have now been rebuilt. While most have incorporated fire protection designs that they did not have originally, the area remains, in many ways, much as it was before the fire. It continues to be a mix of flammable plant material, a hot and dry climate, and a mass of human-provided fuel with which to start another blaze. *See also* INTERMIX FIRE.

Further Reading: Parker, Donald P. "The Oakland-Berkeley Hills Fire: An Overview," <http://www.sfmuseum.org/oakfire/overview.html>, 07/19/99; Sullivan, Margaret. *Firestorm! The Story of the 1991 East Bay Fire In Berkeley*. Berkeley: City of Berkeley, 1993.

BESSEMER, SIR HENRY (1813–1898)

Sir Henry Bessemer was an English inventor who developed an efficient, inexpensive method for producing steel, changing that metal's status from a nearly precious metal to one that could be used for a host of everyday purposes.

Personal History

Bessemer was born in Charlton, Hertfordshire, England on 19 January 1813, the son of an engineer. He inherited his father's interests in making new kinds of devices. One of his first inventions was a device for stamping deeds that saved the British government more than £100,000 a year. The government failed to compensate him for the invention, however, much to Bessemer's chagrin.

In spite of this disappointment, Bessemer continued to develop new inventions, eventually receiving 110 patents for his ideas. In recognition of his accomplishments, he was made a Fellow of the Royal Society in 1879 and was knighted in the same year. He died in London on 15 March 1898.

Invention of the Blast Furnace

Bessemer's most famous invention by far was the blast furnace, a device for making steel. That invention was prompted by an earlier invention he had made in the 1850s, a new type of cannon ball. The cannon ball he designed was made so that it would spin as it traveled through the air, giving it a longer and more accurate projectile. The British government was not interested in the cannon ball, but Bessemer was able to convince the French army to try it out.

The cannon ball worked well, but the French reported that their cannons were not strong enough to fire the new type of missile. The fit between cannon ball and cannon had to be so tight that the cannon often blew up when it fired the ball. Bessemer realized that he had to develop a new and stronger material from which to make cannons if his cannon ball were ever to be used.

The problem was that cannons were then made with cast iron, a form of iron that contains relatively large amounts of carbon. Cast iron is very hard, but it breaks very easily. The only substitute available for cast iron at the time was wrought iron, which is nearly pure iron. Wrought iron was not suitable for making cannons (or almost anything else) because it was too soft.

The ideal middle-ground material was steel, which has less carbon than cast iron, but more carbon than wrought iron. It combines the hardness and strength of cast iron with the durability of wrought iron. The problem was that no one had found an inexpensive and effective way to make steel with just the right amount of carbon.

Bessemer's idea was to blow air through molten cast iron. He knew that oxygen in the air would react with carbon in the cast iron to form carbon monoxide and carbon dioxide. If the blast of air could be carefully controlled, he could burn off as much or as little carbon as he wanted. He could produce a type of steel that had the correct balance of carbon between cast iron and wrought iron.

Bessemer's colleagues had little faith in his idea. They warned him that blowing cold air through molten cast iron would cool the iron and make it solidify. In that case, the whole process would come to an end with no result.

When Bessemer actually tried the process, however, he made a fascinating discovery. Instead of cooling off the molten cast iron, a blast of air increased the temperature of the mixture. His critics had neglected the fact that heat would be released when oxygen combines with carbon, making it unnecessary to add heat from an outside source once the reaction had begun. Bessemer's blast furnace turned out to be a great success because it made possible the production of steel with a very exact amount of carbon in it at a much lower cost than had been possible before. *See also* FURNACES; METALLURGY.

Further Reading: Asimov, Isaac. *Asimov's Biographical Encyclopedia of Science & Technology.* 2nd revised edition. Garden City, NY: Doubleday, 1982, pp. 574–575; Hart-Davis, Adam. "Henry Bessemer, Man of Steel," < http://www. exnet.com/1995/09/27/science/science.html>, 02/23/99; Porter, Roy, ed. *The Biographical Dictionary of Scientists*, 2nd edition. New York: Oxford University Press, 1994, pp. 67–68; "Sir Henry Bessemer F.R.S. An Autobiography," <wysiwyg://27/http://www.bibliomania.com/NonFiction/Bessemer/Autobiography/>, 02/23/99.

BIBLICAL ALLUSIONS TO FIRE

Fire is mentioned in the Bible in a number of places. Some of the most important of those references are the following.

The Burning Bush

The third chapter of the book of Exodus in the Bible tells of an extraordinary event experienced by Moses. In verse 1, Moses is said to be tending a flock of sheep owned by his father-in-law Jethro. He suddenly sees a bush break into flames. That event in itself is not unusual in the desert, where the land is arid and hot. What *is* unusual is that the bush is not consumed by the flame. Moreover, Moses sees an angel of the Lord in the midst of the

St. Catherine's Monastery was built on the site where the burning bush is thought to have appeared to Moses. *Courtesy of the Egypt Tourist Authority.*

"Burning Bush." And the angel brings an astounding message to Moses.

In any case, God, speaking through the voice of the angel in the fire, explains to Moses His plan for the future. He tells Moses that He is aware of the unhappy state of affairs of his Chosen People (the Jews). He has decided to release them from their misery and to deliver them to their Promised Land in Canaan. He warns Moses that he (Moses) will not find it easy to convince the Jews of their new message. But Moses' commission is to lead his people out of Egypt and into their Promised Land.

That message sets the stage for an important period in the history of the Jewish people. For 400 years they have been wandering in the Egyptian desert, scratching out a meager living. In fact, one interpretation of the Burning Bush story is that the bush itself represents the downtrodden state of the Jewish people and the fire, the persecution they have suffered under their Egyptian overlords. The Burning Bush itself is also symbolic. Fire in nearly all cultures is regarded as a cleans-

ing event, a form in which the highest level of purity is attained. It is in this form that God appears and speaks to His creation.

The spot on which the Burning Bush was thought to have been located is now occupied by St. Catherine's Monastery, built by order of Emperor Justinian between 527 and 565. The monastery holds one of the world's finest collections of illuminated manuscripts. It is operated today by the Greek Orthodox Church.

There has been considerable discussion as to precisely what kind of bush it was that is described in Exodus. Some scholars believe that it was a member of the species *Rubus sanctus* that grows near oases in the desert. Other authorities believe the bush was a member of the species *Cassia senna*. Still others think the bush may be *Dictamnus fraxinella*, a not uncommon garden plant. This bush secretes a flammable vapor on hot days, a vapor that can easily be lit and burns without destroying the bush itself.

Today, *burning bush* is the common name for a plant whose systematic name is *Euony-*

mus alata. The plant gets its name from the fact that its leaves turn a brilliant red color in the fall. Except for its name, the plant has no connection with the Burning Bush story in the Bible.

Chariots of Fire

Chariots of fire are mentioned in various parts of the Bible as a symbol of God's presence and power. In II Kings, Chapter 2, for example, chariots of fire and horses of fire appear in the sky in the moments before the prophet Elijah is taken away into heaven. In Chapter 6 of the same book, chariots of fire also appear to Elijah's follower, Elisha, at the moment that the king of Syria and his forces are attacking the Israelites.

Pillar of Fire

A pillar of fire is mentioned more than a dozen times in the Bible. It first appears in the book of Exodus, when God leads the Israelites out of Egypt by means of a "pillar of cloud by day" and a "pillar of fire by night" (Exodus 14:24). As described here, the "pillar" seems to be a large vertical shaft of light that acts as God's beacon leading the Israelites through darkness.

But the reference has other connotations also. Later in Exodus, the pillar is described as a sign of God's protecting presence with the Israelites. In Chapter 33, verses 9 and 10, the writer tells that "the Lord looked down from the pillar of fire and cloud at the Egyptian army and threw them into a panic."

In some cases, the pillar of cloud and pillar of fire actually represented God himself, or at least it provided a conduit by which He could speak to humans. In Numbers 14:14, for example, God appears at the tent of Aaron and Miriam in the form of a pillar of cloud and fire. In Deuteronomy 1:33, God appears again to the Israelites as a pillar of cloud and fire when they have assembled to protest their lack of progress to Moses and Aaron.

In other parts of the Bible, the pillars of cloud and fire are also offered by God as His symbol of protection for the Israelites. In Isaiah 19:1, for example, the prophet tells that God will send down a "cloud in the daytime and smoke and a bright flame at night" that will "cover and protect the whole city."

Shadrach, Meshach, and Abednego

Shadrach, Meshach, and Abednego were three Jews who had been raised to positions of authority under King Nebuchadnezzar after the Jewish nation had been subjugated by the Babylonians. The three men had reputedly been given authority because they were wiser, healthier, and more attractive than other Jews and most Babylonians. Their story is told in the Bible in the book of Daniel, Chapter 3.

Problems developed for the three men when they refused to bow down to a huge golden idol that the king had ordered made of himself. When Nebuchadnezzar heard of their refusal, he became angry and ordered them to be cast into a very hot furnace. The furnace was heated to seven times its usual temperature and was so hot that it burned to death the men who cast Shadrach, Meshach, and Abednego into the flames.

But in spite of the heat, the king was able to see four men walking about in the hot flames of the furnace. The four men were Shadrach, Meshach, and Abednego and an angel of God who was protecting them from the fire. Nebuchadnezzar was so impressed with the protection offered by the Jewish God that he acknowledged God's authority, bowed down before Him, and returned Shadrach, Meshach, and Abednego to their positions of authority. *See also* OLD BELIEVERS; RELIGIOUS ALLUSIONS TO FIRE.

Further Reading: "David Guzik Study Guide for Exodus Chapter 3," <http://www.khouse.org/blueletter/Comm/david_guzik/sg/Exd_3.html>, 12/22/98; "The Men in the Furnace," <http://www.virtualchurch.org/fire.htm>, 08/25/99; Rossotti, Hazel. *Fire*. Oxford: Oxford University Press, 1993, p. 246; Sarna, Nahum, "Theophany at the Burning Bush," <http://www.jewishheritage.com/topics/fire/bush.html>, 12/22/98.

BONFIRE OF THE VANITIES

The term "bonfire of the vanities" refers to two huge bonfires held in the Piazza della Signoria, in Florence, Italy, in 1497 and

1498. The bonfires were inspired by the preachings of the Dominican monk, Girolamo Savonarola, who had been warning the citizens of Florence that their love of art, commerce, and other worldly pursuits was to bring destruction to their city.

Savonarola arrived in Florence just as the last of the Medici family, Piero, was being driven out of the city. For a period of four years, between 1494 and 1498, Savonarola was the de facto ruler of the city, although he never held official office. His influence resulted from his ability to move masses of people with compelling and powerful rhetoric. In particular, he taught that the Roman Catholic Church had been taken over by hypocrites who kept donations for themselves and ignored the needs of the poor. He called the Church "a monster of abomination," and saved his most vitriolic criticism for Pope Alexander VI.

As part of his message, Savonarola taught that salvation was available only to those who gave up the worldly goods on which they doted. He told women to throw out their jewelry, men to discard their finest books, and artists to destroy paintings of anything other than sacred subjects. The final result of this preaching was the two "bonfires of the vanities." It is said that one reason that only pictures with sacred themes remain from that period is that art with secular subjects was largely destroyed in the bonfires. We do know that Sandro Botticelli, then a young artist, contributed many of his paintings to the bonfire at the Piazza della Signoria on that day.

Savonarola's dominance in Florence was to continue for only a few months after the second bonfire. It appears that many citizens began to have second thoughts about total abandonment of "the good life." Merchants, in particular, found that they suffered serious economic losses as they attempted to follow the monk's teachings. In addition, Savonarola was drawn into the ongoing intrigue between members of his own Dominican order and their hated enemies, the Franciscans.

By May of 1498, the citizens of Florence had disavowed Savonarola. He was taken from his convent by an angry mob, subjected to torture for two months, and then hanged and burned. His ashes were thrown into the River Arno. *See also* PROTEST AND DEFIANCE BURNINGS.

Further Reading: Oliver, Revilo P. "Evangelical Democracy." *Liberty Bell*, September 1973. Also at <http://www.stormfront.org/rpo/SAVONARO.htm>, 5/11/99; Polizzotto, Lorenzo. *The Elect Nation: The Savonarolan Movement in Florence 1494–1545*. Oxford: Clarendon Press, 1995; Ridolfi, Roberto. *The Life of Girolamo Savonarola*. London: Routledge & Kegan Paul, 1959; Weber, Erwin. "Girolamo Savonarola." *The Lutheran Journal*, 53, no. 3 (1986).

BONFIRES

A bonfire is a large fire in the open air that is often, but not always, made of logs and other forms of wood. Bonfires are made for a number of different purposes, but they are usually part of a celebration of some kind or another. For this reason, bonfires are sometimes also called *festival fires*.

Researchers have found that bonfires are traditional in a great many different cultures over very long periods of time. In many cases, modern-day traditions appear to have evolved from celebrations and festivals that extend well back into the culture's earliest mythology. For example, the Nordic tradition of lighting fires to celebrate certain festivals is thought to be descended from the cremation fire of one of the most famous Nordic gods, Balder.

Characteristics of Bonfires

No matter where or when they are held, most bonfires appear to have a number of characteristics in common. For example, the fuel needed for the fire is often collected by young children or adults who go from door to door begging for pieces of wood or coal. In Belgium, for example, the tradition was for children to go from farm to farm in the week before the fire, collecting fuel. Anyone who refused to give something for the fire could be expected to be revisited after the fire to have their faces blackened with ashes of the fire.

Festival fires could be built at any time of the year, but were most common during certain parts of the season, such as the summer or winter solstice or during a religious holiday, such as Easter or Christmas. Sir James Frazer claims that the most common period for festival fires in Europe was Midsummer Eve (23 June) or Midsummer Day (24 June). He points out that these festivals were eventually associated with some holy Christian holiday, such as St. John the Baptist's day. But it is not difficult to see that the celebrations existed many centuries before Christianity had come to dominate European culture.

A fascinating characteristic of many festival fires was the burning of some type of effigy in the fire. In many areas, people claimed to be "burning the witch," "burning the old wife," "burning winter's grandmother," "burning the Easter Man," or otherwise exorcising some evil person or spirit.

In this regard, bonfires were thought to provide protection for a community from ill luck and evil in the coming year. In parts of Germany and Lithuania, for example, fires were supposed to protect against natural disasters, such as thunder and hail, as well as against illness among people and cattle.

Bonfires were also thought to have positive effects on the natural world. In many parts of Europe, bonfires are explicitly associated with improving the success of crops in the coming year. In Bavaria, among other regions, celebrants believed that crops would grow to the same height as the flames in the bonfire. In many cases, people would take charred sticks from the fire to plant in their fields, ensuring that crops would grow as tall as possible.

Many fire festivals often involved people jumping over the fires also. The purpose of this tradition is not always clear, but it appears to have some connection with maintaining the moral purity of the individuals who do the jumping and/or protecting them from physical disease. In many cases, traditions are combined, as in parts of Germany where young men and women jump through the smoke of the fire and pray that their crops will grow as tall as possible in the coming season.

The Bonfire Tradition in Great Britain

The tradition of festival bonfires survived in most parts of Europe at least to the end of the nineteenth century, and, in some places, much longer than that. There is probably no country in the world in which bonfires remain such an important form of annual celebration, however, than in Great Britain.

From the early fifteenth century, bonfires became more and more popular as a way of celebrating important local or national events at which a whole community would gather. Historical records from many English towns make note, for example, of sums of money set aside to provide for wood, tar, coal, and other materials from which to make a bonfire. Bonfires were set across the nation to celebrate events such as the inauguration of a new monarch, the victory in a war or battle, and an important religious or secular holiday.

The one event that is still celebrated most often by bonfires in Great Britain is Guy Fawkes Day, which occurs on 5 November each year. That celebration commemorates the reputed effort by Guy Fawkes to blow up the Parliament building and kill all of its members in the famous Gunpowder Plot of 1605. In memory of that event, 5 November has also become known in England as *Bonfire Night*.

Today, many English towns continue to support "bonfire societies," whose purpose it is to organize and conduct celebrations of Guy Fawkes Day and other memorable occasions. Links to the websites for some of these organizations can be found at <http://www.users.globalnet.co.uk/~bonboy/bfl.htm>.

Etymology

Etymologists have been unable to agree as to the source of the English word *bonfire*. The favored explanation appears to be that the word was derived sometime in the fifteenth century from the two words "bone fire," with the first "e" eventually having been lost. That explanation is derived from numerous tales of pagan societies that built fires of human bones to frighten off dragons, demons, and evil spirits. The problem with this explanation is that human bones burn very poorly, if at all.

Other scholars suggest that the term *bonfire* originally derived from two terms, "bon" and "fire." In this case, the "bon" part of the word is assumed to have come from the French word *bon*, or the Latin term *bonus*, both meaning "good." In this explanation, a bonfire is a fire lit to celebrate some special occasion. One problem with this explanation is that the French term *bon* was usually translated in English as "boon," a form that does not occur in *bonfire*. *See also* BONFIRE OF THE VANITIES; CHRISTIANITY AND THE BURNING OF HERETICS; RITUAL FIRE; RELIGIOUS ALLUSIONS TO FIRE.

Further Reading: Frazer, J.G. *Balder the Beautiful: The Fire-Festivals of Europe and the Doctrine of the External Soul*, 2 vols. London: Macmillan, 1913, Chapters 4–6; Frazer, Sir James George. *The Golden Bough: A Study in Magic and Religion*. New York: The Macmillan Company, 1958, Chapters 62–63; Goudsblom, Johan. *Fire and Civilization*. London: Allen Lane, 1992, pp. 132–134.

BUILDING AND FIRE RESEARCH LABORATORY

The Building and Fire Research Laboratory (BFRL) is a division of the National Institute of Standards and Technology, formerly the National Bureau of Standards. The goals of BFRL are twofold: First, to improve the productivity of the U.S. construction industry, and second, to reduce the human and economic losses resulting from fires, earthquakes, winds, and other disasters.

The laboratory is primarily a research center that studies fire science and fire safety engineering practices, building materials, and construction-related problems. The results of the laboratory's research are often incorporated into building codes, fire safety standards, and other rules and regulations dealing with construction and fire safety issues. Two major fire-related projects on which the Bureau has recently been working are an industrial fire simulation system and a program for the development of fire-safe materials.

For further information, contact:
Building and Fire Research Laboratory
National Institute of Standards and Technology
Gaithersburg, MD 20899
(301) 975–6850
http://www.bfrl.nist.gov/

BUNSEN BURNER

A Bunsen burner (also bunsen burner) is a device for producing very hot flames. The burner was invented by the German chemist Robert Bunsen (1811–1899). It is the primary source of heat used in most chemical laboratories. Over the years, a number of modifications of Bunsen's original design have been made. These variations are known as Tirrill, Fletcher, and Meeker burners after their respective inventors.

Method of Operation

All forms of the Bunsen burner operate on the same principles. The fuel is supplied to the burner from a source of natural gas, such as an ordinary gas line. The gas enters through an intake valve at the base of the burner. It then rises upward in the barrel to the mouth of the burner.

Air also enters the burner through openings at its base. The air also rises in the burner barrel, mixing with the gas as it does so. The combination of air and gas is ignited at the mouth of the burner producing the flame.

The character of the flame produced depends on the relative amounts of air and gas fed into the burner. With an excess of gas, the flame produced is yellow and relatively cool. It contains some unburned gas that escapes from the barrel before it can be ignited. With an excess of air, the flame will not light at all because there is insufficient gas to burn.

The correct balance of gas and air can be obtained in the burner by adjustments on the burner barrel or at its base. In the original form of the burner, for example, the amount of air that enters the barrel is controlled by rotating a sleeve with an inlet hole at the base of the burner. There is no way to adjust the flow of gas in this design.

The Tirrill burner, by contrast, has adjustments for both air and gas. The amount of gas that enters the barrel is controlled by a needle valve at the base of the burner. Turning this valve allows more or less gas to enter the burner. The amount of air that enters the burner is controlled by turning the barrel of the burner itself. Twisting the barrel allows more or less air to enter the barrel.

The Meeker burner produces one of the hottest flames used in a chemistry laboratory. It has large openings at the base to permit a copious supply of air, and its mouth is very wide and covered with a fine mesh. The mesh divides the burning gas into dozens of smaller sections, each of which acts like a single Bunsen burner flame.

Nature of the Flame

The flame produced at the top of the burner consists of two parts, usually known as the outer cone and the inner cone. When the mixture of gas and air has been adjusted to produce maximum temperature, the outer cone is blue and the inner cone a very pale blue, nearly invisible. The interior of the inner cone is actually quite cool. A piece of wood laid across the top of the burner will develop a circular burn, indicating that the gas is burning along the outside of the mouth of the burner, but not within the inner cone.

The hottest part of the flame is at the tip of the inner blue cone. The temperature at this point can reach about 1,800°C.

The two parts of a burner flame have different chemical properties. Within the inner cone, combustion is incomplete. The fuel is being broken down into carbon, hydrogen, carbon monoxide, and other products. These substances behave chemically as reducing agents; that is, they tend to take oxygen away from other materials. This portion of the flame is known, therefore, as the reducing cone. Within the outer cone, however, the gas and its by-products are being completely oxidized. This portion of the flame is known as the oxidizing zone.

Modifications of the Bunsen burner also have everyday applications. For example, the flame produced in a gas stove used for cooking has essentially the same design as that of a laboratory Bunsen burner. A Welsbach lantern is another adaptation of the Bunsen burner. In a Welsbach lantern a mantle is attached to the mouth of the burner. When heated, the mantle glows and gives off light. A Welsbach lantern is, therefore, a Bunsen burner adapted to produce light primarily rather than heat.

Further Reading: Rossotti, Hazel. *Fire*. Oxford: Oxford University Press, 1993, pp. 12–16; Shugar, Gershon J. et al. *Chemical Technicians' Ready Reference Handbook*. New York: McGraw-Hill, 1981, pp. 47–50.

"BURN, BABY, BURN!" *See* PROTEST AND DEFIANCE BURNINGS

BURNING MAN

"Burning Man" is a festival held annually in the Black Rock Desert in northwestern Nevada. The festival had its beginning in 1986 when Larry Harvey and Jerry James constructed an eight-foot tall wooden figure at Baker Beach in San Francisco. A crowd of about twenty people watched as the two men set fire to the statue. The event was said to have been staged in honor of the summer solstice.

The event was repeated at Baker Beach again from 1987 through 1990 with the wooden figure increasing in size to twenty feet in 1987, thirty feet in 1988, and forty feet in 1989 and 1990. Attendance at the event also increased over the years, from eighty in 1987 to about 200 in 1988, 300 in 1989, and 800 in 1990.

Park officials and police eventually became concerned about plans to continue holding the festival at Baker Beach, and organizers decided to move the burning event to federal land in the Black Rock Desert. In 1990, the figure was assembled at Baker Beach, but then transported to the desert for burning. About 90 people from the San Francisco group traveled to the desert to witness the burning of the figure.

From 1991 through 1999, with only one exception, the celebration of the Burning Man festival continued to be held at Black Rock. The one exception came in 1997 when participants attempted to hold the event on private land at Hualapi Playa. The following year, they returned to their original location at Black Rock.

By the end of the twentieth century, the Burning Man festival had become a large event that, for one week of the year at least, supported the largest community in Nevada's Pershing County. Attendance for the event was estimated at 10,000 in 1997, 15,000 in 1998, and 18,000 in 1999. The festival had also begun to take on many of the features of a permanent community, with a daily newspaper, radio station, ongoing Internet web site, organized street system, and governing body. Media representatives from virtually every part of the world have visited the festival to write feature stories and make documentary films recording the festivities.

Observers have found it difficult to characterize the nature of the Burning Man festival. It includes a fair amount of performance art, along with many special theme productions, such as "Christmas Camp," "Exploding Man," and "Spirit Tree." The event closes each year with a dramatic burning of the wooden figure, now festooned with neon lights and a hat that contains fireworks. Participants are charged an entry fee to attend the festival, but then money is not used again during the week, except at an on-site coffee shop.

Further Reading: Sullivan, James. "Eccentri-City," *San Francisco Chronicle*, 4 September 1999, B1+; The "Burning Man," <http://www.burningman.com/>.

BURNING OF ROME

A huge fire that swept through Rome in 64 A.D. is one of the greatest fires of antiquity. The fire began on 19 July of that year in the midst of a shopping area near the Circus (a large area used for chariot racing), where a variety of combustible materials were being sold. It spread quickly, partly because of a strong wind that was blowing that day. It quickly ran through the lower part of the city, then up into the hills, and finally back into the lower regions again. The fire lasted for six days and essentially destroyed ten of the city's fourteen districts. Even the shrine of the Vestal Virgins was consumed by the fire. Ironically, the date of the fire corresponds with the sacking of Rome by the Gauls in 390 B.C.

Historians have long debated the supposed role of the Emperor Nero in the fire. A popular legend claims that the ruler started the fire and then climbed to the Tower of Maecenas, where he played his violin while the fire raged. From this legend came the expression "fiddling while Rome burns" that is used to describe a person who carries out unimportant acts while some major catastrophe is occurring. Both Suetonius and Diodorus Cassius, historians living at the time of the conflagration, accused Nero of giving orders that the fire be started. His purpose, they said, was to clear land so that the city could be completely rebuilt with many new parks and palaces.

By contrast, the historian Tacitus argues that Nero had no part in starting the fire, even it if did eventually serve his purposes. The emperor was at his villa in Antium at the time of the fire, according to Tacitus, and did not return until his own palace was threatened by the fire. When he did come back to Rome, he opened his own grounds to those fleeing the fire and provided temporary shelter for those who had lost their homes in the blaze. If arson was involved in the fire, as it appears to have been, it was probably carried out by political foes of the emperor, according to this version of events.

In any case, the fire made possible the reconstruction of Rome that Nero had been planning. Streets and buildings were planned not only to make the city more magnificent than before, but also to provide thoroughfares through which fire equipment could be carried to deal with future fires. Nero himself lived only four more years before being assassinated on 9 June, 68 A.D.

Further Reading: Hadas, Moses and the editors of Time-Life Books. *Imperial Rome.* New York: Time Incorporated, 1965, pp. 61–62; "Nero Didn't Fiddle While Rome Burned," in David Wallechinsky and Irving Wallace, *The People's Almanac, #2.* New York: Bantam Books, 1978, pp. 1239–1240; Tacitus. *The Annals of Imperial Rome,* revised edition. Translated by Michael Grant. London: Penguin Books, 1989.

BURNS AND SCALDS

A burn is an injury caused by fire, intense heat, chemicals, or an electrical current. A scald is a burn-like injury caused by hot water, steam, or other hot liquids. Burns and scalds are among the most common form of injuries to the body. They vary in severity from the very mild to the extremely serious. At least 200,000 Americans are hospitalized each year for burn and scald injuries of which about 5,000 lose their lives. Such injuries are the third leading cause of accidental death in the United States. The nature and treatment of burns and scalds are similar. The following discussion focuses on burns, but applies in general to scald injuries also.

The Nature of Burn and Scald Injuries

Skin that has been exposed to intense heat, to chemicals, or to an electrical current undergoes both biological and chemical changes. Two of these changes are of primary importance. First, proteins in the skin are destroyed, causing destruction of skin tissue. Second, blood vessels become more permeable, allowing the escape of fluids into tissues.

The most serious consequences of a burn injury result from these two changes. In the former case, burn wounds are ideal situations for the growth of disease-causing microorganisms, such as fungi and bacteria. Without care, infections develop quickly that spread through the body. In the latter case, the loss of blood volume can result in shock. In the most serious cases, these conditions can cause the death of a person. As painful as a burn may be, then, the most serious risk that burn victims face is not the burn itself, but the consequences that may develop for other parts of the body and for one's overall medical condition.

Burn wounds tend to heal naturally, depending on the severity of the damage. In the mildest cases, the epidermal (outer) layer of the skin begins to regrow fairly soon. Little or no scarring results as a consequence of the wound.

In more serious cases, the wound heals only when both epidermal and dermal (inner) skin tissue regenerates. In such cases, the opening in the skin caused by the burn is pulled together to form an irregular, scarred surface that can be very disfiguring. The most severe case of burn scarring may require reconstructive ("plastic") surgery to improve the victim's physical appearance.

Burn Severity

The severity of a burn is measured by two criteria: the amount of skin area damaged by the burn and the depth to which the damage occurs. The severity of a burn is said to be

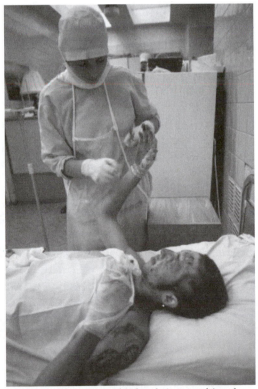

A patient with burns on his hands is treated in a burn unit in St. Petersburg, Russia. © *Steve Raymer/CORBIS.*

small if less than 15 percent of the total body surface area (BSA) has been damaged. It is said to be moderate if 15 to 50 percent BSA has been affected; large, if 50 to 70 percent BSA has been damaged; and massive, if more than 70 percent BSA has been destroyed.

Burns are also defined as being first, second, or third degree in severity depending on the depth to which damage occurs. A first degree burn is also called a *superficial burn* because it affects only the outermost layer of skin. A first degree burn is generally red, very sensitive to the touch, and often somewhat moist. The burn blanches (turns white) if pressure is applied to it. First degree burns generally do not result in the formation of blisters. They tend to heal in less than a week.

A second degree burn is one in which both the epidermis and at least part of the dermis is damaged. Blisters may or may not form, and the skin in the area around the burn is very red and moist. Second degree burns usually heal in less than three weeks without leaving any noticeable scarring.

Third degree burns are the most serious type of burn. The burn injury destroys both epidermal and dermal layers of the skin. For this reason, a third degree burn is also called a *full thickness burn.*

Third degree burns may result in the formation of blisters, but they usually do not. The burned area may be bright red, white, or even black. In some cases, the skin actually appears to be quite normal. This level of damage is, however, the most serious of all burn injuries and requires immediate and aggressive treatment.

Short-Term Treatment

The type of treatment required for burn injuries depends on the severity of the wound.[1] In general, the first steps that need to be taken are (1) to stop the burning process, (2) to exclude air from the burned area, (3) to prevent the onset of infection, and (4) to prevent shock.

Most people who have had some experience with burns know that a simple first step after receiving a burn is to immerse the burned area in cold water. The cold water absorbs heat from the wound and prevents

further damage. A second step involves covering the burn with a loose, clean bandage. The bandage protects the wound from infection, but allows air to aid in the healing process.

The wound must also be cleaned to remove any bacteria or other foreign objects on the wound. The cleaned wound can then be treated with an antibiotic to prevent infection. In some cases, the antibiotic can be administered orally to avoid having to cover the wound directly.

In simple burn cases, it may not be necessary to deal with the possibility of shock. If that does appear to be a problem, the burn victim should begin to receive fluids as soon as possible. If there is evidence that the burn is a serious second degree or third degree burn, the patient should be taken to a hospital immediately.

Long-Term Treatment

The long-term treatment of patients with severe third degree burns over much of their body has become an important field of medicine. Such individuals often face a serious risk of death and, if they survive, many months or years of treatment. When large areas of skin are destroyed, the body's natural regenerative processes are not sufficient to permit healing to occur on its own.

Historically, the most common method of treating extensive third degree burns is with skin grafting. A skin graft involves the removal of a section of skin from a healthy region of the patient's body and its reattachment on the damaged area of skin. For example, suppose that a person receives a serious third degree burn on the arm. A physician may treat this injury by taking a section of skin from the patient's upper leg and graft it over the damaged area of the arm.

Under the best of circumstances, skin grafts are difficult, complex, and dangerous procedures. The skin may or may not begin to grow normally in its new position. Before and after the grafting procedure, the patient is at risk for infections in both the donor and recipient regions of the skin. In cases of extensive damage, more than one graft may be necessary over a period of time.

The treatment of burn victims after grafting is also difficult and lengthy. The skin graft may take many weeks or months to begin growing. During that time, the damaged region must be immobilized to prevent the graft from breaking loose. Patients may be required to wear casts, compression garments, positioning devices, splints, and other mechanical devices to prevent damage to the graft.

Once the graft has begun to heal, patients typically are assigned to a program of physical therapy. The damaged area is likely to be stiff and sore and may require a carefully designed program of exercise in order to make it flexible and functional once again.

Research

The search for new and more effective methods of treating burns is an active field of medical research. One problem is that patients who have extensive burns may not have enough healthy skin to permit grafts on the damaged areas. Grafts from other individuals are generally not possible because of autoimmune responses in the recipients.

One of the approaches that has been developed is to use some form of "synthetic skin" for a graft. For example, an artificial skin made of collagen (a protein that occurs naturally in skin and other body parts) has been used in place of a patient's own skin to form the graft. In another approach, collagen impregnated with clusters of the patient's own skin cells has been used for grafts. The presence of the body's own cells may increase the rate and efficiency with which the graft is accepted by the body.

Research is aimed at finding ways to increase the rate of healing also. In 1998, for example, a group of researchers from China reported that they had had success in using hormone treatment with burn victims. Patients treated with a genetically engineered product, recombinant bovine basic fibroblast growth factor, experienced healing of their wounds significantly more rapidly than did those without the treatment.

Given the pain, potential for disfigurement, incapacitation and other serious long-term problems associated with burn injuries, it is not unreasonable to anticipate that research on burn treatments will continue to be a major field of study in the medical sciences for the foreseeable future.

Note

1. The treatment procedures described here are adapted from *The Merck Manual* (15th Edition), Section 254.

Further Reading: American Burn Association. "ABA Education Resource Manual," <http://www.ameriburn.org/pub/edres/tocedres.htm>, 12/21/98; Berkow, Robert, ed. *The Merck Manual*, 15th edition. Rahway, NJ: Merck & Co., 1987, pp. 2350–2359; Cook, Allan R, ed. *Burns Sourcebook: Basic Consumer Health Information about Various Types of Burns and Scalds, Including Flame, Heat, Cold, Electrical, Chemical, and So Forth*. Detroit: Omnigraphics, 1999; "Cool the Burn," <http://www.cooltheburn.com/learn/about/>, 12/21/98; Munster, Andrew M. *Severe Burns: A Family Guide to Medical and Emotional Recovery*. Baltimore: Johns Hopkins University Press, 1993; Podolsky, Doug M. *Skin: The Human Fabric*. Washington, DC: U.S. News Books, 1982, pp. 150–154; Wardrope, Jim, and June A. Edhouse. *The Management of Wounds and Burns*. New York: Oxford University Press, 1999.

C

CAMP FIRE BOYS AND GIRLS

The Camp Fire Boys and Girls is a coeducational organization providing a variety of programs for about 630,000 participants in forty-one states of the United States and the District of Columbia. It was founded in 1910 as Camp Fire, an organization designed for girls exclusively. The founders, Luther and Charlotte Gulick, believed that campfires were the first human communities, they asserted that young women who learned how to make campfires discovered how to control fire, which was the first step in developing community. In 1975, the organization rescinded its "girls-only" policy and began to accept boys also.

Today, the organization consists of about 46 percent males and 54 percent females organized into 125 councils. The organization's primary programs focus on self-reliance and service-learning classes, camping and environmental education, child care, clubs and mentoring opportunities, and leadership development.

National headquarters for the organization are at:

Camp Fire Boys and Girls
4601 Madison Avenue
Kansas City, MO 64112–1278
Tel: (816) 756–1950
Fax: (816) 756–0258
http://www.campfire.org/
e-mail: info@campfire.org

CAMPFIRE

A campfire (or camp fire) is a fire built out-of-doors, often for the purpose of cooking or providing warmth. Campfires have a long tradition in human history, especially among groups of nomadic people, such as Native Americans and American cowboys. Among such peoples, campfires served more than utilitarian functions. They provided an opportunity for individuals to bond with each other, share common experiences, and pass on their cultural heritage.

The use of campfires for food and warmth has declined significantly as the use of permanent houses has increased. Today, campfires are used primarily for recreational purposes, such as during camping or hiking expeditions. In such cases, campfires still serve many of the functions for which they were used by nomadic peoples.

Fire safety experts suggest a number of cautions in the construction and use of a campfire. For example, they point out that such fires should be built on nonflammable spaces, such as bare ground, on a rocky surface, or in a metal or stone grill. The area surrounding the campfire should be clear of combustible material to a radius of about 10 feet. Any tents in particular should be placed far enough from the fire to prevent sparks from the fire reaching them. The fire should never be left unattended and should be completely extinguished with water or dirt when breaking camp.

For people on hiking or camping trips, campfires still provide an opportunity to share stories, songs, and personal experiences. Indeed, the reason that many people take part in such outdoor experiences is often to experience the bonding provided by a campfire.

Although the utilitarian uses of campfires have decreased significantly, their symbolic uses have become much more widespread and popular. The term *campfire* has come to mean any setting in which groups of individuals can interact with each other verbally or through music. The Internet has provided a new and promising environment in which many different kinds of campfires have been able to develop.

As an example, The Global Campfire Home Page (http://www.indiana.edu/~eric_rec/fl/pcto/campfire.html) is a website containing a number of stories "in development." Anyone can read any one of the stories on the website and make his or her addition to the storyline. The webmaster says that he had three goals in creating the website: (1) to promote the use of the Internet to facilitate learning; (2) to promote language and keyboarding skills; and (3) to promote interaction between geographically separated schools.

Some Internet campfires are focused on specific topics, such as the Great Books of literature (http://killdevilhill.com/faulknerchat/live/chat.cgi); Native American stories, legends, and myths (http://www.geocities.com/RainForest/5292/stories.htm); and astronomical topics (http://killdevilhill.com/astronomychat/wwwboard.html).

CANDLE

A candle is a piece of wax or tallow containing a wick. When the wick is lit, the wax or tallow melts. The melted wax or tallow then burns at the tip of the wick to produce light.

History

The earliest candles of which we know were made by the Egyptians in about 3,000 B.C. These candles were often made by soaking the pith (central column) of a reed in animal fats. The pith served as the wick for such candles. The Egyptians also made the first beeswax candles. These candles were usually made in a cone shape with a wick running through the center of the cone.

Candles were also used widely in other early civilizations. Early Chinese candles, for example, were made from oil extracted from the seeds of the tallow tree. In India, the wax needed for candles was produced by boiling the bark of the cinnamon tree. Native Americans used a somewhat different process. They impaled a species of fish known as the *candlefish* on a long stick. The candlefish contains a high percentage of oil and can be burned to produce light.

Candles have often been made of other natural objects also. In Colonial America, for example, a branch from the tree known as *candlewood* (in New England) and *fatwood* or *lightwood* (in the South) was often used as a source of light. The branch was held in a simple iron holder and the pitch inside the wood was then ignited to produce light. One of the most widely used candles during this period was the *rushlight*. The rushlight consisted of a piece of meadow rush (*juncus effusus*) soaked in tallow and held in a simple iron holder.

One important function of candles, of course, is for illumination. For many hundreds of years, they were one of the few ways of providing light in dark places and at night. But candles took on religious and ceremonial functions early in human history. Hieroglyphic drawings found in Egyptian pyramids suggest that they were used during special events, such as funerals or weddings. Among ancient Greeks, candles were often used during funerals to help guide the dead on their journey to the next world.

By the Middle Ages, candlemaking had become an important industry. The craft was usually divided into two parts, one of which made only tallow candles and the other only beeswax candles. Beeswax candles were generally more desirable since they burned with a cleaner flame and had a more pleasant smell. In contrast, tallow candles were made from the rendered fat of slaughtered animals and were less desirable. Because of their cost, however, beeswax candles were generally available only to the wealthy.

A minor revolution in candlemaking occurred in Colonial America when women discovered that a sweet-smelling wax could be obtained by boiling bayberries. For a short time, bayberry candles became very popular as a source of lighting in Colonial homes. That situation changed, however, largely because of the laborious process required for the extraction of bayberry wax.

By the eighteenth century, yet another source of candle wax had been found: the spermaceti wax obtained from whales. Like beeswax, spermaceti wax burns cleanly and with relatively little odor. By the end of the century, the vast majority of candles made for everyday use were being produced from spermaceti wax.

The modern candlemaking industry reached its peak during the first half of the nineteenth century. The discovery of petroleum and methods for refining the new fuel produced an important by-product: paraffin. Candlemakers soon found that paraffin was a nearly ideal substance for the manufacture of their product. In 1834, the American inventor Joseph Morgan had produced the first automated device for the manufacture of candles.

The era of lighting by candle largely came to an end in 1879 when Thomas Alva Edison invented the electric light. Candles are still used on rare occasions, however, when electrical power is lost and for secular and religious ceremonial occasions.

Candle Flames

A candle flame is a deceptively simple phenomenon. Some authors of chemistry textbooks have made a study of the candle flame an introductory exercise in the laboratory portion of their books. They use the exercise because what appears to be an uncomplicated spot of light can actually be analyzed to yield a number of different physical and chemical changes.

The first step in producing a candle flame is to light the wick. When the wick is lighted, it begins to burn. Heat from the burning wick rather quickly causes wax at the top surface of the candle to melt. Melted candle wax then moves upward through the wick by means of capillary action. When the melted wax reaches the outer surface of the wick, it vaporizes. The vapor is then ignited by the burning wick. At this point, the wick itself ceases to burn. Its only function beyond this point is to serve as a conduit by which melted candle wax can reach the outer surface of the wick.

Combustion of the vaporized candle wax is a complex chemical and physical process. For example, the shape of the candle flame itself is affected by gravitational forces. The flame usually takes a pear-shaped form that we associate with most flames. In the absence of gravity, however, a candle flame has a perfectly spherical shape.

The tallow, wax, or other material of which a candle is made contains complex organic compounds consisting of carbon, hydrogen, and, sometimes, oxygen. For example, the chemical formula for cetyl palmitate, the primary component of spermaceti wax, is $C_{32}H_{64}O_2$, and paraffin contains a mixture of many solid hydrocarbons with formulas such

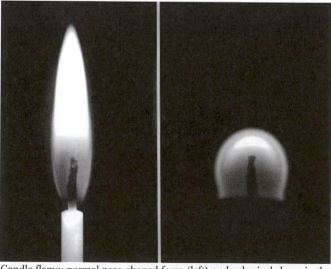

Candle flame: normal pear-shaped form (left) and spherical shape in the absence of gravity (right). *Courtesy of NASA Glen Research Photo Archives.*

as $C_{32}H_{66}$ and $C_{34}H_{70}$. When compounds such as these burn, they form a mixture of products, including carbon dioxide (CO_2), water (H_2O), and unburned element carbon (C). It is the unburned carbon that is responsible for the appearance of the candle flame itself. The unburned carbon exists in the form of minute particles, each only a few microns (thousandths of a meter) in size. Heat from the candle flame causes these particles to glow, giving off the characteristic yellow candle flame, although the particles do not themselves become hot enough to burn.

Further Reading: "Candle Information," <http://www.genwax.com/candle_instructions/>; "Candles," <http://www.ywh.com/Tips/Table/candles/candles-history.html>; Cooke, Lawrence S., ed. *Lighting in America: From Colonial Rushlights to Victorian Chandeliers*. New York: Main Street/Universe Books, 1975; "History of Candlemaking," <http://www.candles.org/history.htm>; "The Magick of Wicks and Wax Part I: The History of Candle Making," <http://www.triplemoon.com/wickswax.html>; Newman, Jon. *Candles*. San Diego: Thunder Bay Press, 2000.

CARBON

Carbon is a chemical element with atomic number 6 and atomic mass 12.011. Carbon is the sixth most abundant element in the universe, the fourth most common element in the Solar System, and the seventeenth most common element in the Earth's crust. Its abundance in the crust is estimated to be 200–300 parts per million. Compounds of carbon also occur in living organisms. About 18 percent of the human body consists of carbon.

Carbon occurs in nearly all of the compounds that make up living organisms, including proteins, carbohydrates, fats and oils, and nucleic acids. It also occurs in many rocks, primarily in the form of carbonates, such as calcium carbonate ($CaCO_3$; aragonite, calcite, chalk, limestone, marble, and travertine) and magnesium carbonate ($MgCO_3$; magnesite). It is also a constituent of carbon dioxide (CO_2), the fourth most abundant gas in the atmosphere.

Physical and Chemical Properties

Carbon is one of the most unusual and interesting chemical elements. It occurs in a number of allotropic forms. Allotropes are forms of an element with different physical and chemical properties. Two of carbon's allotropes are crystalline, diamond, and graphite. Diamond is the hardest natural substance known. It has a Mohs rating of 10. The Mohs scale is used to rank the hardness of materials and ranges from 0 to 10. By contrast, graphite is a very soft material, with a Mohs rating of about 2. It is so soft that it leaves a mark when rubbed on most materials. It is because of this property that graphite is used as the "lead" in lead pencils.

A number of amorphous (noncrystalline) forms of carbon are also known. These include coal, lampblack, charcoal, carbon black, and coke. In 1987, a new allotropic form of carbon was discovered. The molecules of which this allotrope consists are made of sixty carbon atoms joined to each other in a soccer-ball-like formation. These molecules were given the name of *buckminsterfullerenes* after the American architect Buckminster Fuller (1895–1983). Fuller used closed, spherical structures consisting of hexagons and heptagons in many of his architectural designs.

Carbon-Based Fuels

The vast majority of fuels used today for residential, industrial, or other applications contain carbon in one form or another. These fuels include wood, the fossil fuels (coal, petroleum, and natural gas), as well as many synthetic fuels.

Coal, for example, contains both elemental carbon and hydrocarbons, compounds of carbon and hydrogen. The relative proportion of elemental and combined carbon varies for different types of coal. Anthracite, or "hard" coal, consists of higher proportions of carbon and lower proportions of hydrocarbons than does bituminous, or "soft" coal. By contrast, petroleum and natural gas consist primarily of hydrocarbons. Wood also contains carbon compounds, although these are more complex in chemical structure than are the hydrocarbons.

Many thousands of years ago, humans learned how to make naturally occurring fuels, such as wood and coal, into more efficient fuels, such as charcoal and coke. Both of these fuels are made by the process known as *destructive distillation*. Destructive distillation is the process of heating wood or coal in the absence of air. When oxygen is not present, these fuels are not able to burn. Instead, they break down into simpler components, which are better fuels than the original materials.

The destructive distillation of wood, for example, produces charcoal, which is nearly pure carbon. The destructive distillation of soft coal produces coke, which is also nearly pure carbon. In the latter case, coal gas and coal tar are also produced. Coal gas consists of gaseous hydrocarbons and can also be used as a fuel. Coal tar is viscous black liquid consisting of a complex mixture of hydrocarbons. It is used for waterproofing, pipe coating, and insulation, although its most important use may be as a source for the raw materials from which plastics, drugs, dyes, and many other synthetic materials are produced.

Combustion of Carbon-Based Fuels

When carbon-containing fuels burn, they always produce carbon dioxide, almost always carbon monoxide, and, often, water. Pure carbon, of course, can yield the first two compounds only:

$$C + O_2 = CO_2 + heat$$

$$2C + O_2 = 2CO + heat$$

The relative amounts of the two gases formed depends on a number of factors, including the availability of oxygen during the combustion process.

In the case of hydrocarbons and more complex carbon-containing compounds, water is also formed. For example, the combustion of natural gas (methane; CH_4) produces carbon dioxide, carbon monoxide, and water vapor:

$$CH_4 + O_2 = CO_2 + CO + H_2O + heat$$

Again, the relative amounts of carbon dioxide and carbon monoxide formed depend on the amount of oxygen available, as well as other factors. *See also* CARBON DIOXIDE; CARBON MONOXIDE; COMBUSTION; FUELS.

Further Reading: Grayson, Martin, ed. *Kirk-Othmer Encyclopedia of Chemical Technology,* 3rd edition. New York: John Wiley, 1978. Volume 4, pp. 556–709; Greenwood, N.N., and A. Earnshaw. *Chemistry of the Elements.* Oxford: Pergamon Press, 1984, Chapter 8; Newton, David E. *Chemical Elements: From Carbon to Krypton.* Detroit: Gale Research, 1999. Volume 1, pp. 101–112; Radovic, Ljubisa R. *Chemistry and Physics of Carbon.* New York: Marcel Dekker, 2000; Sparrow, Giles. *Carbon (The Elements: Groups 1).* Buckinghamshire, England: Marshall Cavendish, 1999.

CARBON DIOXIDE

Carbon dioxide is a chemical compound with the formula CO_2. It is a clear, colorless, odorless gas with a density of 1.97 grams per liter. Its melting point is -78.5°C, at which temperature it sublimes directly from a solid to a gas. Solid carbon dioxide is also known as dry ice. Carbon dioxide is moderately soluble in water. At 0°C, 171 milliliters of the gas dissolve in one liter of water. At 20°C, its solubility drops to 88 milliliters per liter and, at 60°C, to 36 milliliters per liter of water.

Carbon dioxide is the fourth most abundant gas in the atmosphere, with a concentration of about 330 parts per million. The gas is derived from both natural and anthropogenic sources, primarily from combustion and decay of organic matter. Over the past two hundred years, the concentration of carbon dioxide in the atmosphere appears to have been increasing at a fairly steady rate, probably due to the combustion of fossil fuels by humans.

Carbon dioxide is formed when most fuels are burned. Most fuels contain carbon in elemental or combined form, as in fossil fuels (oil and natural gas). When such fuels are ignited, the carbon in the fuel combines with oxygen to form carbon dioxide:

$$C + O_2 = CO_2 + heat$$

or

$$C_xH_yO_z \text{ (as in wood)} + O_2 = CO_2 + H_2O + \text{other products} + heat$$

For example, in the case of natural gas (primarily methane, CH_4), the reaction is as follows:

$$CH_4 + 2O_2 = CO_2 + 2H_2O + heat$$

Carbon dioxide itself is noncombustible. Because of this fact, it is often used as a fire extinguishing agent. In some types of fire extinguishers, carbon dioxide is produced by the chemical reaction between two other chemicals and then released onto a fire. For example, the familiar metal-tank extinguisher once found in many homes and offices generated carbon dioxide by the reaction between sodium carbonate (Na_2CO_3) and hydrochloric acid (HCl):

$$Na_2CO_3 + 2HCl = 2NaCl + CO_2 + H_2O$$

Some fire extinguishers contain pure carbon dioxide gas under high pressure. When a valve on the extinguisher is opened, the gas is released and can be sprayed on a fire.

Carbon dioxide is generally regarded as a harmless by-product of the combustion of fuels. Since it occurs naturally in the atmosphere, its release causes no problems for plants and animals. In fact, its production can be considered to be a benefit since it is one of the raw materials (along with water) used by plants in the process of photosynthesis.

In recent decades, however, some concerns have been raised about the effects on the Earth's climate by the increasing amounts of carbon dioxide released by anthropogenic actions. Carbon dioxide molecules trap heat energy reflected from the Earth's surface, tending to cause an increase in the atmosphere's average annual temperature. Considerable disagreement exists as to the severity of this problem and the long-term effects it may have for life on the Earth. *See also* CARBON MONOXIDE; CLIMATE CHANGE; COMBUSTION; FUELS.

Further Reading: Budavari, Susan, ed. *The Merck Index*. Rahway, NJ: Merck & Company, 1989, p. 1820; Grayson, Martin, ed. *Kirk-Othmer Encyclopedia of Chemical Technology,* 3rd edition. New York: John Wiley, 1978. Volume 4, pp. 725–742; Lewis, Richard J., Sr., ed. *Hawley's Condensed Chemical Dictionary*, 12th edition. New York: Van Nostrand Reinhold Company, 1993, pp. 219–220.

CARBON MONOXIDE

Carbon monoxide is a chemical compound with the formula CO. It is a colorless gas with a very slight odor, almost impossible to recognize. It has a freezing point of -207°C and a boiling point of -190°C. It has a density of 1.250 grams per liter and a solubility of about 33 milliliters per liter in water at 0°C and about 23 milliliters per liter at 20°C.

Carbon monoxide is about the thirteenth most abundant gas in the Earth's atmosphere, with an abundance of about ten parts per billion. Its occurrence is low because it reacts readily with oxygen in the air to form carbon dioxide.

About the only natural source of carbon monoxide is the combustion of carbon-containing fuels, such as the fossil fuels (coal, oil, and natural gas). When such fuels burn, some carbon monoxide is often formed. The gas is produced because carbon present in the fuel does not have sufficient time to react completely with the oxygen around it. The process that occurs is known as *incomplete combustion:*

$$2C + O_2 = 2CO + heat$$

Carbon monoxide burns readily with a violet flame, producing carbon dioxide:

$$2CO + O_2 = 2CO_2 + heat$$

Carbon monoxide is produced routinely during many combustion processes used by humans. For example, the gas is a major constituent of the exhaust gases escaping from automobile, truck, airplane, and other types of engines. It is also a product of the combustion that occurs in home furnaces, space heaters, and other heating devices.

Carbon monoxide is an important raw material in many industrial operations. For example, it is used to remove oxygen from ores and impure forms of metals such as lead oxide (PbO) by reacting with oxygen present in the ore or metal. For example:

$$PbO + CO = Pb + CO_2$$

The compound is also used in the preparation of many important commercial products. For example, methanol (methyl alcohol; CH_3OH) can be made by combining carbon monoxide with hydrogen gas in the presence of a catalyst:

$$CO + 2H_2 = CH_3OH$$

Carbon monoxide poses an extreme health risk for humans and other animals. When taken into the body, it combines with hemoglobin molecules that carry oxygen through the circulatory system. It bonds with hemoglobin 200 times as efficiently as does oxygen, preventing the blood from carrying oxygen to cells where it is needed for anabolic and catabolic reactions. The gas presents a special threat because it can not be seen, smelled, or tasted except in very high concentrations. Moderate exposure to carbon monoxide gas can produce nausea, disorientation, and headaches. In larger doses, it can lead to loss of consciousness and death.
See also CARBON DIOXIDE; CLIMATE CHANGE; COMBUSTION; FURNACES.

Further Reading: Berkow, Robert, ed. *The Merck Manual of Diagnosis and Therapy*. Rahway, NJ: Merck & Company, 1987, pp. 2404–2405; Budavari, Susan, ed. *The Merck Index*. Rahway, NJ: Merck & Company, 1989, p. 1821; Grayson, Martin, ed. *Kirk-Othmer Encyclopedia of Chemical Technology*, 3rd edition. New York: John Wiley, 1978. Volume 4, pp. 772–793; Lewis, Richard J., Sr., ed. *Hawley's Condensed Chemical Dictionary*, 12th edition. New York: Van Nostrand Reinhold Company, 1993, p. 221.

CERAMICS

Ceramics are compounds or mixtures that are usually processed at room temperature and then heated to relatively high temperatures. The chemical and physical structures of ceramics may be very complex and include substances such as glass, cements, plaster, bricks, tiles, and porcelain. Ceramics are now divided into two general categories: traditional and modern ceramics. Most of the materials listed above are traditional ceramics. Modern ceramics include substances such as silicon nitride, silicon carbide, aluminum nitride, and many other materials that have been strengthened by the addition of zirconium, niobium, lead, lanthanum, and other metals.

History

Ceramics are among the oldest manufactured products known to humans. The earliest ceramics were probably made out of wet clay that had been shaped, dried in the sun, and then baked in an oven (or *kiln*). Some of the earliest pottery known was produced in about 25,000 B.C. in an area now known as the Czech Republic. It consists of clay animals that were fired by some means and then painted. The first pottery wheels and kilns date to about 3000 B.C. in the region of Mesopotamia. Superb examples of pottery remain from the pre-Christian era from regions throughout the world, including the Middle East, China, India, and Greece.

Properties

The properties of ceramic materials are in sharp contrast to those of metals, another group of important structural materials. Metals can generally be heated and shaped a number of times. By contrast, ceramics are typically worked and shaped at moderate temperatures and then "fired" (heated to a high temperature). Once they have been formed, they can not be reheated and reshaped. Indeed, they tend to maintain their integrity to very high temperatures, probably the major property that determines their practical applications. As an illustration of this property, the maximum temperature at which the best steel alloys can be made is about 1,000°C (1,800°F). By contrast, many ceramics can easily be used at temperatures as high as 1,500°C (2,700°F).

Processing

The production of ceramic materials typically involves five steps. Step one involves the selection of powders to be used in the product. Choice of powders is based on a number of criteria that determine the property of the final product—criteria such as particle size, purity of the powder, and reactivity of the material. Additives may also be mixed with the primary powder to improve characteristics such as strength and heat resistance.

In step two, generally known as *green forming*, the powder mixture is formed into some useful shape, such as a cup, a rod, or a plate. Many procedures are available for green forming, including pressing, casting, extrusion, and injection molding. In all of these procedures, the dry powder mixture, a slurry, or some other form of the powder is forced into some desired shape. For example, a slurry of the powder may simply be poured into a mold with the shape desired for the product.

Step three involves removal of the water or other liquid used in the original powder slurry. In the most primitive methods used in making ceramics, simply exposing the product to the sun would achieve this objective. Today, the product may simply be allowed to air dry, or it may be exposed to heat.

Step four involves densification (increasing the density) of the ceramic material. Pressure is applied to the material to remove air trapped between particles and make the product more compact. A common procedure used to achieve densification is sintering, in which heat and/or pressure may be applied to the ceramic material.

The final step in ceramic production is finishing, a process in which the product is polished, buffed, ground, or otherwise treated so as to give it exactly the correct dimensions and shape.

Applications

Traditional and modern ceramics have a very large number of applications. Some examples of the former are bricks, pottery, dishes, cements, glass, fiberglass, tile, concrete, plumbing materials, abrasives, and refractories. Modern ceramics are often subdivided into two categories, based on their applications: structural and electronic ceramics. Some uses of structural ceramics include components of engines, bearings, valves, cutting tools, corrosion-resistant parts, heat-exchange components, and biological implants. Electronic uses include sensors, superconductors, capacitors, transducers, magnets, waveguides, and display units.

For further information:
American Ceramic Society
P.O. Box 6136
Westerville, OH 43086–6136
Tel: (614) 890–4700
Fax: (614) 899–6109
http://www.ceramics.org/
e-mail: info@acers.org

Further Reading: Chavarria, Joaquim. *The Big Book of Ceramics: A Guide to the History, Materials, Equipment, and Techniques of Hand-Building, Molding, Throwing, Kiln-Firing, and Glazing.* New York: Watson-Guptill Publishing, 1994; Richerson, David W. *Modern Ceramic Engineering.* New York: Marcel Dekker, 1982.

CHARCOAL

Charcoal is the primary product formed when carbonaceous (carbon-containing) materials are heated in the absence of air. By far the most common material used to make charcoal is wood, in which case the product is known as *wood charcoal*. But other plant materials, such as nut shells, are also used in the manufacture of charcoal as are animal bones under some circumstances. In such cases, the charcoal-like product may be called a *char*.

From a chemical standpoint, the process of making charcoal is known as *destructive distillation*. During destructive distillation, wood or another raw material is heated to temperatures in the range of 500–600°C. Air must be excluded from the operation to prevent the raw material from catching fire. At these temperatures, volatile materials, such as water, are driven off, leaving behind nearly pure carbon. On average, charcoal contains about 85–90 percent carbon. Charcoal also tends to be much more porous than wood or other materials from which it is made.

The origins of charcoal-making are unknown. Evidence suggests that humans understood the process for making charcoal very early in history, at least a few thousand years before the birth of Christ. A number of passages from the Bible refer to a material used in making fires that is almost certainly charcoal.

One reason for the early discovery of charcoal is that the product forms naturally when carbonaceous materials are allowed to smolder in the absence of air. It required little imagination for early humans to notice and copy this process.

Probably the first method for producing charcoal was one that duplicates the natural process, the *pit kiln process*. In this process, wood is set afire in a shallow pit and then covered loosely with sand or dirt. After a period of time, the wood has been converted to charcoal. This process was later improved by constructing an aboveground structure in which to hold the smoldering wood and dirt covering. The structure might still be called a *pit* or, alternatively, a *pile,* or a *kiln.*

For many centuries, the primary use of charcoal was as a fuel. It burned more hotly and with fewer waste products than wood, the fuel it usually replaced. With the dawn of the Industrial Revolution, the demand for charcoal as a fuel in the manufacture of iron and steel skyrocketed. In some steel-making operations, up to half of the workforce was involved in the production of charcoal for the plant.

The production of charcoal began to drop during the second half of the eighteenth century. Two factors were responsible for this change. First, the raw material needed for charcoal production—trees—was rapidly being depleted. In some areas, whole forests had been cut down to supply the wood needed to make charcoal for industrial operations. It was not unusual for an average-sized steel mill to use all the trees on an acre of land to fuel its operations for a single day. As steel mills grew in size and number, they soon placed an enormous demand on the wood resources of surrounding areas.

Second, a superior fuel had been found to replace charcoal: coke. Coke is made by a process similar to that used in the production of charcoal: destructive distillation. In the case of coke, it is bituminous (soft) coal that is heated in the absence of air to produce the final product. Coke burns with an even hotter flame than charcoal. As the nineteenth century came to a close, coke largely replaced charcoal as the fuel of choice in the manufacture of steel and iron, and in many other metallurgical processes.

Charcoal is still used today in much smaller quantities for a variety of other applications. Its most important commercial application today is as a filtering medium. In a finely divided form (known as *activated charcoal*), charcoal has the ability to adsorb very large amounts of other materials onto its surface. Because of this property, it is used to remove undesirable colors and odors from processed foods. For example, solutions of raw sugar are run through charcoal-lined filters to remove the product's natural brownish color. Activated charcoal is also used for the same reason in gas masks, where it adsorbs toxic fumes. Probably one of the best known applications of charcoal today, although one of relatively modest economic significance, is in outdoor grills for barbecuing.

CHARIOTS OF FIRE. *See* BIBLICAL ALLUSIONS TO FIRE

THE CHEMICAL HISTORY OF A CANDLE

The Chemical History of a Candle is a book based on a series of lectures given in December of 1860 by Michael Faraday at London's Royal Institution. The lectures have become famous in the history of chemistry because of the clear and fascinating way in which Faraday uses a simple candle to illustrate so many fundamental principles in chemistry and physics. In fact, chemical educators of the second half of the twentieth century adopted Faraday's strategy and designed lessons for the new curricula arising out of the revolution in science education of the 1960s around the burning candle.

Faraday designed his Christmas Lectures of 1860 for children of the members of the Royal Institution because he thought they had little opportunity to use the resources of that great institution as did their parents. By all accounts, his lectures were received with great enthusiasm by his young audiences.

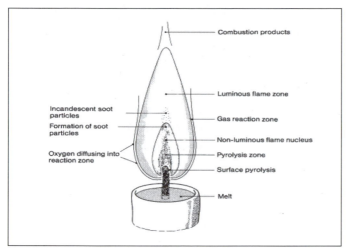

The mechanics of combustion as seen in a candle flame. *Courtesy of Troitzch, J.,* International Plastics Flammability Handbook, *2nd ed. Carl Hanser Publishers, Munich (1990).*

Faraday's presentation consisted of six lectures dealing with the structure and properties of a candle flame, the nature of combustion, water, hydrogen, oxygen, and carbon. The original lectures proved to be so successful that they became a regular offering of the Royal Institution. Faraday himself gave a total of nineteen other lectures on similar topics, such as chemical reactions, magnetism, electricity, and gravitation.

Faraday made no written record of his own lectures, but they were transcribed verbatim by a reporter and then prepared for publication by William Crookes, later to become one of the greatest scientists of the early twentieth centuries. It is Crookes' edition that eventually saw the light of day as *The Chemical History of a Candle*.

Original copies of the book are rare, but an inexpensive paperback edition was produced by The Viking Press in 1960.

Further Reading: Faraday, Michael. *The Chemical History of a Candle*, edited by William Crookes. New York: The Viking Press, 1960.

CHERNOBYL NUCLEAR REACTOR FIRE

The Chernobyl nuclear reactor fire broke out on 26 April 1986 in Unit 4 of the V. I. Lenin Atomic Power Station at Chernobyl, then in the USSR's Ukrainian Soviet Socialist Republic (now Ukraine). Ironically, the disaster occurred during a routine safety test of the reactor. In preparation for their tests, plant operators disconnected certain components of the plant's automatic safety system. When power production began to drop to levels too low for tests to continue, operators overreacted and increased the power level to unacceptably high levels. Safety systems designed to prevent the reactor from overheating at such levels were no longer in operation, and energy production in the reactor core went out of control. Two explosions occurred, the first of which destroyed the reactor core and blew the roof off the building. The second explosion was even more powerful than the first, sending lumps of burning graphite into the air.

The first fire crew to arrive on the scene consisted of three fire engines from the town of Chernobyl. The Chernobyl fire fighters quickly realized that they required substantial aid in fighting the plant fires and called for help from fire companies throughout the region of Kiev, the nearest large city. Eventually, more than 37 fire crews, 81 fire engines, and 186 firefighters responded to this call. Unfortunately, none of the fire fighters had had any training in dealing with fires involving radioactive materials, and they had none of the protective clothing, special breathing apparatus, or other equipment needed for such fires.

Firefighters were faced with two kinds of fires at the nuclear facility. The first was the burning core of the reactor itself. The core was made largely of graphite, a highly flammable material. After the second explosion, the graphite caught fire and began to burn at temperatures as high as 1,500°C (2,800°F). At the same time, burning pieces of graphite thrown out of the reactor core had fallen on other parts of Unit 4, on parts of the Unit 3 building, and on surrounding structures. Eventually, thirty separate fires developed.

Many of these fires started because the material used for roofing on reactor buildings was bitumen, a highly flammable petroleum product.

Firefighters were able to bring the thirty outside fires under control in about four and a half hours. They were able to use traditional firefighting techniques successfully on these fires. The burning reactor core presented a different challenge. Water from fire hoses was essentially useless against the intense heat of this fire. Instead, firefighters flew over the core in helicopters and dropped more than 5,000 tons of boron, lead, sand, dolomite, clay, and other materials in an attempt to smother the fire and absorb radiation released by the flames. The graphite core continued to burn for another twelve days before it was extinguished.

One of the worst aspects of the Chernobyl accident was the release of radioactive materials into the atmosphere. A huge plume of hot gases and radioactive debris was carried into the upper atmosphere, where it was carried northwest across the USSR's Belorussian Soviet Socialist Republic (now Belarus) and then across Europe. Increases in atmospheric radiation were detected as far away as Great Britain.

The precise health and environmental effects of the Chernobyl disaster will probably never be known. About thirty people were killed immediately, and perhaps 500 more died shortly after the explosion. Some Ukrainians claim that 150,000 died in the decade after the accident and another 55,000 became invalids. Other health authorities question the accuracy of these estimates. They say that those numbers may have been exaggerated in order to qualify for more international assistance. One statistic that is suggestive is the rate of thyroid cancer in children. In the decade prior to the Chernobyl accident, there were seven such cases in nearby Belarus. Between 1990 and 1996, more than 300 cases of thyroid cancer were reported.

The Chernobyl explosion also devastated the environment around the plant. Thousands of villages were abandoned and their inhabitants relocated to new homes. Whole herds of cattle were slaughtered and farmland was declared unusable because of high levels of radioactivity. More than ten years later, 20 of 21 farming districts in southern Belarus are still abandoned. The area within an 18-mile radius of the Chernobyl plant has been declared uninhabitable.

As soon as it was safe, a huge concrete and steel sarcophagus was built over the damaged power plant. The purpose of the sarcophagus was to prevent the further release of radioactive gases into the atmosphere. Topsoil around the plant was scooped off and buried in underground "graves," some of which were covered, and some not. Radioactive materials continue to leach out of the soil, drain into groundwater, and eventually work their way into rivers, streams, and lakes in the region.

The Chernobyl disaster remains an ongoing problem for the governments of Ukraine and Belarus. Both nations are still spending significant portions of their national budgets to pay for the long-term effects of the explosion. Most experts agree that the efforts made so far to deal with health and environmental problems resulting from the disaster are inadequate. But it is difficult to see how finances will be found to improve those efforts.

Further Reading: Cheney, Glen Alan, and Glenne Alan Cheney. *Chernobyl: The Ongoing Story of the World's Deadliest Nuclear Disaster*. New York: New Discovery Books, 1993; Chernousenko, V.M. *Chernobyl: Insight from the Inside*. New York: Springer Verlag, 1991; Medvedev, Zhores A. *The Legacy of Chernobyl*. New York: W. W. Norton, 1992; Yaroshinska, Alla et al. *Chernobyl: The Forbidden Truth*. Lincoln: University of Nebraska Press, 1995.

CHICAGO FIRE OF 1871. *See* THE GREAT CHICAGO FIRE OF 1871

CHIMNEY SAFETY INSTITUTE OF AMERICA

The Chimney Safety Institute of America is an organization that provides information on chimney, firewood, gas logs, and related issues. They provide referrals of Certified Chimney Sweeps who have been approved and certified by the Institute.

Chimney Safety Institute of America
8752 Robbins Road
Indianapolis, IN 46268
(800) 536–0118
e-mail: CSIA@CSIA.ORG

CHRISTIANITY AND THE BURNING OF HERETICS

The early Christian church was profoundly concerned about the issue of heresy. There was much debate as to what constituted true church dogma, and a number of councils were called to decide on many points regarding Church teachings. Any individual who rejected the decisions of these councils and of the popes was regarded as a heretic and was subject to exclusion from the "true" church.

According to ecclesiastical law, heretics were subject to a number of penalties and punishments, the most serious of which was excommunication. Under the Christian Roman emperors, however, the secular state became more involved in the punishment of heretics. Constantine I (280–337), Theodosius I (347–395), and Valentinian III (313–424), for example, all published edicts that assessed civil penalties for those convicted of heresy under ecclesiastical courts. The most severe of these penalties was first announced by Valentinian in 382.

The first use of burning to punish heretics appears to have been authorized in 1184 by the Synod of Verona. Later synods, including the Fourth Lateran Council in 1215 and the Synod of Toulouse in 1229 reinforced this policy and established the use of investigative committees in every parish to search out heretics. The use of burning as a final punishment for those convicted of heresy grew over the next 300 years, reaching a climax in the sixteenth and seventeenth centuries. Two of the most famous examples of the burning of heretics were the auto-da-fé and the witch-hunts.

Auto-da-fé

The *auto-da-fé* means, literally, an act of faith. It was a ceremony first used during the Inquisition that was instituted in the thirteenth century by the Roman Catholic Church. The purpose of the Inquisition was to seek out heretics and win them back to the holy faith or, if unsuccessful in that effort, to burn them to death. The auto-da-fé was often the last episode in a series of events that included the accusation, testing, judgment, and imprisonment of the guilty. Among those sentenced to suffer death by burning were Jews, Muslims, unbelievers, Protestants, gypsies, bigamists, and others regarded as evildoers by the inquisitors.

Over the centuries, the auto-da-fé, with its burning of heretics and others outside society became major social events in some parts of Europe. They drew very large crowds of people for whom the event was a break from a routine and otherwise dull existence. The auto-da-fé also had broader uses than the punishment of heretics. For example, Voltaire tells in Candide of an auto-da-fé held in Portugal to prevent earthquakes of the type that had destroyed much of the city of Lisbon. He wrote that "the sages of that country could think of no means more effectual to prevent the kingdom from utter ruin than to entertain the people with an auto-da-fé, it having been decided by the University of Coimbra, that the burning of a few people alive by a slow fire, and with a great ceremony, is an infallible preventative of earthquakes."

Witch Burning

Witch burning refers to the form of death prescribed for certain individuals convicted of being witches during the great witch-hunts of the sixteenth through eighteenth centuries.

Characteristics of Witches. The first witch-hunt took place in Switzerland in 1427. The movement gained force a few decades later with the publication of a "textbook" on the subject, *The Malleus Maleficarum (The Witch Hammer)*. The book was written by two Dominican monks, Heinrich Kramer and James Sprenger. It came two years after the same monks had obtained approval from

Pope Innocent VIII to search out and punish witches.

The Malleus Maleficarum outlined the methods by which one could recognize a witch, the kinds of activities in which they engaged, and the punishments that could be meted out to them. The guidelines provided by the book, and that developed in greater detail later, made it possible to accuse almost anyone of being a witch. Even young children were encouraged to be alert for witches and report them to the authorities. Women were especially likely to be identified as witches, especially those who pursued "unusual" occupations for women of the day. Midwives, herb gatherers, widows, and spinsters were especially likely to be accused of being witches.

Witches were greatly feared because they were thought to be responsible for causing disease and death among both humans and animals. They were also accused of bringing bad weather and causing natural disasters. They were thought to have these talents because of their allegiance to Satan, who gave them special abilities not possessed by ordinary people.

The Search for Witches. Witch-hunting reached its peak between about 1580 and 1660 and was most common in Germany, Austria, and Switzerland. The number of individuals accused of witchcraft is in serious dispute. One authority, Dr. Marija Gimbutas, of the University of California, has estimated that up to 9 million people were convicted of being witches and killed for the offense. Other authorities put the number at no more than a few tens of thousands.

The punishment for being convicted of witchcraft varied considerably from country to country. In Scotland, the usual procedure was to torture and then burn anyone convicted of being a witch. In England, torture was prohibited, and witches were typically executed by hanging. In Spain and Italy, witch-hunts were often carried out by officers of the Inquisition, but burning was rarely used as a method of execution.

The execution of witches, whether by burning or other means, was often an event of great community interest. People came from miles around to take part in an occasion that had the feeling of a carnival. The Church sometimes imposed a period of fast-

Burning of witches in a German marketplace c. sixteenth century. © *Bettmann/CORBIS*.

ing and religious rites before the actual execution took place.

The hysteria over witches spread to the New World after 1650, and about forty people were eventually convicted of the crime and executed. Half that number died in the Salem Witch Trials of 1692.

Further Reading: Baroja, Julio C. *The World of Witches*. Chicago: University of Chicago Press, 1964; Canetti, Elias. *Auto-da-Fé*. Translated from German. London: Johnathan Cape, 1946; *The Malleus Maleficarum*, <http://www.paganteahouse. com/malleus_maleficarum/>, 07/25/99; Monter, E.W., ed. *European Witchcraft*. New York: Alfred Knopf, 1969; Thomas, Keith. *Religion and the Decline of Magic*. Oxford: Oxford University Press, 1971.

CHURCH FIRES. *See* INTIMIDATION FIRES

CLEMENTS, FREDERIC EDWARD (1874–1945)

Frederic Clements is sometimes credited as providing the first comprehensive analysis of the role of fire in ecosystems. This work came in his 1910 study, *The Life History of Lodgepole Burn Forests* (USDA Forest Service Bulletin 79).

Clements was born in Lincoln, Nebraska on 16 September 1874. He entered the University of Nebraska at the age of sixteen and was awarded his Bachelor of Science degree four years later. He then stayed on at Nebraska to earn his M.A. in 1896 and Ph.D. in 1898. While still a student at Nebraska, he worked as part of a research team studying the vegetation of Nebraska. That work later led to his first book, *The Phytogeography of Nebraska* (1898), written with Roscoe Pound, later to become a renowned jurist.

Clements served on the faculty at Nebraska from 1897 to 1907, and then accepted an appointment as chair of the botany department at the University of Minnesota. He left Minnesota in 1917 to become a research associate at the Carnegie Institution of Washington, a post he held until his retirement in 1941.

Throughout his lifetime, Clements took a holistic view of ecology, arguing that plants, animals, and the physical environment had an intimate and inescapable interrelationship with each other. He was among the first ecologists to recognize that fire was one of the variables involved in this relationship, and no study of an ecosystem was complete without an analysis of the role that fire played in that ecosystem. *See also* FIRE ECOLOGY.

Further Reading: Hagen, Joel B., "Clements, Frederic Edward." In John A. Garraty and Mark C. Carnes, eds, *American National Biography*. New York: Oxford University Press, 1999, pp. 53–54.

CLIMATE CHANGE

The term *climate change* refers to modifications in long-term weather patterns that may occur as the result of a variety of factors. In recent times, it has been used frequently to describe changes that may be occurring as the result of human activities that result in the release of carbon dioxide and other so-called greenhouse gases into the atmosphere.

The term *global warming* is also used to describe such phenomena because the anticipated, long-term, overall planetary changes are likely to be a warmer atmosphere. However, this pattern is not expected to be manifested in exactly the same way—as a warming trend—in all parts of the planet. Therefore, climate change is often preferred as a way of expressing significant changes, of whatever type that may be, that may result as a consequence of the release of carbon dioxide and greenhouse gases into the atmosphere.

Historian of fire, Stephen J. Pyne, has suggested that recent concerns about climate change represent a whole new era in the relationship between fire and human culture. The possibility that the Earth's climate may be undergoing a significant change as the result of fossil fuel combustion has changed our focus on fire issues out of the specific and immediate—such as a particular forest fire or urban conflagration—to the planetary level. He writes that the possibility of climate change has "thrown into confusion the combustion calculus of the Earth. How much combustion can the Earth absorb?" He asks, "What is the relationship between fire and combustion? How much industrial combustion can be added, and how much wildland burning withdrawn, without ecological damage?" (Pyne 1997, 322–323).

Greenhouse Effect

Understanding the way in which climate changes may be occurring depends on an appreciation of the greenhouse effect that occurs in the Earth's atmosphere. The term *greenhouse effect* was coined because heating of the Earth's atmosphere occurs by a process similar to that by which air inside a greenhouse is heated. Energy from the Sun that reaches the Earth's atmosphere experiences one of three outcomes. Some of that energy is reflected off molecules in the atmosphere back into space. Another portion of the Sun's energy is absorbed by molecules in the atmosphere. Absorption of solar energy by molecules causes them to move faster and, therefore, to become warmer. Finally, some solar energy passes through the Earth's atmosphere without absorption to strike the Earth's surface.

The solar energy that reaches the Earth's surface may experience one of two fates. Some of that energy is absorbed by water, soil, plants, and other materials on the surface. The rest is reflected back into the Earth's atmosphere.

The process of reflection causes a change in the nature of the solar energy, however. A large fraction of the solar radiation that strikes the Earth's surface occurs in the form of ultraviolet (UV) radiation. Upon reflection, however, UV radiation returns to the atmosphere in the form of infrared (IR) radiation (heat). Some of the reflected IR radiation passes through the atmosphere and out into space. But some of it is absorbed by gaseous molecules in the atmosphere. One of the substances that absorbs infrared radiation most efficiently is carbon dioxide. From a simplistic standpoint, then, the more carbon dioxide in the Earth's atmosphere, the more IR radiation (heat) is absorbed.

Carbon dioxide is by far the most important but by no means the only substance capable of absorbing IR radiation. Other substances with the same capacity are known as greenhouse gases because they contribute to the warming effect that takes place in the Earth's atmosphere. Some other greenhouse gases include methane, nitrous oxide, chlorofluorocarbons (CFCs), and ozone. All of these gases except CFCs are produced by both natural and anthropogenic mechanisms.

Trends in Atmospheric Carbon Dioxide

Concern about climate changes are based fundamentally on the fact that humans have vastly increased the amount of carbon dioxide released to the atmosphere over the past 200 years. The dawn of the Industrial Revolution in the eighteenth century resulted in a rapid increase in the amount of wood, coal, oil, natural gas, and other fuels burned to operate machinery and transportation vehicles.

Scientists in general have been aware of this trend for many decades. However, concrete data to support this fact has become available only recently. Scientists have been measuring the amount of carbon dioxide in the atmosphere above an observatory on top of Mt. Mauna Loa in Hawaii since the late 1950s. The data show a very clear and steady increase in the level of carbon dioxide over the observatory. That level has increased from about 315 ppm (parts per million) in the late 1950s to more than 355 ppm in the late 1990s. The graph (known as the Keeling Curve), that shows this change has become one of the most famous scientific illustrations in history.

Data about the amount of carbon dioxide in the atmosphere is available from other sources also. For example, it is possible to extract ice cores from glacial ice sheets, such as those that cover Greenland Island. Analysis of air bubbles trapped in these ice cores show that levels of carbon dioxide in the atmosphere have been increasing since the late 1700s. This finding confirms the simple prediction that increasing use of fossil fuels on Earth corresponds to increasing levels of carbon dioxide in the atmosphere.

Implications of These Trends

What is the significance of an increased concentration of carbon dioxide in the atmosphere? One simplistic answer should be obvious from the description of the greenhouse effect given above. One might expect that an increase in the amount of carbon dioxide would result in an increase in the temperature of the Earth's atmosphere. That is, the planet should be getting warmer as the result of more carbon dioxide in the air.

Scientists have tried to estimate how much any given increase in carbon dioxide would affect the annual average global temperature. According to one calculation that is frequently cited, a doubling in the amount of carbon dioxide in the atmosphere might result in an increase of 1.5°C to 4.5°C in the planet's average annual temperature. This calculation is subject to considerable uncertainty, however, and immediately raises a number of other questions. For example, what changes could be expected in the Earth's hydrosphere (oceans, glaciers, lakes, and other bodies of liquid and frozen water)? Again, answering this question requires that one make a number of assumptions and some calculations involving high degrees of uncertainty. The most common estimate seems to be that an increase of 1°C in the Earth's annual average temperature would result in a rise of the sea level worldwide of about 25 cm (10 in). The increase in ocean volumes would result not only from the melting of glaciers and ice sheets, such as those at the Poles, but also from the expansion of water as it is warmed.

The consequences of a higher sea level would be more serious for some parts of the world than for others. Many of the world's largest cities, such as Tokyo, New York City, London, and Rio de Janeiro, lie on the ocean. As the sea level rises, significant parts of such cities would be expected to be swallowed up by the oceans. In the most extreme cases, some South Pacific islands, whose highest point is no more than a few meters above sea level, might be completely inundated by the rising oceans.

Another effect of a warmer atmosphere would be an increase in the rate of evaporation of water from the oceans, seas, lakes, and other bodies of water. With more water vapor in the air, one might expect an increase in the volume of clouds formed in the atmosphere. More clouds might, in turn, result in an increase of rainfall and more severe meteorological phenomena. These changes would almost certainly not occur uniformly across the planet, but would be more or less significant depending on topological, geographic, and other features. Ultimately, an increase in cloud cover might change dramatically agricultural patterns in various parts of the world. Regions that are now unsuitable for agriculture, such as the Sahara Desert, might become fertile areas, while currently prosperous agricultural regions might become arid or water-soaked.

Scholars who imagine scenarios such as these have called upon humans to begin thinking seriously about our current patterns of releasing ever-increasingly larger amounts of carbon dioxide into the atmosphere.

Contrary Opinions

Not everyone agrees with the scenario presented thus far. In fact, one can find reasons to disagree with this outline at almost every point. It is probably more appropriate to describe each of the steps above as being "almost certain," "highly probable," or "less probable." For example, almost no one disagrees with the observation that the amount of carbon dioxide in the Earth's atmosphere is increasing significantly. Not everyone agrees, however, that this increase will cause an increase in the Earth's annual average temperature, that sea levels will rise as a result, or that such changes will bring about significant climatic changes.

For example, no one is quite sure what the ultimate fate of carbon dioxide released to the atmosphere will be. Carbon dioxide is soluble in water and a certain fraction of the gas present in the atmosphere eventually dissolves in the oceans. To what extent and in what way will this phenomenon affect any future warming of the atmosphere? Also, an increase in the volume of clouds in the atmosphere might have a moderating effect on the climate. With more clouds in the air, a greater fraction of solar energy might be reflected back into space, causing a reduction in the temperature of the atmosphere.

The Role of Fire

Human use of fire lies, of course, at the root of the whole climate change controversy. On the one hand, our increasing dependence on combustion for heating homes and offices, generating electricity, operating cars and other forms of transportation, and powering all manner of industrial operations means that increasing amounts of carbon dioxide will be released into the atmosphere in the future. The only way to change that scenario, it would seem, is to cut back on the combustion of fossil fuels; that is, to adopt a somewhat simpler lifestyle.

Widespread forest fires are also an important factor in possible future changes in climate. Over the past few decades, increasingly large amounts of tropical forests have been destroyed by intentional burning. The purpose of these fires is usually to make available land for agricultural, commercial, industrial, residential, or other purposes. When tropical forests are burned, however, large amounts of carbon dioxide are released to the atmosphere. In addition, one of the most important natural sinks for carbon dioxide—plants that use carbon dioxide in photosynthesis—is diminished.

Policy Decisions

After three decades of debating this issue, very little action has been taken to deal with the possibilities of climate change. International conferences have been held at which most nations have appeared to accept the need for cutting back on carbon dioxide emissions. But so far, little concrete change has come about as a result of those conferences. Most political and industrial leaders seem to feel that the costs of cutting back on carbon dioxide emissions is too great for governments, the general public, and industry to accept. *See also* CARBON DIOXIDE; COMBUSTION; FUELS; INTERNAL COMBUSTION ENGINES.

Further Reading: Cushman, John H., Jr. "Industrial Groups Plan to Battle Climate Treaty," *New York Times*, 26 April 1998, A1; Gerstenzang, James.

"Campaign to Halt Global Warming Moves Glacially," *The Sunday Oregonian*, 14 June 1998, A6; Mahlman, J.D. "Uncertainties in Projections of Human-Caused Climate Warming," *Science*, 21 November 1997, 1416–1417; Manning, Anita. "'90s Contain Hottest Years of Hot Century," *USA Today*, 23 April 1998, D1; Miller, G. Tyler, Jr. *Environmental Science: Sustaining the Earth*. Belmont, CA: Wadsworth Publishing Company, 1991, Chapter 10; Newton, David E. *Global Warming: A Reference Handbook*. Santa Barbara, CA: ABC-CLIO, 1993; Pickering, Kevin T., and Lewis A. Owen. *An Introduction to Global Environmental Issues*. London: Routledge, 1994, Chapters 2 and 3; Pyne, Stephen J. *World Fire: The Culture of Fire on Earth*. Seattle: University of Washington Press, 1997; Schneider, Stephen. *Global Warming: Are We Entering the Greenhouse Century?* New York: Random House, 1989; Warrick, Joby. "5 Years Later, Leaders Gather for Second Earth Summit," *The Oregonian*, 23 June 1997, A4.

COAL COMBUSTION BY-PRODUCTS. *See* ASH

COCONUT GROVE NIGHTCLUB FIRE OF 1942

The Coconut Grove Nightclub fire occurred in Boston nearly a year to the day after the outbreak of World War II. The Coconut Grove was one of the largest and finest nightclubs in "nightclub row" in downtown Boston. The fire occurred on the evening of 28 November 1942 and resulted in the death of about 490 people. It was the largest fire, in terms of human fatalities, in American history since the Iroquois Theater fire in Chicago in 1903.

The Coconut Grove consisted of two levels, an upstairs dining room and a downstairs "Melody Lounge." On the night of the 28th it was packed to capacity with tired war workers and off-duty servicemen. Among the estimated 800–1,000 guests was film star Buck Jones and twenty-four of his Hollywood friends. Conspicuous by their absence were members of the Boston College football team. The team had planned to celebrate their anticipated victory over traditional rivals Holy Cross that evening. After suffering a humiliating 55–12 defeat, however, the Boston

College team had chosen to forego their plans, saving their lives in the process.

The fire started in the Melody Lounge at about 10:15 P.M. when a patron decided to unscrew a light bulb to cut down on the light in his seating area. The bartender noticed the patron's action and sent busboy Stanley Tomaszewski to replace the bulb. Unable to see what he was doing, Tomaszewski lit a match. The match immediately set fire to an artificial palm tree made of supposedly fireproof material. The fire quickly spread to silk draperies and the cloth-covered ceiling. Within minutes, it had spread to the upper-floor restaurant.

The Coconut Grove was nearly devoid of fire safety facilities. The main entrance was a revolving door, that almost immediately became jammed with bodies. Nine additional exit doors were available, but none carried a lighted sign and one was locked. In addition, all windows in the nightclub had been covered with heavy black fabric as an air-raid precaution. Finally, curious spectators outside the nightclub blocked access to the building, preventing firefighters and rescue workers from reaching those trapped inside.

The Boston fire occurred nearly forty years after the great Iroquois Theater fire in Chicago. That fire, probably the worst human-made fire in American history, had occurred because of contractors' refusal to incorporate fireproof systems that had already been developed and were widely available. These same omissions were responsible for the Coconut Grove fire.

For example, most of the electrical wiring in the building had been done without permit. Most of the materials used in decorating the club were either flammable or smoldered with the release of suffocating and toxic gases. The owners of the club carried no liability insurance on their building. In spite of a host of obvious code violations, the club had been inspected by a member of Boston's Fire Prevention Bureau only a week before the fire and had passed with flying colors.

One consequence of the Coconut Grove fire was a renewed concern about the implementation of fire codes both in Boston and throughout the United States. As the fire demonstrated, however, the presence of codes is never in and of itself an adequate measure of protection when professional inspectors do not see that they are enforced.

Further Reading: Benzaquin, Paul. *Fire in Boston's Coconut Grove: Holocaust*. Boston: Branden Press, 1967; DeVoto, Bernard. "The Easy Chair," *Harper's*, February 1943, 333–337; Keyes, Edward. *Coconut Grove*. New York: Atheneum, 1984; Veltfort, Helene Rank, and George E. Lee, "The Coconut Grove Fire: A Study in Scapegoating." *Journal of Abnormal and Social Psychology* 38 (1943): 138–154.

COKE

Coke is the primary product formed when coal or some related natural material is heated in the absence of air. By far the most common type of coke is that produced from bituminous (soft) coal. Some coke is also made from coal tar (*pitch coke*) and some from petroleum (*petroleum coke*).

The process by which coke is made is called *destructive distillation*. Destructive distillation is the process by which the raw material, such as soft coal, is heated in the absence of air. During this process, volatile materials, such as water and most hydrocarbons, are driven off. A very pure form of carbon remains behind as the major product. In addition, small amounts of ash and other products may be mixed with the coke.

Various forms of coke can be made depending on the temperature used in the heating process. At temperatures of less than about 750°C, a relatively impure form of carbon is produced. This form of coke has limited use, primarily as a residential fuel in Great Britain. Medium-temperature coke is produced when soft coal is heated to temperatures of 750–900°C, and high-temperature coke is produced at temperatures of about 900°C. One of the purest forms of coke is *calcined coke*, produced by heating petroleum coke to temperatures in excess of 1,300°C.

Coke was first manufactured in England at the close of the sixteenth century. The process used was similar to that employed in the manufacture of charcoal. Piles of bituminous coal were set afire and then covered with a thin layer of soil to prevent air from reaching the fuel. After a period of time, the coal was converted to coke. At the time, the process was too inefficient to make coke competitive with charcoal. As supplies of wood were depleted during the nineteenth century, however, coke became a more attractive alternative to charcoal. By the beginning of the twentieth century, coke had largely replaced charcoal for nearly all commercial and industrial applications.

Coke is produced today by two methods: the beehive-oven process and in by-product ovens. The beehive-oven process operates on essentially the same principle as that used four centuries ago. Soft coal is piled into a large, beehive-shaped oven. The lower layer is then set afire, and air is excluded from the oven. As the lower layer burns, it provides sufficient heat to convert the remaining coal to coke. In the by-product oven, heat from an external source beneath the closed oven is used to bring about the same kind of change.

By far the most important use of coke today is in the production of iron and steel and in other metallurgical operations. About 90 percent of all coke is now used in blast furnaces for the production of unrefined pig iron. The coke serves two functions in this process. First, its combustion provides the heat needed to bring about the chemical conversion of iron ore to elemental iron. Second, as coke burns, it forms carbon monoxide, which acts as a reducing agent that converts iron oxide to elemental iron:

$$2C + O_2 = 2CO$$

$$3CO + 2Fe_2O_3 = 3CO_2 + 4Fe$$

Very small amounts of coke are also used in foundries, in the manufacture of water gas, and in other industrial operations.

Further Reading: Anderson, Nils, and Mark W. Delawyer. *Chemicals, Metals and Men: Gas, Chemicals and Coke: A Bird's Eye View of the Materials that Make the World Go Around*. New York: Vantage Press, 1995.

COMBUSTION

Combustion is the scientific term for the process more commonly known as *burning*.[1] Combustion is one of the first chemical processes to be used by humans. The discovery of methods by which combustion can be controlled to cook foods, warm shelters, and bring about chemical changes was an important breakthrough in the history of the human species.

Scientific Definition

Today, combustion is defined more precisely as an oxidation reaction in which noticeable heat and light are produced. Such reactions are known as *exothermic* ("heat out") reactions.

Oxidation has two meanings to the chemist. The simpler definition of oxidation refers to any process in which some material reacts with oxygen. For example, when charcoal is burned, the carbon of which charcoal is made combines with oxygen to produce carbon dioxide. Energy is released in this reaction and appears as heat and light.

According to a more scientific definition, oxidation is the loss and gain of electrons between two substances. An example of this form of oxidation can be seen when iron metal is dropped into a container of chloride gas. The iron and chlorine react with each other vigorously, giving off heat, light, sparks, and smoke. The reaction looks like the vigorous burning of wood or charcoal. In addition, the chemical change that occurs during the process has certain fundamental similarities to that of the oxidation of charcoal. In the case of the combustion of charcoal, carbon atoms give up electrons to oxygen atoms. Carbon is said to be oxidized because it loses electrons. In the reaction between iron and chlorine, iron atoms give up electrons to chlorine atoms. Iron is said to be oxidized because it loses electrons.

Necessary Conditions

In order for combustion to occur, three requirements must be met:

1. There must be a *fuel*, a substance that will undergo oxidation. Some typical fuels are wood, paper, coal, petroleum, and natural gas. But many other substances can behave as fuels also. The power needed to lift rockets into the atmosphere, for example, often uses hydrogen or hydrazine as a fuel.

2. An oxidizing agent must be present to cause the fuel to burn. The most familiar oxidizing agent is oxygen itself. But many other oxidizing agents are known. In the reaction between iron and chlorine above, for example, chlorine is the oxidizing agent because it causes the combustion of iron.

3. The fuel and oxidizing agent must be brought to some minimum temperature at which oxidation (combustion) will occur. That temperature is known as the *ignition point* for that material. Ignition points vary dramatically for various substances. Some substances are so difficult to ignite that they have no clearly defined ignition point. Nylon and rubber are examples of such materials.

The ignition points for fuels vary considerably because of the many forms in which they occur. Examples of approximate ignition points of some fuels that are relatively pure chemically are the following:

coke or charcoal: 400°C

methane: 630°C

hydrogen: 580°C

carbon monoxide: 610°C

carbon disulfide: 100°C

phenol: 715°C

History

The earliest scientific attempt to explain combustion was offered by the Flemish physician Johann Baptista van Helmont (1580–1644). Van Helmont based his explanation on some obvious characteristics of a combustion reaction. He noted that a burning substance appears to release something (smoke and/or heat) to the air. He suggested that the substance must contain something that escapes into the air when it is burned. He called that substance *spiritus silvestre*, or "wild spirit."

Van Helmont's ideas were later reworked by the German chemist Johann Joachim Becher (1635–1682) and his student, Georg Ernst Stahl (1660–1734). Becher and Stahl argued that combustible materials contain a gas-like material that escapes into the air upon combustion. Becher called the material *terra pinguis*, or "fatty earth," while Stahl adopted the term *phlogiston* (from the Greek meaning "to set on fire") for the material. For well over a century, the phlogiston theory was widely accepted among scientists as the proper explanation for the process of combustion.

The phlogiston theory was a satisfactory explanation for combustion in many ways. But it had one serious drawback: It did not make sense quantitatively. When many objects burn, they weigh more than they did before combustion. If something (phlogiston) is *lost* during combustion, how can the object *gain* weight?

The resolution of this dilemma was provided in the late 1770s by the French chemist Antoine Laurent Lavoisier (1743–1794), often called the "Father of Modern Chemistry." Lavoisier's analysis of combustion was based on the discovery of oxygen by the Swedish chemist Karl Wilhelm Scheele (1742–1786) and the English chemist and physicist Joseph Priestley (1733–1804) in the early 1770s. Lavoisier realized that oxygen played a crucial role in the combustion process. He suggested that combustion occurs when a material combines or reacts with oxygen gas. This explanation of combustion is essentially the one that chemists accept today. Lavoisier presented his ideas so clearly and so convincingly that his "oxidation theory" of combustion soon replaced the older phlogiston theory.

Combustion Mechanics

Today, chemists have a very clear idea about the physical and chemical changes that take place during combustion. Essentially, that process involves the breaking of chemical bonds between the original combustible material and oxygen and the formation of new chemical bonds in the products of this reaction. For example, imagine that a piece of charcoal is ignited. The heat of the match used to start the burning process is sufficient to break chemical bonds between carbon atoms that make up the charcoal. At the same time, the chemical bonds that hold oxygen molecules together begin to break. An illustrative, if chemically inaccurate, way to depict these processes is as follows:

$$-C-C-C-C-C-C-C-C-C- \text{ —heat energy—} = -C- + -C-C- + -C- \text{ (etc.)}$$

$$O-O \text{—heat energy—} = O + O$$

The breaking of these bonds in the fuel and in oxygen itself releases heat energy. That heat energy is then used to break more chemical bonds in wood and oxygen molecules.

As bonds break in both fuel and oxygen, the fragments of molecules formed are free to recombine with each other in new ways. For example, free oxygen atoms formed by the destruction of oxygen molecules may combine with free carbon atoms from charcoal to make carbon dioxide:

$$-C- + O + O = CO_2$$

In such cases, the total amount of energy in the new chemical bonds (such as those that hold CO_2) together is *less* than the total amount of energy in the original chemical bonds (such as those that hold carbon together in charcoal and oxygen together in oxygen molecules). The excess energy is released in the form of heat and light, the evidence we associate with the burning process.

Applications

The earliest use of combustion by humans for cooking foods and heating shelters is lost in antiquity. Archaeologists have found evidence that primitive humans may have used fire as early as 1.5 million years ago in areas that are now part of South Africa and Kenya. But they are uncertain as to whether those fires were wildfires or built by humans. The oldest date for building and using fire about which all archaeologists appear to agree is about 230,000 years ago in a site called Terra Amata along the French Riviera.

The use of fire for metallurgy occurred much later in history. The first human-made metal products formed with fire date to about the fifth millennium B.C. in India. The use of fire for glassmaking and other ceramic arts can be traced to about 25,000 years ago in an area now part of the Czech Republic.

Arguably the greatest single breakthrough in the use of combustion for industrial purposes occurred during the eighteenth century in the period now known as the Industrial Revolution. The Industrial Revolution was made possible by the ingenuity of a number of inventors who discovered how to make productive use of the energy released during combustion. The most important of the devices invented during this period was the steam engine.

In the steam engine, a combustible material, such as wood or coal, is burned in order to boil water. Steam formed from the boiling water can then be used to move a piston or drive machinery. By the mid-nineteenth century, steam-powered engines were being used to mine coal, operate threshing and other agricultural machines, clean cotton, drive trains and boats, and perform countless other tasks. The world of work was revolutionized as steam engines took over many of the jobs that had previously been performed by human or animal power.

The full realization of the Industrial Revolution depended, however, on the conversion of combustion engines from wood and coal to gasoline and its chemical relatives in the internal combustion engine. Today, one can argue that modern society really operates on the power produced by the combustion of hydrocarbon fuels such as gasoline, diesel oil, and kerosene.

Environmental Issues

No one can adequately gauge the benefits that combustion has provided human society. On the other hand, the extended use of combustion-based devices has also produced its share of problems for the world. For example, the combustion of coal produces not only the heat needed to operate a variety of machines, but also pollutants, such as carbon monoxide, sulfur dioxide, oxides of nitrogen, and particulates. These pollutants became social problems within a matter of years after the discovery of the steam engine, and they have grown in severity as human use of combustion reactions has grown and expanded. Today, the nations of the world face the constant dilemma as to how they will be able to improve their standards of living by expanding their use of combustion-based machines while maintaining an environment congenial to human safety and happiness. *See also* BECHER, JOHANN JOACHIM; FLASH POINT; FUELS; INTERNAL COMBUSTION ENGINES; LAVOISIER, ANTOINE LAURENT; PHLOGISTON THEORY; PYROPHORIC MATERIAL; SPONTANEOUS COMBUSTION; STAHL, GEORG ERNST; STEAM ENGINES.

Note

1. Many introductory high school and college chemistry textbooks provide a general introduction to the topic of combustion.

Further Reading: Bradley, John N. *Flame and Combustion Phenomena*. London: Methuen, 1969; "Combustion." In Bridget Travers, ed. *Gale Encyclopedia of Science*. Detroit: Gale Research, 1997, volume 2, pp. 885–887; Combustion and Fire." In *Macmillan Encyclopedia of Chemistry*, New York: The Macmillan Company, 1977, volume 2, pp. 358–387; Glassman, Irvin. *Combustion*. New York: Academic Press, 1996; Turns, Stephen R. *An Introduction to Combustion: Concepts and Applications*. New York: McGraw-Hill, 1995.

CONGRESSIONAL FIRE SERVICES INSTITUTE

The Congressional Fire Services Institute (CFSI) is a nonprofit, nonpartisan policy institute organized in 1989 to educate members of Congress on issues of fire and fire safety. The institute's motto is: "So That Fire Responders Never Stand Alone." In order to achieve that goal, CFSI works with 45 national fire service organizations to keep issues of fire and fire safety constantly before the attention of the Congress. These organizations include firefighting departments, manufacturers of fire safety equipment, emergency service organizations, and other agencies involved in firefighting and fire prevention.

CFSI is especially concerned with issues such as threats of terrorism, aging equipment, and reduced funding for firefighting systems. Another function of CFSI is to keep individual and group members aware of actions taken by the Congress on fire fighting issues.

Each year, CFSI sponsors an Annual National Fire and Emergency Services Dinner, held in Washington, DC, and awards recognition to a Legislator of the Year, Fire Service Organization of the Year, and Mason Lankford Fire Service Leadership Award, cosponsored by the Institute and Motorola.

Congressional Fire Services Institute
900 Second Street, N.E., Suite 303
Washington, D.C.20002
(202) 371–1277
http://www.cfsi.org/
e-mail: info@cfsi.org

CONTROLLED BURN. *See* PRESCRIBED BURN

COOKING

Cooking is the process of preparing food for consumption by heating. Food can be heated in a number of ways: by placing it directly over a fire, as in barbecuing a piece of meat over an open flame; by warming it in a closed space, as in roasting meat in an oven; by exposing it to a hot vapor, as in steaming vegetables; by exposing the food to a hot surface, as in frying potatoes; or by placing it in a warm or hot liquid, such as boiling a chicken in a pot. All forms of cooking change the physical, chemical, and biological properties of food.

History of Cooking

No one knows when humans first started cooking. Some authorities have hypothesized the circumstances under which this discovery might have been made. Even before early humans knew how to make fire, they witnessed wildfires started by lightning or other natural forces. In these wildfires, some animals would be killed and plants would be cooked. Dutch scholar Johan Goudsblom describes such a scenario in his book *Fire and Civilization*:

> At the outbreak of the fire they [early humans] would have seen game fleeing. Later they would have basked in the glow of the dying embers, and picked up partly charred animals and fruit from the ashes, savouring them. In this manner they would have learned to appreciate the advantages of broiling and roasting . . .

Some authorities believe that the first evidence of cooking by humans can be found in caves near the city of Zhoukoudian (Choukoutien), China. It was in these caves that the famous fossil known as *Peking Man* was discovered in 1927. Among the many human fossils uncovered in the Zhoukoudian caves were roasted bones that could be construed to have been formed during the roasting of animals. Those bones are thought to be about 400,000 years old. Other authorities question this claim. They point out that the roasted bones from Zhoukoudian may have come from animals killed in wildfires. Many archaeologists believe that humans may not have begun cooking until about 380,000 to 465,000 years ago. The cooked remains of a rhinoceros found at the Menez-Dregan archaeological site in southern France from this period are taken by many researchers as the first evidence for cooking by humans.

The earliest methods used for cooking are also unknown, but not difficult to imagine. Perhaps the earliest cooking technique of all was simply to hold a plant or piece of meat over a fire, allowing the flames to cook the material directly. The modern process of barbecuing is essentially the same process, indicating that some cooking techniques remain largely unchanged over thousands of years.

Baking foods inside an oven was probably another early technique. A common method was to heat the inside of a stone oven with burning wood or charcoal. The hot ashes from this fire were then removed and the food to be cooked would be placed inside the oven. This method is still used in some parts of the world where modern ovens are not available.

The kitchen oven found in nearly every home today is only a more sophisticated version of this technique.

Boiling foods may have developed later, since devices for holding boiling water had to be invented first. The first devices for holding boiling water were woven baskets and ceramic pots. Cooking in such devices is generally more difficult than cooking over an open campfire or in an oven, however, and boiling did not become widely popular as a method of cooking until the invention of metal pots and pans.

Benefits of Cooking

Cooking is an activity with many benefits for individuals and groups. Some benefits are largely aesthetic. Most people are familiar with the sensuous pleasures aroused by the sight and smell of a cooked meal. Reading cooking magazines and watching cooking programs on television are popular activities, not just because people learn more about nutrition, but also because they are appealing experiences. Also, the preparation of food and sharing of a meal have long been important community activities which help families and groups bond to each other. Thanksgiving dinner, Sunday meals, and anniversary parties are as much social activities as they are opportunities to nourish the body.

Cooking brings about many beneficial physical and chemical changes in foods. For example, rice, wheat, corn, and other grains occur naturally in the form of individual particles surrounded by a hard cellulose coating. If these particles are eaten raw, the cellulose coating protects them from being digested in the human body. Cooking makes the particles swell, however, causing the cellulose coating to burst. The exposed material within the particle can then be digested more easily.

Perhaps the most beneficial effect of cooking from a physiological standpoint is the predigestion that it makes possible. Most nutrients (carbohydrates, proteins, and lipids) consist of large, complex molecules. When these molecules are ingested, enzymes in the human digestive system begin to break them down. In some cases, digestion does not occur quickly enough for the nutrients to enter the blood stream. They are excreted in a partially undigested form. Cooking begins the process of digestion before the nutrients enter the digestive system.

As an example, a protein molecule consists of a number of subunits known as amino acids joined to each other by chemical bonds. If A_1, A_2, A_3, and so on, represent different amino acids, then a protein molecule can be represented as:

$$[-A_1-A_5-A_7-A_2-A_2-A_5-]_n$$

with the subscript n indicating that this chain continues to a very great length.

The human digestive system contains enzymes that break the bonds between amino acids, thereby reducing a single protein molecule into smaller and smaller subunits. For example:

$$A_1-A_5-A_7-A_2 + \text{enzyme} = A_1-A_5-A_7 + A_2-A_2-A_5$$

But heating a protein molecule achieves the same results. Thus, cooked meat contains protein molecules that are already partially digested.

Food Preservation

As humans engaged in cooking, they discovered a subsidiary benefit: food preservation. In the first place, high heat kills many of the bacteria that are responsible for food spoilage. As a result, cooked foods tend to take longer to decay than raw foods. In the second place, certain aspects of the cooking process may contribute to the preservation of foods. Humans learned early on that smoke from burning wood or charcoal is an excellent preservative. Components of the smoke itself, we now know, kill bacteria that tend to promote decay. Even today, smoking is a widely used and effective method of food preservation.

Dehydration is another method of food preservation that makes use of heat. When foods are heated, the water they contain tends to be driven off as water vapor or steam. This process helps decrease the rate of decay because most microorganisms require water in

order to survive and remain active. As foods dry out, they are less likely to be affected by such organisms.

Disadvantages of Cooking

Cooking also has some deleterious effects on foods. Some vitamins are destroyed by cooking. For example, vegetables are a good source of vitamin A, needed for good eyesight. But cooking destroys vitamin A. So vegetables that have been cooked have lost an important part of their nutritional value.

Cooking may also result in the formation of harmful compounds. One of the most thoroughly studied cases involves the broiling, grilling, frying, or roasting of meat. When meat is heated, researchers have found, some of the creatinine in muscle tissue is converted to a group of compounds known as heterocyclic amines. Heterocyclic amines are known mutagens and carcinogens. They have been implicated in the development of breast and other forms of cancer. According to one research study, people who eat beef at least once a day are at twice the risk for stomach cancer as those who had it only once a week ("For Cancer Risk").

Further Reading: Barham, P. The Science of Cooking. New York: Springer Verlag, 2000; "Beyond Vegetarianism," <http://www.beyondveg.com/nicholson-w/hb/hb-interview2c.shtml>, 07/27/99; Derry, T.K., and Tevor I. Williams. *A Short History of Technology from the Earliest Times to A.D. 1900*. New York: Dover Publications, 1993, passim; "For Cancer Risk, Rare Steak Better than Well Done," <http://usatoday/com/life/health/lhs476.htm>; Francis, F.J., ed. *Encyclopedia of Food Science and Technology*, 2nd edition. New York: John Wiley, 2000; Goudsblom, Johan. *Fire and Civilization*. London: Allen Lane, 1992, Chapters 1–3, passim; Rossotti, Hazel. *Fire*. Oxford: Oxford University Press, 1993, Chapter 6; Scholliers, Peter, ed. *Food, Drink and Identity: Cooking, Eating and Drinking in Europe since the Middle Ages*. Oxford, UK: Berg Publishing, 2001; Selinger, Ben. *Chemistry in the Marketplace*, 4th edition. Sydney: Harcourt Brace Jovanovich, 1988, Chapter 3.

CREMATION

Cremation is a method of disposing of a dead body. The practice has been followed for many thousands of years by many different cultures. Over the past few decades, there has been an increased interest in the use of cremation rather than burial in the United States, Canada, and other parts of the world. Today, about one-fifth of all deaths in the United States and about one-third of all deaths in Canada are followed by cremation. The rate of cremations varies widely in both nations, from a low of less than 10 percent in most southern states and Atlantic provinces to nearly 50 percent in most western states and over 60 percent in British Columbia.

Cremation rates vary widely worldwide, depending on a number of factors, including religious limitations, custom, and land area available for burial. For example, in England and Japan, cremation rates may be as high as 90 percent because of the lack of land that can be used for burial.

History

Historians believe that humans first started using cremation during the early Stone Age, around 3,000 B.C. probably in the Near East and/or Europe. The practice seems to have spread across Europe and, by about 800 B.C., it had become the most common method of disposing of human bodies in Greece. The practice then appears to have spread to Rome, where it eventually became so prevalent that an imperial decree was issued prohibiting the use of cremation within the city itself. With the fall of the Roman Empire, cremation appears to have become less popular, at least partly because of prohibitions against the practice in the Christian religion. Between about the fifth and twentieth centuries, burial was by far the most popular form of disposing of bodies after death in the Western world.

Interest in cremation began to grow in Great Britain and the United States after 1873 when the first modern, efficient cremating chamber was invented by the Italian, Brunetti. The first crematory was opened in the United States in Washington, Pennsylvania in 1876. By 1900, the practice of cremation had spread throughout the United States and much of Europe. The Cremation Association of America was founded in 1913 and later changed its name, in 1975, to the Cremation Association of North America.

Reasons for Cremation

Although cremation is practiced in some regions for religious reasons, this method of disposing of a body is usually chosen for strictly personal reasons. An individual might, for example, have some personal objection to burying a dead loved one in the ground. Many people find solace, also, in the practice of spreading a dead person's ashes over a sacred or familiar spot, or a location with special meaning.

Process

Cremation is performed in a large oven, called a retort, designed especially for the purpose. The oven is large enough to hold the body enclosed within a casket or some other container. The temperature in the oven is raised to 1,600°F, at which point, all body parts ex-

A Hindu funeral pyre on the River Ganges, India. Cremation is used because, for Indian Hindus, the body is an empty shell after the soul departs.© *Bojan Breceli/CORBIS.*

cept bones are incinerated. The incineration process takes about two hours. The remaining bones are then processed mechanically to reduce them to fine particles. The total time required for this step is another three hours. On average, the remains from this process weigh between four and eight pounds, which are then transferred to an urn given to relatives or loved ones for final disposal. *See also* FUNERAL PYRE; NAZI DEATH CAMPS.

Further Reading: Cremation, <http://www.nfda.org/resources/marketplace/brochures/cremation.html>, 07/15/99; Cremation Information, <http://www.cremation-tx.com/crematio.htm>, 07/15/99; Goudsblom, Johan. *Fire and Civilization*. London: Allen Lane, 1992, pp. 101–102; Phipps, William E. *Cremation Concerns*. Springfield, IL: Charles C. Thomas Publishing, 1989; Prothero, Stephen. *Purified by Fire: A History of Cremation in America*. Berkeley: University of California Press, 2000.

CUYAHOGA RIVER FIRES

The Cuyahoga River fires were a series of fires that occurred on the Cuyahoga River between 1959 and 1969. They took place because the river had become so fouled with flammable industrial pollutants that the surface of the river itself was actually set ablaze.

The Cuyahoga River is a 166-kilometer (103-mile) long waterway that rises near the town of Huntsburg in eastern Ohio. It then flows southwest through Akron, and then northward through Cleveland and into Lake Erie. It passes through the Youngstown-Akron-Cleveland area that has traditionally been one of the most industrialized regions in the Midwest.

As industries grew up along the river during the twentieth century, they used it as a waste disposal reservoir. They dumped oil, organic wastes and by-products, and other toxic and flammable liquids directly into the river. The volume of wastes was so large that the river's natural flow was not able to dilute and discharge them into the lake rapidly enough. These wastes began to collect on the surface of the river's water and, in November 1959 for the first time, they caught fire. Local firefighting teams were unfamiliar with the task of putting out fires on the river. The more they sprayed water from their fire hoses on the fire, the more they spread the flammable liquids and the fire itself.

At first, local residents regarded the river fires with bemusement and even some amount of pride. The huge accumulation of industrial wastes were, after all, visible evidence of a hugely successful economic growth in the area. A November 1959 article in the business magazine *Fortune* seemed to put just this slant on the event.

By the 1960s, pollution of the Cuyahoga had gotten out of control. Industrial plants were dumping up to 155 tons of waste per day into the river. Lower stretches of the river were totally devoid of any form of life, even such organisms as leeches and sludge worms that normally thrive in polluted waters. Then, on 22 June 1969, the most serious river fire of all broke out. This fire developed as floating tree trunks and a large accumulation of trash formed a natural dam in a section of the river known as "The Flats." When the ever-present oil and organic waste on top of the river caught fire, it spread quickly several miles along the river. It nearly destroyed two railroad bridges passing over the river along "The Flats."

The fire appeared to be the "last straw," not only for the residents of Ohio, but also for environmentally concerned citizens throughout the United States. The city of Cleveland had already acted nearly a year earlier to pay for an improved sewage control system. But the 1969 fires convinced a number of companies to institute their own pollution-control systems. Three major steel companies announced plans to install cooling towers, eliminate dumping procedures, and institute other pollution-control systems.

Nationwide, the Cuyahoga River fires were an important factor in the creation of the first Earth Day, celebrated on 22 April 1970. Earth Day was a nationwide celebration in which more than 20 million people lobbied for better environmental controls. Experts have also credited the fires as being a strong motivating force in passage of the Clean Water Act of 1972.

Further Reading: "Cuyahoga River Area of Concern," <http://www.epa.gov/glnpo/aoc/cuyahoga.html>; "Cuyahoga River Remedial Action Plan Program," <http://chagrin.epa.state.oh.us/programs/rap/cuyahog.html>; "The Crooked River Project," <http://www.lerc.nasa.gov/WWW/K-12/fenlewis/Cuyahoga_River.html>.

D

DANTE'S *INFERNO*. *See Inferno*

DAVY SAFETY LAMP

The Davy safety lamp is a device invented by the English chemist and physicist, Sir Humphry Davy (1778–1829), in about 1815. The lamp was used for illumination by coal miners and revolutionized the safety with which mining could then be done.

The Industrial Revolution of the late eighteenth century greatly increased the demand for coal to operate the many new machines using steam power that had been invented. At first, much of the needed coal was mined from seams very near the Earth's surface. There was relatively little danger to human life involved in the process.

Dangers of Deep Coal Mining

As near-surface sources played out, however, deeper mines had to be dug. Deeper mines exposed valuable new sources of coal, but vastly increased the danger faced by miners working hundreds of feet underground. One of the most serious of those dangers was the possibility of explosions caused when underground gases were ignited by miners' lamps.

Lamps of some kind were, of course, needed to see in underground tunnels. But the flames they produced often caused the ignition of carbon monoxide, methane, and other gases and mixtures of gases found in connection with coal. A number of solutions were suggested for dealing with this problem. For example, one individual, known as the *fireman*, was sometimes used to go ahead of the miners themselves to check for combustible gases. The fireman was covered with damp blankets, and he carried a candle at the end of a long pole. A change in the appearance of the candle flame could often be used to indicate the presence of dangerous gases.

As one might expect, this system had both its advantages and disadvantages. Potentially dangerous explosions were avoided when combustible gases were discovered by the fireman. But the candle flame itself was sometimes enough to set fire to these gases, and the fireman himself was often the cause of a mine explosion.

Davy's Invention

A possible solution to the mine explosion problem was suggested by Davy in the mid-1810s. It occurred to him that a lamp might be produced in which the amount of heat produced was less than that needed to set fire to the mine gases. He designed such a lamp in which the candle flame was surrounded by a fine wire gauze. The flame produced enough light for miners to work by, but the wire gauze absorbed much of the heat produced by the candle flame.

The Davy lamp was also helpful in another way. As previously noted, the shape and color of a candle flame tends to change when the composition of the surrounding atmosphere changes. In the presence of firedamp, for example, the flame tends to become longer and develops a distinctive blue cap. A miner carrying the lamp would be forewarned of the presence of an explosive gas and would adjust his behavior accordingly. The Davy safety lamp turned out to be a lifesaving device for miners. By one account, one of the first men to use the lamp in a mine exclaimed that, "We have at last subdued this monster" (Derry and Williams, 472).

Other workers made further refinements in the Davy safety lamp. For example, Davy's original lamp often became so hot that the wire gauze was not able to remove enough heat. A wind passing through a mine tunnel might, for instance, cause the flame to burn brightly enough for sufficient heat to escape and cause an explosion. A simple solution for this problem was the installation of a second wire gauze to absorb even more heat produced by the candle.

Further Reading: Derry, T.K., and Trevor I. Williams. *A Short History of Technology from the Earliest Times to A.D. 1900*. New York: Dover Publications, 1993, pp. 470–472; Pohs, Henry A. *The Miner's Flame Light Book: The Story of Man's Development of Underground Light*. Denver: Flame Publishing, 1995; Travers, Bridget. *World of Invention*. Detroit: Gale Research, 1994, pp. 189–190.

DEATH BY FIRE

Fire is a leading cause of death and injury in all developed countries. In the United States between 1994 and 1998, an average of 4,400 people died from fire-related factors annually. Another 25,000 people were injured in fire-related accidents. About 80 percent of all fire deaths in the United States occur in the home. Fire is the third leading cause of accidental death in the American home, after poisoning and falls. Children under the age of five and adults over 65 represent a disproportionately large fraction of fire fatalities in the United States. Blacks are twice as likely to be killed in fires than whites, and men, twice as likely as women ("Facts on Fire").

Pattern of Fire-related Deaths

Deaths from fire have been described in literature as far back as Greek and Roman times. Except for large conflagrations such deaths were rare, however, largely because most homes were furnished simply. Fire represents a much greater threat today to families who live in developed countries, where homes often contain an abundance of combustible materials, many of which produce toxic gases when they burn.

Even in developed countries, the rate of deaths from fire can vary significantly. The rate of fire-related deaths in Canada is about the same as that in the United States, but it is only about half as large in France, Japan, Sweden, and the United Kingdom, and only about one-third as large in Australia, Switzerland, and Germany. These differences can not easily be explained, although fire safety education and regulations may be one possible factor.

Causes of Fire-related Deaths

Cooking and heating accidents are the two most common causes of residential fires. However, the major cause of fire-related fatalities is cigarette smoking. A smoker may fall asleep and drop a cigarette or accidentally leave a still-smoldering cigarette on a combustible surface. In such cases, materials may smolder for a considerable period of time before breaking suddenly into flame and spreading rapidly throughout a room. Such fires annually cause 30–45 percent of all fire-related deaths in the United States. Faulty heating and electrical systems constitute the second and third leading causes of death, respectively. The most important secondary factor that can contribute to death by fire is alcohol consumption. People who have been drinking too heavily may not be able to respond to a fire as quickly as they might without having had alcohol in their system.

Burns from heat are an obvious cause of death in many fires. A typical residential fire can quickly reach temperatures of about 600°C (1,100°F), and the presence of certain types of polymer materials can increase that intensity to 1,200°C (2,200°F). No living organism can survive at these temperatures.

Nonetheless, heat itself is not the major cause of death in fires. Instead, it is the products of combustion that are more likely to cause death. Most important among these products, and the single most common cause of fire-related deaths, is carbon monoxide (CO). Carbon monoxide is produced in any fire, but fires in confined spaces (such as a home), where oxygen supplies are limited, produce even greater concentrations of carbon monoxide.

Carbon monoxide is toxic because it combines with hemoglobin in red blood cells, preventing cells in the body from getting the oxygen they need to function. At concentrations of 200 ppm (parts per million), as may be experienced at a busy downtown traffic intersection, carbon monoxide can cause headaches, nausea, and disorientation. At concentrations of 1,000 ppm, a person may quickly become confused and will die in about two hours without immediate treatment. When the concentration of carbon monoxide exceeds 12,000 ppm, unconsciousness may occur after only 2–3 breaths, and death follows in 1–3 minutes.

Carbon monoxide is not the only toxic gas produced during a fire. Some other toxic gases and the burning materials from which they are produced include:

- acrolein (CH_2CHCHO): fats, oils, and petroleum products;
- hydrogen chloride (HCl): plastics;
- hydrogen cyanide (HCN): wool, silk, and certain plastics;
- hydrogen fluoride (H_2F_2): plastics, especially certain types of glassy polymers;
- hydrogen sulfide (H_2S): wood, meat, sulfur-containing organic compounds;
- sulfur dioxide (SO_2): organic compounds containing sulfur, such as certain types of rubber.

The toxicity of these gases ranges widely. The most dangerous of those listed is hydrogen cyanide, which is immediately fatal in concentrations of 280 ppm or higher, and fatal within ten minutes at concentrations of 180 ppm or more.

Oxygen deprivation itself may be a contributing factor to death by fire. As a fire burns, it uses up oxygen in a closed space more rapidly than it can be replaced through cracks around doors, windows, and other openings. As the concentration of oxygen decreases, a person becomes disoriented. At an oxygen concentration of about 15 percent (compared to a normal concentration of about 21 percent), most people lose their ability to make intelligent decisions. Even if they would normally know what to do in case of a fire, they quickly lose the ability to act on this knowledge when a fire breaks out. If the oxygen concentration drops to less than 10 percent, a person becomes unconscious and can be restored only if given a supply of oxygen quickly.

Smoke inhalation is also a contributing factor in fire-related fatalities. One reason is that smoke particles clog alveoli in the lungs, making it more difficult for oxygen to reach the bloodstream. In addition, smoke particles may irritate cells in the respiratory system, increasing their susceptibility to toxic gases. Finally, smoke particles may serve as nuclei on which solid toxic particles produced in a fire may collect and concentrate their effects on cells.

A small number of fire-related deaths are caused during efforts to escape from a blaze. Such cases are more common in large conflagrations, such as the Iroquois Theater fire in 1903, when the majority of victims died as they were crushed at exits or while attempting to jump from fire escapes.

Further Reading: "Facts On Fire," <http://www.usfa.fema.gov/safety/facts.htm>; "Fire-Related Injury and Death Among U.S. Residents," <http://www.cdc.gov/ncipc/duip/fire.htm>; "Fire Safety & Education," <http://www.usfa.fema.gov/safety/facts.html>; "Fire Statistics in the United States," <http://www.execpc.com/~fireinv/safety.htm>; Rossotti, Hazel. *Fire*. Oxford: Oxford University Press, 1993, Chapter 18; Runyan, C.W. et al. "Risk Factors for Fatal Residential Fires." *New England Journal of Medicine*, 49 no. 43 (1972): pp. 859–863.

DRAFT-CARD BURNING. *See*
PROTEST AND DEFIANCE BURNINGS

DRAGONS

Dragons are mythical creatures that have been mentioned in the folklore of societies all around the world. In many cases, dragons are pictured as breathing fire, steam, hot air, or smoke.

Pictorial images and stories of dragons can be traced as far back as the fourth millennium B.C. to cultures as diverse as those in Egypt, Mesopotamia, India, and China. Dragons often played major roles in the creation stories and other myths of ancient cultures. For example, dragons played an essential role in the Babylonian epic of creation in which Apsu (god of waters beneath the earth) and Tiamat (goddess of the oceans) battle with each other.

Similar stories can be found in mythological tales from nearly every part of the world. For example, in Dahomey, a dragon-like creature known as Aido Hwedo was thought to have assisted the god Mawn as he created the world. In the New World, another dragon-like creature, the feathered serpent Quetzalcoatl, was said to be the creator of the universe.

The physical appearance of dragons has differed widely in various parts of the world and at different times. For example, one description of Chinese dragons says that they have, "Head of a camel, horns of stag, the eyes of a demon, ears of a cow, neck of a snake, belly of a clam, scales of a carp, claws of an eagle, and the paws of a tiger. Breathes smoke, not fire," (Dragons). The primary difference among Asian dragons, according to one authority, is that Chinese dragons have five toes, Korean dragons have four toes, and Japanese dragons have three toes.

The description of dragons in the Western World has also varied widely. In general, they are usually said to have a snake- or lizard-like body that is covered with fireproof scales and the wings of a bat or bird. Most Western dragons were fire-breathers who killed their opponents by breathing on them. One of the most familiar dragons from Western mythology, for example, was the firedrake, frequently mentioned in Germanic literature. The firedrake was usually pictured as a fire-breathing, cave-dwelling monster responsible for protecting a great treasure.

The major difference between Asian and European dragons was that the former were generally kind and benevolent creatures responsible for many of the benefits enjoyed by humans. In the West, they were thought to be malevolent creatures that had to be destroyed. In both Asia and Europe, dragons took on important symbolic meanings. Until late in the Middle Ages, for example, Christians regarded dragons as the symbol of Satan that had to be slain in much the way that human sin had to be overcome.

Dragons, mythical creatures from folklore of societies around the world, are often pictured as breathing fire. This one decorated a nobleman's badge and is from the Ming dynasty, seventeenth century. *Courtesy of Art Resource, NY.*

Further Reading: "The Dragon as an Archetype," <wysiwyg://138/http://www.geocities.com/Area51/Lair/6252/ds.html>; "Dragon Myths from Vorarlberg," <http://members.tripod.com/~gfriebe/v-dragon.htm>, 08/26/99; *Encyclopedia Mythica*, <http://www.pantheon.org/mythica/articles/>; Greenberg,

Martin H., ed. *Dragons: The Greatest Stories*. New York: Fine Communications, 1997; Hague, Michael. *The Book of Dragons*. New York: William Morrow & Company, 1995; McCaffrey, Anne. *A Diversity of Dragons*. New York: Harper Prism, 1997; "Morgana's Observatory: Dragons," <wysiwyg://50/http://www.dreamscape. com/ morgana/galatea2.htm>, 08/26/99; "Odin's Castle of Dreams & Legends," <http:// www.odinscastle.org/odin3.html>, 08/26/99;

Rose, Carol. *Giants, Monsters, and Dragons: An Encyclopedia of Folklore, Legend, and Myth*. Santa Barbara, CA: ABC-CLIO, 2000; Shuker, Karl P.N. *Dragons: A Natural History*. New York: Simon & Schuster, 1995; Victoria and Albert Museum. *A Book of Dragons & Monsters*. New York: Abbeville Press, 1992; Weis, Margaret. *A Dragon-lover's Treasury of the Fantastic*. New York: Warner Books, 1994.

E

EARTH, AIR, FIRE, AND WATER.
See ELEMENTS OF ANCIENT GREECE

ELEMENTS OF ANCIENT GREECE

Earth, air, fire, and water were generally regarded by early natural philosophers to be the four elemental substances of which all materials in the natural world are composed. The idea of an element as some basic and fundamental substance was first enunciated by Greek philosophers during the period 600 to 300 B.C. Similar ideas can also be found in ancient manuscripts from other cultures. For example, the *Shu Chings*, or "Book of Records," from the Chou dynasty (722–221 B.C.) mentions "five things" or "five movers" thought to be responsible for the composition of all objects in the natural world. Those "five things" were water, fire, wood, metal, and earth. However, the idea of elements was not as fully developed anywhere in the world outside of ancient Greece.

Early Greek Science

The period between the sixth and third centuries B.C. in Greece is generally regarded as the beginning of modern science. It is during this period that humans first began to think about the fundamental nature of the world, exclusive of any practical applications or religious significance that knowledge might have. Although the Greeks made little or no effort to relate their theoretical ponderings

with empirical data about the world, they developed some profoundly sophisticated musings as to what the true nature of the physical world—beyond its superficial appearances—was like.

Most early natural philosophers seem to have had a strong desire to find simplicity and order in a physical world that seemed to be characterized by complexity and chaos. That seemingly almost innate drive still dominates much of the theorizing about nature that goes on in science today.

In any case, one of the handful of fundamental questions Greek philosophers asked was what it was (and they assumed there *was* such a thing) that was fundamental about matter. They could not believe that iron and blood and water and linen and hair . . . and all other forms of matter they encountered in the natural world were truly and uniquely different from one another. Instead, they chose to think that such materials are modifications or combinations of one, two, or a small number of truly fundamental substances. Today, we would give the name *elements* to those substances.

The Nature of Elements

The oldest expression we have of this view can be found in the writings of Thales of Miletus (640–546 B.C.). Thales thought that there was but a single element, water. Every material object we see, he said, can be made by the different ways in which water evaporates and condenses. By contrast Anaximenes (560–500 B.C.) made air the one and only elemental substance. He believed that air could condense to form water or earth or rarefy to make fire.

Heraclitis (536–470 B.C.) taught that fire was the one fundamental element. He made his choice based on a more general philosophy that the only reality in the world is change. Nothing remains constant, he taught, and all is in a state of flux. He often used a river to illustrate his point that a seemingly concrete physical reality was, in fact, constantly changing and being renewed at every instant. In such a case, what could be more representative of and fundamental to the real world than fire, which is always being born, changing, and dying?

In order to understand Heraclitus's theory, one must realize that he and other Greek philosophers were not referring to concrete, physical materials when they spoke of "water," "air," and "fire" as fundamental elements. It was not the flame of a candle or a burning piece of wood to which Heraclitus alluded when he described "fire" as the most basic thing in the world. Instead, he was referring to fundamental properties of fire, the ability of a flame to grow brighter, to flicker, and to become weaker. It was this constant transformation of shape and properties that was important about fire and, for Heraclitus, about everything in the world.

The Four Elements

Over time, philosophers developed more complex and sophisticated views of the nature of elements. Matter, they said, was probably made of two, three, or more elementary substances which could be combined in a variety of ways. For example, Empedocles (490–430 B.C.) was among the first scholars to list earth, air, fire, and water as the four basic elements of matter. He said that these four materials could combine and separate in an endless number of ways, but they always remained fundamentally the same. "Only commingling takes place," he wrote, "and the separation of the commingled." Out of this commingling and separation, he said,

arose the "illusion" of a physical world (Van Melsen 1952, 23).

The doctrine of the four elements was brought to its highest level of development by Aristotle (384–322 B.C.). He took the analysis of matter one step further back by saying that earth, air, fire, and water were not, themselves, fundamental, but were, instead, manifestations of various combinations of four other fundamental properties: heat, cold, dryness, and moistness. Thus, he taught that heat and moisture combined to produce air; heat and dryness, to produce fire; dryness and cold to produce earth; and cold and moistness to produce water.

With the spread of Christianity and the disparagement of worldly knowledge in the first few centuries after the death of Christ, the development of scientific thought largely came to an end in the Western world and was not revived until the early Renaissance. When the new breed of scientists who then appeared began to think about the nature of elements, earth, air, fire, and water were no longer part of their theories.

Further Reading: Bynum, W.F., E.J. Browne, and Roy Porter. *Dictionary of the History of Science*. Princeton, NJ: Princeton University Press, 1981, pp. 117–119; Partington, J.R. *A Short History of Chemistry*. London: Macmillan & Company, 1937; Rossotti, Hazel. *Fire*. Oxford: Oxford University Press, 1993, pp. 256–259; Van Melsen, Andrew G. *From Atomos to Atom*. Pittsburgh: Duquesne University Press, 1952.

ETERNAL FLAMES

An eternal flame is a fire that is lit to commemorate some important person or event. For most Americans, the best known eternal flame may be the one that burns at the gravesite of President John F. Kennedy at the Arlington National Cemetery. The eternal flame at the grave is located at the center of a circular granite stone, five feet in diameter. The flame was originally lit by Mrs. Kennedy during the funeral for her husband held on 25 November 1963. The flame is fed by a mix of natural gas and air through a special burner designed by the Institute of Gas Technology in Chicago. It is kept lit by an electric spark at the nozzle of the burner that reig-

The Eternal Flame at the grave of John F. Kennedy, Arlington National Cemetery, Virginia. *Courtesy of U.S. Army.*

nites the flame in case it is extinguished by wind or rain.

An equally famous eternal flame burns at the Tomb of the Unknown Soldier in the Arc de Triomphe in Paris. Construction of the Arc was begun in 1806 by Emperor Napoleon, but was discontinued at his death in 1814, after the emperor abdicated. The work was finally completed and dedicated in 1836. The eternal flame honoring the unknown dead in World War I was put into place and first ignited on 11 November 1920. Each year, the President of the Republic revisits the Arc which also memorializes the dead who fought for France in World War II.

Sacred Eternal Flames

Some eternal flames have profound religious, historical, and cultural meanings also. For example, the Cherokee Indians maintained an eternal sacred flame long before Europeans arrived in North America. The sacred fire was the center of important religious ceremonies and was used to light the fires of every Cherokee household.

When the Cherokees were expelled from their homelands in 1838, they went into exile in the hills of Tennessee and took their sacred fire with them. Shamans of the tribe were assigned the responsibility of keeping the fire alive. In 1951, representatives of the Cherokee Historical Association set out on an expedition to locate the sacred fire and found that it was still being kept in the Blackgum Mountains of Oklahoma. They were allowed to collect live coals from the fire and return them to the original Cherokee homelands in Tennessee. Fire generated from those coals has been used for more than fifty years in a Cherokee dramatic production, "Unto These Hills," performed in Cherokee, North Carolina. The flame also took one more trip in 1984, this time to the ancient capital of the Cherokee Nation, Red Clay, Tennessee. There, a torch that was to become an Eternal Flame of the Cherokee Nation was lit from the coals of the sacred fire.

Eternal flames have been a part of many cultures throughout history. For example, Moses is said to have lit a holy fire that continued to burn in a secret hiding place for the seventy years of the Babylonian exile. After that fire was lost, it was replaced in 165 B.C. in the restored Temple at Jerusalem. The fire was said to have issued forth from the stones of the new altar as the result of prayers offered by the Maccabees, the family responsible for the restoration of the Jewish political and religious life. The date on which this event occurred is now taken as the beginning of the Jewish holiday Hanukkah.

Perhaps the most famous eternal flame in ancient history was that maintained at the Temple of Vesta in Rome. According to tradition, that flame was first ignited during the reign of King Numa Pompilius (715-673 B.C.), after which it burned until 394 A.D.

Further Reading: Goudsblom, Johan. *Fire and Civilization.* London: Allen Lane, 1992, pp. 120–121; "Paris Digest Data on the Arc de Triomphe," <http://www.parisdigest.com/monument/arcdetriomphe-theflag.htm>, 07/21/99; "President John Fitzgerald Kennedy and Arlington National Cemetery," <http://www.mdw.army.mil/fs-m01.htm>, 01/17/99; Pyne, Stephen J. *Vestal Fire: An Environmental History, Told through Fire, of Europe and Europe's Encounter with the World.* Seattle: University of Washington Press, 1997, pp. 81–146.

EUROTUNNEL FIRE

On 18 November 1996, a fire broke out in the Channel Tunnel (Eurotunnel, or "Chunnel") that joins England and France

A view of the Eurotunnel showing some of the damage caused by the 1996 fire. *Courtesy of Archive Photos © 6635/ Gamma.*

beneath the English Channel. The fire occurred about one-third of the way into the tunnel from the French end. Five truck drivers were taken to hospitals and treated for smoke inhalation, while another twenty-five drivers were evacuated safely from the tunnel.

The long-term effects on the tunnel were enormous. More than $350 million dollars in damage was sustained inside the structure, and the tunnel had to be closed for more than six months. It was not able to reopen until June of 1997. Perhaps the most serious damage of all was to the tunnel's reputation. Fire had been the major safety concern in the structure since it had first been designed. Architects were convinced that every possible safety precaution had been taken to protect passengers, train crews, and equipment. But the extent of the November 1996 fire raised questions anew about fire safety within the tunnel.

Nearly eighteen months after the fire, French authorities announced that the fire had been intentionally set. They had found traces of a flammable liquid in the portion of the tunnel where the fire had occurred. The French planned no further action on the arson since vandalism is not a criminal act under French law.

Further Reading: Department of the Environment, Transport and the Regions, "Inquiry into the Fire on Heavy Goods Vehicle shuttle 7539 on 18 November 1996," <http://www.railways.detr. gov. uk;ctsa/18nov96/ctsa.htm#exec>; "Eurotunnel Fire 'Deliberate'," <http://news. bbc.co.uk/low/ english/world/europe/newsid_67000/ 67970.stm>, 12/21/98; "Fire Breaks Out in Channel Tunnel." *The Independent*, 19 November 1996, 11.

EXPLOSIVES

Explosives are compounds or mixtures that undergo rapid chemical or nuclear reactions that produce large volumes of gas, usually at very high temperatures. The gases expand outward very rapidly, producing a shock wave. The shock wave is generally responsible for most of the immediate damage caused by the explosion, such as the splitting of rock or the destruction of buildings. However, the heat that accompanies an explosion usually ignites flammable materials in the area, causing fires. The composition of explosive devices can be modified to alter the relative amount of shock and fire damage produced.

History

Until the 1940s, the only type of explosive known was a chemical explosive. The first such material was invented in China in the tenth century. That material was gunpowder, or black powder, a mixture of carbon (charcoal), sulfur, and potassium nitrate (nitre). When gunpowder is ignited, the compounds of which it is made react with each other to produce carbon monoxide (CO), carbon dioxide (CO_2), sulfur dioxide (SO_2), and nitric oxide (NO) (all gases), as well as other products. The heat released in the reaction causes the gases to expand outward very rapidly, forming shock waves.

It appears that the Chinese used gunpowder primarily for fireworks displays and never appreciated or took advantage of its potential military applications. By about the thirteenth century, the recipe for making gunpowder had become known to Islamic tribes in Asia, who used the new weapon against Europeans. Europeans, in turn, brought that recipe home, where they used it against each other in warfare.

The age of modern explosives can be traced to 1846 and the invention of nitroglycerine by the Italian chemist Ascanio Sobrero (1812–1888). Sobrero treated glycerine with nitric acid in the presence of sulfuric acid as a catalyst to produce the new compound. Nitroglycerine is an enormously powerful explosive, but very unstable. It can be detonated by a spark or the slightest blow. It was considered too dangerous to use for most purposes until, two decades after its invention, the Swedish inventor Alfred Nobel (1833–1896) invented dynamite. Dynamite is nitroglycerine deposited in an absorbent material. The absorbent material makes the nitroglycerine much safer to use. With the invention of dynamite, both the construction trades and mining, as well as military warfare itself, were dramatically changed.

The next important breakthrough in explosive technology was the invention of trinitrotoluene (TNT). The compound was first prepared in 1863 by the German chemist J. Wilbrand while he was searching for a new dye. However, no use for the material was found for about forty years. Then, chemists realized that TNT was a powerful explosive that had important military (and, later, industrial) applications. TNT differed from all previous explosives, however, in that it did not detonate of its own accord, but required some other material, such as lead azide or mercury fulminate, to make it explode. For many years, TNT was the most powerful explosive known.

Over the last half century, many new explosive compounds and mixtures have been invented. Some are entirely new materials, while others are new formulations of TNT or other traditional explosives with new compounds. One of the most common industrial explosives today is called ANFO, an acronym for ammonium nitrate–fuel oil. ANFO consists of tiny pellets of a 94:6 mixture of ammonium nitrate and fuel oil. Another common explosive is SBS, an acronym for slurry blasting agent. SBS also consists primarily of ammonium nitrate in a gelatinous slurry that may also contain dynamite, aluminum, or some other material.

The nature of explosive technology changed dramatically in the 1940s with the invention of *fission* and *fusion* weapons ("atomic bombs" and "hydrogen bombs"). These weapons produce their explosive effects by the process of nuclear fission, in which uranium or plutonium atoms are broken apart, or the process of nuclear fusion, in which hydrogen and helium atoms are joined with each other. The power produced by fission and fusion weapons is many magnitudes greater than that of chemical weapons. Their destructive power is commonly expressed by comparing the damage they do by a comparable amount of TNT. For example, most fission weapons have a destructive power of a few tens or hundreds of kilotons (thousands of tons) of TNT. Fusion weapons may have a force equivalent to a few megatons (millions of tons) of TNT.

The heat produced by such weapons is also far beyond that associated with chemical explosives. At the instant of ignition, both fission and fusion weapons produce temperatures of the order of a few million degrees Celsius, roughly comparable to that found at the center of most stars.

Types of Explosives

Explosives are usually categorized according to their chemical composition, rate of decomposition, use, or other criterion. Some common types of explosives are the following:

An explosion at Atlantic Richfield's Black Thunder strip mine in Gillette, Wyoming breaks coal in the mine into pieces small enough to be transported by truck. The coal is burned to generate electricity in cities around the country. © *Jonathon Blair/CORBIS.*

Low Explosives. Low explosives burn rather than explode. The damage they cause results more from the heat and flames they produce than from any shock waves. Low explosives usually begin burning at one surface, after which a flame moves relatively slowly through the mass of the material.

High Explosives. High explosives are compounds that detonate throughout every part of their mass very quickly. The chemical reactions that cause the explosion are often completed within a few millionths of a second after ignition. Some high explosives, called *primary explosives,* are very sensitive and will detonate easily when heated or shocked. Other high explosives, known as *secondary explosives*, require a detonator for an explosion to occur.

Initiating or Detonating Explosives. Explosives in this category are compounds that are very sensitive to heat or shock and will detonate easily and quickly. They are a kind of primary explosive used not so much for their own destructive potential, but because of their ability to ignite other explosives.

Blasting Agents. Blasting agents are high explosives used primarily in construction work and mining. They are used to blow apart rock in order that roads can be constructed or to gain access to ores and minerals. Blasting agents have also become popular among terrorists, who combine the explosives with clay or soft plastic. These materials can then be attached to surfaces or molded into almost any shape before being detonated.

Military Explosives. Military explosives tend to be of two general types, propellant (or impulsive) explosives and disrupting (or bursting) explosives. The former are used to propel shells, rockets, missiles, depth charges, and other objects from guns and cannons. The latter are the actual objects—artillery shells, bombs, and hand grenades, for examples—designed to cause damage to the enemy.

Further Reading: Akhavan, J. *The Chemistry of Explosives*. New York: Springer-Verlag, 1998; Brown, George I. *The Big Bang: A History of Explosives*. Dover: Sutton Publishing, 1998; "Explosives," in Bridget Travers, ed., *Gale Encyclopedia of Science*. Detroit: Gale Research, 1996, Volume 3, pp. 1425–1430; Partington, J.P. *History of Greek Fire and Gunpowder*. London: W. Heffer and Sons, 1961; Pickett, Mike. *Explosives Identification Guide*. Albany, NY: Delmar Publishers, 1998; Rossotti, Hazel. *Fire*. Oxford: Oxford University Press, 1993, Chapter 12.

F

FARADAY, MICHAEL. *See* CHEMICAL
HISTORY OF A CANDLE

FESTIVAL FIRES. S*ee* BONFIRES

FIRE ALARM SYSTEMS

A fire alarm system is a method by which reports of a fire can be forwarded to firefighting units. Until the early nineteenth century, only primitive methods were available for signalling about the start of a fire. In Imperial Rome, for example, sentinels (or *nocturnes*) were stationed throughout the city on towers. When they saw that a fire was under way, they passed on that information by blowing a trumpet to notify the nearest unit of firefighters (known as *vigeles*).

Today, fires can be reported to neighborhood fire stations by telephone calls or by means of a *fire alarm box*. The first fire alarm system was invented by the English inventor Ithiel Richardson in 1830. Richardson's system consisted of strings running through the rooms of a building. In case of a fire, the strings burned through, setting off a bell in a central station.

The first self-contained fire alarm box of modern design was invented by William F. Channing in 1839. The alarm box contained an automated telegraph that was activated when a person turned a crank on the outside of the box. The telegraph transmitted a coded message to a central station that indicated the location of the box. A dispatcher at the central station could then notify the nearest fire station.

Channing's idea met with little enthusiasm, and he sold the rights to his invention to the John N. Gamewell Company of South Carolina. Gamewell was much more successful in marketing Channing's invention, and the company became the premier producer of fire alarms for many years.

Modern Alarm Box Systems

Variations of Channing's original alarm system are still in use. Alarm boxes may use telephones or radio signals to notify a central station about a fire. Telephone systems have the advantage of being easy to use and familiar to citizens who wish to report a fire. They require a tie-in, of course, with an existing telephone system. Radio systems have the advantage that they do not require such systems, and can be put into operation in remote areas where no telephone or electric wires are available.

Perhaps the best known fire alarm systems today are those painted a distinctive red color containing a glass front with the words, "Break in case of fire." When the glass is broken, a person wishing to report a fire pulls down a small handle inside the box. The handle activates a telegraphic system that sends out a distinctive signal through an electric wire connected to a central station. The signal consists of a certain number of distinctive sounds, such as two short and three long sounds, that ring at the central station. At the same time, the signal is recorded on a tape that can be read by a dispatcher at the station.

The signal is repeated by the alarm box telegraphing a certain number of times to permit the central station to confirm the location of the message. After these repetitions, the alarm box automatically resets to its original condition and is ready to be activated again. At the central station, the dispatcher is able to identify the location of the alarm box that has been activated by means of the distinctive code received. The dispatcher is then able to notify the fire station closest to the alarm box to alert firefighters there of the fire. *See also* FIREFIGHTING; LOOKOUT TOWERS; SMOKE DETECTOR.

Further Reading: Bryan, John L. *Fire Suppression and Detection Systems*, 2nd edition. New York: Macmillan, 1982; Ditzel, Paul C. *Fire Alarm!* New Albany, IN: Fire Buff House Publishers, 1994; Rudman, Jack. *Fire Alarm Dispatcher*. Syosett, NY: National Learning Corporation, 1983; Traister, John E. *Security/Fire-Alarm Systems: Design, Installation, Maintenance*. New York: McGraw-Hill, 1995.

FIRE ASSAYING

Fire assaying is a quantitative procedure by which the composition of an ore or some other metallic mixture is determined. The procedure is used primarily to determine the amount of gold and/or silver in a sample, but it can also be used to measure the quantity of platinum group metals, especially platinum and palladium. Fire assaying got its name because the process was originally carried out over an open flame, although an electric furnace is a more efficient device today.

History

The techniques of fire assaying have been known for at least 3,000 years, and probably much longer. Clay tablets found in the Nile Valley mention that Babylonian metallurgists used fire assaying to determine the purity of gold sent to them from the Pharaoh Amenophis (1375–1350 B.C.) and found that

Precious metals (typically gold and silver) are analyzed using a technique called fire assaying. Here a fire assayer pours a molten sample of precious metal ore mixed with reagents and assay grade litharge (PbO). Once cooled, the sample is heated a second time leaving behind a bead of gold and silver, which is then weighed to determine concentration. Fire assaying, a technique that has been practiced for thousands of years, is a procedure by which the composition of an ore or other metallic mixture is determined. *Courtesy of ASARCO Inc., Denver, Colorado.*

they had been shortchanged in the transaction. Methods of assaying were further developed during the Middle Ages by alchemists, for whom the purity of a metal was a critical piece of information.

Assaying is still widely used today, often in much the same form used by crafts people of hundreds or thousands of years ago. Assaying is used to determine the purity of a sample of gold, silver, or other precious metal that has been bought or sold. It can also be used as a quality control technique to make sure that a given product contains the correct amount of a precious metal. Finally, it is an important procedure in determining the quantity of a precious metal found in a given sample of ore.

Technique

The modern process of fire assaying usually begins with a mechanical process in which the material to be analyzed is crushed into a fine granular material. A sample weighing about 29.167 grams is then weighed out into a crucible. This weight is known as an *assay ton*, a unit that has come to be used as a standard in fire assaying. Since there are 29,167 troy ounces in a short ton, the number of milligrams of a precious metal in a sample of one assay ton is numerically equal to the number of troy ounces of that metal in one ton of the raw ore.

To the crucible is also added a flux, a material that lowers the melting point of the mixture and separates pure precious metals in the sample from other materials. The primary component of the flux is litharge (lead oxide). The flux also contains a reducing agent and one or more other substances. The choice of materials to add to the flux depends on the nature of the ore, and may include sodium carbonate, potassium carbonate, borax, silica, organic materials (such as sugar, starch, or flour), and/or potassium nitrate.

The mixture of sample and flux is then heated to a temperature of about 1,000°C (1,900°F). At this point, the reducing agent converts lead oxide to molten lead, which forms as tiny droplets that collect together to form a single "button." Any gold or silver (or other precious metals) present in the sample dissolve in the molten lead and are carried to the bottom of the crucible along with it. The lead button, containing its precious metals, then sinks to the bottom of the crucible. Materials in the ore other than precious metals combine with the other components of the flux to form a slag that floats to the top of the mixture.

After cooling, the lead button is removed and transferred to a small container known as a *cupel*, made of bone ash. The cupel is then placed into another furnace and heated to about 950°C (1,750°F) at which temperature lead on the surface of the button is converted back to lead oxide. The lead oxide is absorbed by the bone ash, leaving pure metallic lead behind in the button. As heating continues, another coating of lead oxide forms on the outside, followed again by the absorption of the lead oxide by the bone ash. This process is repeated over and over again in the cupelling furnace until all of the lead

itself has been converted to lead oxide and absorbed by the bone ash. All that remains behind at that point is gold, silver, and other precious metals. None of these metals is affected by the heating process or the bone ash of the cupel and remain in the bottom of the container as a small bead.

In the final step of the assaying process, gold and silver are separated from each other. If other precious metals are also present, they are also separated out at this point. Gold and silver can easily be separated from each other because the latter is soluble in nitric acid, while the former is not. The pure droplet of gold that remains behind after the separation process can be weighed on an analytical balance. The weight of silver present can be found by subtracting the weight of the gold from the weight of the final gold/silver button. If other precious metals, such as platinum and palladium, are also present in the ore sample, additional steps are necessary to determine their relative fractions.

Fire assaying is a difficult technique that is partly science and partly art. The process almost always results in the loss of small amounts of one or more of the precious metals present. If osmium is present in the mix, for example, it will be almost entirely lost during the heating process. The efficiency of the process also depends on the exact selection of flux components so that nonmetallic compounds will be removed efficiently from the mixture.

Further Reading: Kallman, Silve, "Fire Assaying." In *McGraw-Hill Encyclopedia of Science & Technology*, 8th edition. New York: McGraw-Hill Book Company, 1987. Volume 7, pp. 134–136; Shepard, O.C., and W.F. Dietrich. *Fire Assaying*. Boulder, CO: Met-Chem Publishing, 1989; "What Is a Fire Assay?" <http://www.bc-mining-house.com/prospecting_school/m_fire.htm>, 07/31/99; "What Is Fire Assaying? <http://www.mrsscrap.com/fireassay.htm>, 07/31/99.

"THE FIREBIRD"

"The Firebird" is a ballet written for the great impresario Serge Diaghilev by Igor Stravinsky (1882–1971) in 1910. The ballet is widely regarded as one of Stravinsky's three most important works (along with "Rite of Spring" and "Petrushka"). The ballet, first performed at the Paris Opera on 25 June 1910 with Tamara Karasavina dancing the title role, was an immediate success and has remained an essential part of the ballet repertoire today.

The firebird is a creature popular in Russian folklore. It has very bright feathers that shine like silver and gold and bright eyes that sparkle like crystals. It is said to shine so brightly that it can light up a room at night as brightly as a thousand candles.

The ballet retells a familiar Russian folk story about the hero, Prince Ivan, who finds himself in the kingdom of the demon Kaschchei. Kaschchei has a special fondness for imprisoning young women and turning men into stone. In his wanderings, Ivan notices the beautiful "Firebird," pursues her into Kaschchei's enchanted garden, and finally captures her there. When she promises to give him one of her magic feathers, he allows her to go free.

Sometime later, Ivan comes across thirteen dancing young women who have been imprisoned by the demon in his garden. As dusk falls, the women are forced to return to Kaschchei's castle, and Ivan follows. He is captured by the demon, who threatens to turn him into stone. Only a display of the "Firebird's" magic feather saves him from this horrible fate and allows him to destroy the monster and set free the thirteen young women. As a reward for his valor, Ivan is allowed to marry the leader of the young women, Tsarvena.

FIREBOAT

A fireboat is a marine craft designed to put out fires on other ships, in dock areas, along the shore, or in other areas accessible by water.

History

The first fireboats of which we have any information were put to use in the mid-eighteenth century in various European cities. These fireboats were nothing other than floating versions of land-based firefighting systems. For example, a fire engine might be carried on to a barge and floated or rowed down river to fight a fire. Fireboats of this type were in common use along the Thames River throughout the early nineteenth century, where they were used to fight fires on the docks and in riverside warehouses.

The first fireboat in the United States was *The Floating Engine*, built by volunteer firefighters in New York City in 1800. The boat consisted of a barge on which was mounted a coffee mill–type pump operated by a dozen men. It was rowed to fires where needed on or near the river, but had no fire hoses of its own. Instead, the pump was used to supply water to fire hoses on the shore near the fire.

The mid-nineteenth century saw a revolution in the use of fireboats for two reasons. First, the Industrial Revolution had led to the construction of many large factories, warehouses, docks, and other structures along the waterfront of large cities. The problem of dealing with fires in these structures became one of major importance to municipalities. At the same time, dock areas became much more crowded with ships bringing in raw materials and carrying away finished products. Ship fires became much more common and difficult to fight.

At the same time, the development of steam-powered engines that had led to these developments also made possible a new form of fireboat: one that operated with steam power. The first steam-operated pump installed on a barge for firefighting purposes was built for use on the Thames River in 1852. Three years later, the first fireboat to use steam both for pumping and for propulsion was also put into use on the Thames. The first steam-operated fireboat in the United States was Boston's *William M. Flanders*, put into service in 1872. The *Flanders* was 75 feet long with an iron hull and four pumps capable of delivering a total of 2,500 gallons of water per minute. Probably the most famous of the early fireboats was New York City's *The New Yorker*, launched in 1891. *The New Yorker* had a pumping capacity of 16,000 gallons per minute, equal to that of fourteen land-based fire engines. It remained in service until 1931.

Between 1866 and 1989, 184 fireboats were built for service in New York City, Boston, Chicago, New Orleans, Los Angeles, Tacoma, Seattle, and many other cities with substantial commercial waterfronts. Since the 1960s, there has been an increased interest in the use of smaller fireboats, capable of dealing with a wider variety of fires and other emergencies on and near waterfront areas. Prior to 1960, most fireboats ranged in size from 75 to 140 feet in length. In the last few decades, boats as small as 16 to 24 feet in length have become more popular.

Specifications

The primary features of interest on a fireboat are usually its overall size, method of propulsion, and fire gun capacity. As an example, one of the best known fireboats may be New York City's *Fire Fighter*, commissioned in 1938. *Fire Fighter* has long been a familiar sight in New York harbor, recognizable at one time because of a nearly 50-foot tall fire tower (removed in 1962) located just behind the pilot house. The boat is 90 feet, 6 inches in length and powered by two 1,500 horsepower 750 rpm engines. Its nine fire guns are operated by four 60 horsepower 1,500 rpm motors capable of providing 20,000 gallons of water per minute at a water pressure of 150 pounds per square inch. The latter two numbers are of significance because they indicate the amount of water a boat can supply and the distance its hoses can reach.

The 1960s change in fireboat technology, from larger to smaller ships, reflects at least in part a perpetual debate about the role of fireboats in a city's firefighting arsenal. City officials have long balked at the cost of building large ships that may be used only rarely, even if those rare cases involve major fire disasters. In the last half-decade, then, fire departments have found it easier to get approval for smaller *Boston Whaler*-type boats that can be used for a variety of purposes, including both firefighting and rescue operations. The Los Angeles Fire Department currently owns three small fireboats, for example, each

manned by three specialized firefighters and two firefighter/SCUBA divers. These boats are designed to deal with boat fires, wharf and underwharf fires, emergency medical incidents, drownings, hazardous material spills, oil tanker inspections, underwater search and rescue, stranded boat rescue, salvaging of partially sunken vessels, and many other situations.

Further Reading: Ditzel, Paul. *Fireboats: A Complete History of the Development of Fireboats in America*. New Albany, IN: Fire Buff House, 1989; Fireboat NYC: John J. Harvey, <http://www.fireboatnyc.com/aboutus/index.html>.

FIRE BRICK

Fire brick is a type of brick made of refractory materials. A refractory material is a substance that withstands very high temperatures without melting or decomposing. Refractories are used to line kilns, coke ovens, steel furnaces, and other devices that produce temperatures in the thousands of degrees. Fire bricks are typically made out of fireclay, but they come in a variety of forms for many specialized applications. The differences among different types of fire brick are based primarily on their thermal and physical properties. *See also* FIRECLAY; FURNACES.

Further Reading: Olsen, Frederick L. *The Kiln Book*. Radnor, PA: California Keramos Bassett Books, 1974.

FIREBUG. *See* ARSON

FIRECLAY

Fireclay is a form of clay that can withstand very high temperatures. The term *clay* is used to describe a naturally occurring material that is soft and plastic when moist, but hard when fired. The primary constituent of fireclay is a mineral known as *kaolinite*. Kaolinite is a naturally occurring form of aluminum silicate ($Al_2O_3 \cdot 2SiO_2 \cdot 2H_2O$). It is typically whitish to pale yellow in color and, like other clays, plastic when moist. Some forms of fireclay contain impurities such as quartz that make them nonplastic. In this form, the material is known as a flint clay.

Fireclay's ability to withstand very high temperatures makes it useful in the refractories industry. A refractory is a substance that tolerates very high temperatures without melting or decomposing. Refractories are used to line kilns, coke ovens, steel furnaces, and other devices that produce temperatures in the thousands of degrees. Fireclay is essentially unaffected by temperatures up to about 1,500°C (2,750°F).

Fireclay is also used in the foundry industry to make heat-resistant forms in which metals are cast. The forms consist of about 85–90 percent sand mixed with 10–15 percent fireclay. One advantage of using a mixture of this composition is that it provides a very smooth surface for the mold so that metals cast in the mold also have smooth surfaces.

Further Reading: *Fireclay*. St. Petersburg, FL: Artext Publishing, 1996.

FIRE CLOCK

A fire clock is any device that uses fire to measure time. The principle of using fire to measure the passage of time may have occurred to early humans for two reasons. First, fire was readily available, and, second, there were few alternatives to compete with the use of a fire clock for time measurements.

The general principle of a fire clock is that devices can be built along which a flame can move at a relatively constant rate of speed. In theory, one could simply light one end of a long string and measure the amount of time passed by how long it takes the fire to move along the string. The burning of ten centimeters of string might, for example, be equivalent to a time period of fifteen minutes.

The problem with this general theory is that many factors can affect the rate at which the string (or any other object) will burn. The exact dimensions of the string, air currents around the string, and the care with which observations are made are a few such factors. Fire clocks were of value only to the extent that inventors were clever enough to find ways to minimize these potential sources of error.

For example, candles marked with horizontal lines along their sides were used as measuring devices as early as the time of King Alfred the Great (871–899). The passage of time was measured by the number of lines through which the candle flame had been burned through. The Muslim scholar al-Jazari is credited with having developed a more sophisticated version of this kind of fire candle by attaching a series of weights and pulleys that moved a pointer as the candle burned down.

Fire clocks were also widely used in Asia before the eleventh century. In one version, a long thin piece of wood called a *joss stick* was marked off with horizontal lines indicating the passage of time. In another version, a long rope was knotted at regular intervals and then set to smoldering at one end. The passage of time was noted by the number of knots through which the fire had burned.

A convenient adaptation of the burning joss stick as a timekeeper was the coiled incense burner. The incense burner was very similar in appearance to devices still used today as mosquito repellents. The coiled version of the joss stick was marked at regular intervals along its (coiled) length and then set afire. This device was useful for measuring longer periods of time because of the greater length along which the fire could burn.

Further Reading: Rossotti, Hazel. *Fire*. Oxford: Oxford University Press, 1993, Chapter 15.

FIRE CODE

A fire code is a document that provides a minimum, verifiable, and enforceable level of safety for citizens of a community. Fire codes are usually first developed by groups of professionals with special training and background in the field of fire safety. They are generally developed by consensus; that is, through the process of reaching decisions with which everyone, or the majority of those involved, are satisfied. In many cases, the codes developed by fire professionals later go on to become legal requirements in a community. Those legal requirements determine many aspects of the way a building is constructed, such as the types of materials that can be used, the way the building can be wired, and the width of doorways.

History

The first fire safety codes were developed toward the end of the nineteenth century. An important impetus for the development of these codes was the invention of fire sprinkler systems. Such systems had obvious potential for dramatically reducing the damage to structures caused by fires. The problem that arose, however, was the variety of systems created and installed.

In March 1895, a small group of men representing fire insurance and sprinkler system manufacturers met in Boston to deal with this issue. At the time, members of the group knew of nine totally different sprinkler systems being used within a 100 mile radius of the city. They saw little hope that the potential for fire safety provided by sprinkler systems could be achieved unless and until standards were adopted for those systems.

The Boston meeting had two long-term consequences. The first was the adoption of standards for the design and installation of fire sprinkler systems. The second result was the creation of an association that was to become the National Fire Protection Association (NFPA). Standards have been revised a number of times, but still exist today in a modified form as NFPA 13 (Code 13).

Fire Codes Today

Over the past century, NFPA has greatly expanded the number and scope of codes dealing with fire safety. About 300 such codes and standards now exist. They are reviewed, evaluated, and modified by more than 5,000 volunteers from the field of fire safety working on more than 200 technical committees.

NFPA fire codes are designed to reduce the risk of fire and possible losses from such fire. They cover every conceivable aspect of building design and construction, such as fire extinguishers of every design, installation of sprinkler systems, gas piping, equipment installation, venting of gases, storage and handling of gases, electrical installation, fire alarm design and installation, materials to be used in curtains and drapes, placement of smoke detectors, and the design and use of life safety devices and procedures.

NFPA codes are recommended standards to which the vast majority of individuals and companies conform voluntarily. People involved with fire safety understand that it is to their benefit to follow the advice of professionals in their field. The codes have another important function, however. They are used to write local and state laws (also known as codes) that establish standards that must be met in new construction. The NFPA is active in offering its assistance to governmental bodies in the development and implementation of these codes.

Prescriptive and Performance Codes

Historically, most fire codes have been developed in response to some specific fire disaster. After studying the disaster, officials have recognized certain changes that would have reduced the risk or consequences of the disaster. For example, wider doors or more fire escapes might have been needed in a building. These changes would then later have been adopted as part of fire codes.

Codes of this kind are called prescriptive codes because they dictate very specific structural or procedural methods of fire protection. In recent years, another kind of code has been proposed—a performance code. A performance code is one that requires that certain standards of fire safety and protection be met, no matter how that goal is achieved. For example, a code might require that a certain amount of water be provided from a sprinkler system to cover a given area in a given time. But the code might not state precisely how the system has to be installed to meet this requirement.

Some people argue that performance standards represent a step forward over prescriptive codes because they provide a broader, more general approach to fire safety rather than focusing on specific issues. The Draft British Standard Code of Practice for the Application of Fire Safety Engineering Principles to Fire Safety in Buildings, first developed in the early 1900s, is an example of a performance code. *See also* NATIONAL FIRE PROTECTION ASSOCIATION.

Further Reading: Goudsblom, Johan. *Fire and Civilization*. London: Allen Lane, 1992, pp. 144–150; "History of the NFPA Codes and Standards-Making System," <http://roproc.nfpa.org/code+standard_sys.html>, 01/03/99; "How NFPA Codes and Standards Are Used," <http:roproc.nfpa.org/how.html>, 01/03/99; *International Fire Code 2000*. Whittier, CA: International Conference of Building Officials, 2000; "NFPA Codes & Standards," <http://roproc.nfpa.org/products/listing.html>, 01/03/99; Pyne, Stephen J. *Vestal Fire: An Environmental History, Told through Fire, of Europe and Europe's Encounter with the World*. Seattle: University of Washington Press, 1997, pp. 53, passim; *Uniform Fire Code* (year varies). Whittier, CA: International Conference of Building Officials.

FIREDAMP

Firedamp is a term used to describe natural gas, methane, or an explosive mixture of methane and air. Firedamp is one of two types of "damps" that can occur in a coal mine. *White damp* is the name given to poisonous carbon monoxide, which can collect in coal mines and can cause death. *Black damp* (also called *after damp*) is a mixture of carbon dioxide, nitrogen, and steam that remains after a fire or an explosion has occurred. Black damp is also poisonous and fatal to humans.

When organic matter decays without access to oxygen, it may convert into coal, petroleum, or natural gas—the three so-called fossil fuels. Natural gas is often associated with underground veins of coal and reservoirs of petroleum. This natural gas may seep into coal mines or escape along with petroleum when oil wells are drilled into a reservoir. The presence of natural gas in either of these conditions creates environmental and safety problems. In the case of oil wells, for example, they can be responsible for sudden, dramatic releases of oil known as *gushers*. In the case of coal mines, the natural gas may seep into the mine, forming an explosive and toxic mixture. A spark may ignite the mixture causing extensive damage to the mine and/or mine workers may suffer health damage and death by inhaling the natural gas.

The primary component of natural gas is methane (CH_4). It makes up about 90 percent of a typical sample of natural gas. As far back as the 1670s, the term firedamp was used for the dangerous environment encountered by mine workers. Over time, the term has been used for the natural gas itself, for the methane component of the gas, or for the explosive mixture of natural gas and air.

Further Reading: *The Way Things Work*. New York: Simon and Schuster, 1971. Volume 2, pp. 74–75.

FIRE DEPARTMENT. *See* FIREFIGHTING

FIRE DETECTION. *See* SMOKE DETECTORS

FIRE DRILL

A fire drill is a practice exercise designed to familiarize people with the behaviors to be followed in case of an actual fire. Fire drills may be held in educational buildings, such as schools and colleges; healthcare facilities; certain business establishments; public assembly halls; and private residences. Fire drills are often required by public law or administrative regulation. In such cases, the manner, frequency, and reporting of such exercises is usually established by that law or regulation.

The exact form in which a fire drill takes place is usually determined by the nature of the setting. Fire drills in private homes, for example, are generally less formal than they are in a public school. However, certain general principles tend to apply in any setting. These principles include the following:[1]

- Become familiar with the alarm used to announce a fire in the building being occupied. Remain calm when the fire alarm sounds. If part of a group, follow the directions of the group leader (such as a teacher).
- Be aware of at least two exits from every room. Be prepared to assemble with

other occupants of the building at some prearranged meeting place outside the building.

- Do not open any exit door immediately. Check first to see if the door or door knob is hot or whether smoke is coming in under the door. In either case, leave by an alternative exit if possible. If an alternative exit is not available, leave the room carefully by crawling along the floor. In some cases, conditions outside the room may make it impossible to leave. In such cases, notify rescue workers by telephone (if one is available) or by signalling through a window.
- Never use elevators during a fire drill or actual fire.
- In large buildings, where possible, alert other occupants to a fire or fire drill as you leave the building.
- Do not carry anything with you that will impede your movements as you leave the building.
- If escape ladders or fire escapes are available, make sure you know how they are used.

Individuals responsible for the conduct of fire drills can improve their effectiveness in a number of ways. For example, the time at which drills are held can be staggered so that they do not occur on a regular and predictable schedule. Special emergency conditions can be indicated by placing signs, such as "Exit Blocked," throughout the building. Evaluation sessions can be held following each drill to review the strong and weak points of the exercise.

Note

1. These principles have been adapted from a number of sources providing suggested procedures to be followed in conducting a fire drill. See, for example, Further Reading section below.

Further Reading: "Fire Drills," <http://www.ci.phoenix.az.us/FIRE/firedril.html>; "School Fire Drills," <http://www.firestation24.com/drill.htm>.

FIRE EATING. *See* FIRE RESISTANCE

FIRE ECOLOGY

Fire ecology is the study of the relationship among fire, plants, animals, soil, and other elements in the environment. The science is a relatively new field of research. For much of human history and in many (although certainly not all) societies, fire has been treated as if it is a distinct phenomenon having little or nothing to do with other aspects of the environment.

It took many decades for biologists, fire scientists, and other specialists to understand that fire is a normal and natural factor determining the kinds of plants and animals found in a region and the way those organisms survive, thrive, or die out. The first textbook on ecology in which fire was treated as an ecological variable along with light, water, and other factors was probably Daubenmire's *Plants and Environment* (1947).

Research has now shown that fire has a great variety of effects on the biological and physical components of an environment. Some of those effects are reviewed below.

Germination

Some plants have evolved in such a way that their very survival depends on the regular occurrence of fire in their environment. For example, some plants reproduce by means of seeds stored within a particular type of cone known as a *serotinous* cone. The term serotinous means "late opening," and is used for cones that may hold their seeds for very long periods of time, approaching a hundred years in some cases. Fire is an essential ingredient in the germination of such seeds since it causes cones to break open and release their seeds sooner than they would under nonfire conditions. Some examples of trees with serotinous cones include the knobcone pine (*Pinus attentuata*), monterey pine (*Pinus radiata*), baker cypress (*Cupressus bakeri*), and jack pine (*Pinus banksiana*).

Other plants reproduce by means of seeds covered with very hard coats. These seeds may lie in the ground for years until fire breaks out and provides the heat needed to break open the coating and release the seed. Examples of such plants include species such as *Ceanothus*, *Arctostaphylos*, and *Rhus*.

Plants with special adaptations for the regular occurrence of fire are known as *pyrophytes*. Such plants not only depend on fire for germination and growth but also, by their own life cycle, make fire more likely. Many conifers, for example, release a resinous material when heated that quickly catches fire and increases the intensity of a burn.

Growth and Development

Some plants seem to grow especially rapidly in the period following a fire. These plants have often developed specific adaptations that allow them to survive fire and then take advantage of the postfire conditions. For example, some plants have structures known as *lignotubers* located at the junction of a root and shoot. A lignotuber contains food reserves and a bud that sprouts quickly after a fire has died out. Other plants, without such adaptations, are likely to be killed off entirely by the fire.

Some plants are also able to make unusually good use of nutrients released from organic material in the soil as the result of a fire. This may be especially the case with legumes and certain types of grasses that quickly regenerate after a fire has died out.

One interesting example of the importance of fire in plant development is the fire lily, a common plant in South Africa. The fire lily blossoms only after a fire, and then within twenty-four hours of the fire's dying out. Without fire, the plant never blossoms and can no longer survive in an area.

Nutrient Recycling

Forests, grasslands, and other biomes are gigantic nutrient recycling systems in which older plants die, decay, and return their nutrients to the soil, while younger plants pick up those nutrients and grow to mature organisms. Under most circumstances, especially in climax communities, the rate at which nutrients are removed from the soil is much greater than the rate at which they are returned.

Fire serves an important function, then, in restoring this balance of nutrient recycling. Fire quickly breaks down sticks, branches, leaves, and dead organic matter to their basic components in a matter of hours, a process that would otherwise require many years to occur. In this sense, fire is constantly rejuvenating and restoring the biological productivity of a region.

Insects and Diseases

The interaction among fire, insects, and plant diseases is complex and still not well understood. But some possible connections have already been discovered. For example, older trees are often more susceptible to attack by pests than are younger trees. As balsam fir and lodgepole pine, for example, grow older they become more susceptible to attack by spruce budworm and mountain pine beetle. These pests kill trees, and increase the amount of fuel available for a fire. When fire occurs, it not only burns up the accumulated fuel, but may also destroy the host for the pests as well as the pests themselves.

The presence of fire may, itself, influence the presence of insect pests. Round-headed woodborers are attracted by smoke to the site of a fire, for example, and begin attacking dead trees. The arrival of woodborers is usually followed shortly thereafter by northern three-toed woodpeckers, whose primary food is the woodborer. The occurrence of a fire in an area dramatically alters the predator/prey and other interrelationships among organisms in that area.

Wildlife

Studies show that wildlife tend to be relatively unaffected by even the most serious fires. In the 1988 Yellowstone fires, for example, total casualties included five bison, one black bear, two moose, four deer, and 243 elk, relatively small numbers considering the enormity and duration of those fires.

Fire may, however, have benefits for many types of wildlife by providing new habitats, increasing supplies of certain types of plant and insect food that appear after a fire, and opening up grazing and feeding areas for different animal species.

Community Composition

Fires can also alter the composition of an ecological community. Although some plants are better adapted to fire than others, there is a great deal of variability in survival rates depending on the intensity of a fire, the season at which it occurs, the species present in the area, soil composition, and other factors.

It is clear that some plants have evolved mechanisms that make them more resistant to fire, just as others have evolved mechanisms that use fire for reproduction, growth, and development. For example, a group of plants known as *scleromorphs* have developed a number of adaptations that make them especially resistant to fire. Two of these adaptations are a thick bark and tough, hard leaves. Both of these adaptations allow the plant to retain moisture and to provide a layer of protection against attack by fire.

Perhaps the best known and one of the most widely distributed scleromorphs are those that belong to the genus *Eucalyptus*. More than six hundred species of trees and bushes belong to this genera, and all have evolved a number of mechanisms that make them fire-resistant. Not only do they have thick barks and tough leaves, but they also contain buds, known as *epicormic buds*, just beneath the bark from which new branches can sprout if the bark is damaged by fire.

Some ecologists now classify plants into one of four large categories to describe how they deal with fire: *avoiders*, *evaders*, *invaders*, and *resisters*. The avoiders are plants that have not adapted to the presence of fire, are not present in areas where fire is common, and are destroyed rather quickly when fire strikes. Examples of avoiders include hemlock, western juniper, and subalpine fire. Evaders are plants in which seeds are stored in the soil or in the canopy of trees and that regenerate quickly after fire has passed through an area. Typical evaders are certain varieties of ceanothus, arctostaphylos, gooseberries, and currants. Invaders are plants that need open space to germinate and grow and often move quickly into an area after fire has passed through it. Red alder and black cottonwood are two common invader species. Finally, resisters are plants that are able to survive all but the most serious fires. They include coast redwood, douglas fir, and Pacific madrone.

Fire Ecology and Public Policy

Fire ecology is not simply pure science, research designed to produce a body of abstract knowledge about interactions in nature. Fire ecology has a very direct and significant impact on the way that societies think about and deal with fire in their forests, grasslands, and other wildland areas.

For well over a century, many nations throughout the world practiced a fire management policy that focused primarily (if not exclusively) on fire suppression. As forests became "tree farms" that produced lumber and other human needs, the goal of most fire management programs was to prevent, suppress, eliminate, and otherwise get rid of all forms of natural fire. This policy could probably only have been successful when policymakers knew little or nothing about the importance of fire in the natural environment. Hardly anyone worried about changes in the biodiversity of an area, the loss of plants and animals through extinction, and the widespread damage and injury caused by wildfires that, ultimately, did break out.

For example, in his book *World Fire*, Stephen Pyne reports that the number of plant species found in one national park in Sweden dropped from 208 in 1925 to 122 in 1970. He suggests that this change resulted from the Swedish government's concerted efforts to eliminate the role that fire had played in the park (and elsewhere in the nation) "since the retreat of the ice sheets" (Pyne 246).

In an extended review of U.S. fire management policies and practices over the past century, Pyne also points out the ecological consequences of fire suppression programs carried on in the United States:

> The pine forests changed composition as fire suppression and selective

logging promoted grand fir and Douglas fir reproduction in what became dense thickets. When they reached sapling and pole size, western spruce budworm invaded the fir in what appears, despite extensive spraying with pesticides, to be an unquenchable epidemic. In many stands, kill exceeds 60-70 percent. Simultaneously, bark beetles and root disease infested other sites, and mistletoe swarmed over second-growth ponderosa pine. The bugkill pumped up the region's fuels until, like a speculative binge, the forest threatened to crash, burn, and plunge its dependent human economy into depression. (Pyne 1997, 222)

As our understanding of fire ecology has grown over the past half century, so has our realization that some form of fire—often, a prescribed burn—must be reinstituted into forests, grasslands, and other wild areas. This change in policy and practice is likely to be only the first, if most important, fruit of the new science of fire ecology. *See also* FOREST FIRES; PRAIRIE FIRES; PRESCRIBED BURN; YELLOWSTONE FIRES OF 1988.

Further Reading: Agee, James K. *Fire Ecology of Pacific Northwest Forests*. Washington, DC: Island Press, 1993; Debano, Leonard F., Daniel G. Neary, and Peter F. Folliott. *Fire's Effects on Ecosystems*. New York: John Wiley & Sons, 1998; Hendee, John C., George H. Stankey, and Robert C. Lucas. *Wilderness Management*. Golden, CO: North American Press, 1990, Chapter 12; Kozlowski, T.T., and C.E. Ahlgren, eds. *Fire and Ecosystems*. New York: Academic Press, 1974; "Native Plants Require Fire," <http://www.epa.gov/glnpo/greenacres/ga-fire.html>, 07/05/98; Nodvin, Stephen C., and Thomas A. Waldrop, eds. *Fire in the Environment: Ecological and Cultural Perspectives*. Washington, DC: U.S. Forest Service, 1991; Pyne, Stephen J. *World Fire: The Culture of Life on Earth*. Seattle: University of Washington Press, 1997, passim; "The Role of Fire in Ecosystems," <http://www.blm.gov/education/fire/fire1.html>, 07/05/98; Whelan, Robert J. *The Ecology of Fire*. Cambridge, UK: Cambridge University Press, 1995; Wright, Henry A., and Arthur W. Bailey. *Fire Ecology: United States and Southern Canada*. New York: Wiley-Interscience, 1982.

FIRE ENGINE

A fire engine is a mobile piece of equipment used to fight fires. Modern fire engines (or "fire trucks") come in a variety of models designed for specific firefighting functions. Pumpers, for example, are designed primarily to provide the water needed to fight a fire, while a ladder truck ("hook and ladder") carries the equipment needed to bring firefighters closer to the actual fire.

History

The term *fire engine* refers specifically to a piece of apparatus that can be moved from one place to another with relative ease. In that sense, true fire engines have existed only since the nineteenth century. However, inventors were developing the elements of fire engines as far back as the third century B.C.

During that period, the Greek inventor Ctesibius designed and built a device called a *siphona* for fighting fires. The siphona consisted of a platform that rested in a pool of water and contained two suction pumps on either end. An outlet pipe in the center of the device was activated by a crossbar connected to the two pumps. When two men raised and lowered the crossbar, water was pumped out of the source and into the outlet pipe. It could then be sprayed through a hose on a fire.

The siphona could hardly be considered a fire engine, of course, because it was not readily mobile. However, it did contain the concept of using a force to spray water at a fire from some distance away. Pumps of this design were apparently used widely in the ancient world. They were, however, very inefficient. They required a dependable source of water, of course, which might or might not be close to a fire. They also had to be placed so close to a fire that they ran the risk of catching fire themselves.

The next major advance on the siphona design was the *syringe pump*, first used in Egypt in the first century B.C. The syringe pump consisted of a large container of water capped by a tapering lid. When water was forced out of the container, it squirted through a point in the lid in a stream that could be directed at a fire. The container was placed on a wheeled vehicle that could be pushed to a fire by firefighters. The water was forced out of the container by turning a screw at the bottom of the container.

After the 1600s, the most popular form of the fire engine was a combination of the siphona and the syringe pump. A large barrel holding water was placed on a wheeled vehicle or sled on which it could be rolled or dragged to a fire. The water was forced out of the barrel by means of a pump manned by two or more men. Water expelled from the barrel was replenished by a bucket brigade of men who continuously passed containers full of water to the device.

All siphona- and syringe-pump-type devices shared two major problems in common. First, they were unable to produce very powerful sprays of water and were not, therefore, very effective in putting out fires. Second, they had to be placed so close to a fire that they often were destroyed by flames themselves. These problems were largely solved by a series of improvements made to the siphona/syringe-pump design after 1600.

The first of these improvements was the development of a reliable flexible hose by the Dutch engineer Jan van de Hieden in the early 1670s. Prior to van de Hieden's invention, fire hoses were made of reinforced cloth or other materials. These hoses could not be made to any great length and could not be connected to each other very easily. Van de Hieden's hoses, by contrast, were stronger, sturdier, and easier to attach to each other. With such hoses, fire engines could be stationed several feet from the fire on which they were to be used rather than in such proximity that they ran the risk of catching fire.

By the 1800s, firefighting systems had also been improved so that engines could draw water on municipal water supplies, as they do today, rather than to depend on bucket brigades. A bucket brigade is inherently inefficient, of course, because people tend to drop buckets or spill water. They were often unable to replace a siphon/syringe device as fast as the device was emptied.

Modern Developments

The single most important improvement in fire engines, however, was the development of steam power in the 1800s. At first, steam power was used to power the water pumps on an engine. The engine itself was still drawn to a fire by horses or pulled by humans. But the improvement in efficiency of the engine was remarkable. The first steam-powered

Denver firefighters on an early twentieth century horse-drawn pumper, usually pulled by three horses. *Courtesy of The Denver Firefighters Museum.*

engine was built in London in 1829. It had the capacity to shoot a stream of water thirty meters (about 100 feet) into a fire. This made it the most powerful firefighting device invented at the time.

By the late 1880s, steam power was being replaced by the internal combustion engine. A device introduced in 1888 in Great Britain, for example, used a gasoline-powered pump to operate its fire hoses, although the fire engine on which it rode was still towed to the fire by horses.

Shortly after the turn of the century, however, fire engines were being built in which both the water pumps and the vehicle itself were powered by gasoline. The first device of that kind in the United States was built by the Waterous Fire Engineer Works in St. Paul, Minnesota in the early 1900s.

Ladders

The first use of ladders in firefighting can be traced to the eighteenth century when they were used primarily as a way of helping people escape from fires. During the 1830s, efforts were made to develop ladders constructed of sturdier materials, such as steel, that could be joined to each other in greater lengths. In 1868, a San Francisco firefighter by the name of Daniel Hayes invented the first ladder truck. Hayes' truck carried a ladder that could be raised, extended, and aimed at almost any part of a burning building. In 1906, a German engineering company developed the first telescope ladder capable of rotating through 360°. Over the last century, engineers have created stronger, more flexible ladder systems that allow firefighters to reach heights of up to about fifty meters (160 feet) on portions of a burning building.

The Modern Fire Engine

The two main types of fire engines in use today are the pumper and the ladder truck. The pumper is responsible for providing the water needed to fight a fire. The pumper may carry its own supply of water in tanks mounted on the truck or, more commonly, it has the means to connect to a municipal water supply that provides a virtually constant supply of water. Pumpers also carry a large supply of hoses to bring the water to the fire. It is not unusual for a pumper to carry up to 300 meters (1,000 feet) of hose. The pumper's engine may be capable of supplying a spray of water at a pressure of 150 pounds per square inch at anywhere from 500 to 1,500 gallons per minute.

Pumpers may be modified for dealing with special types of fire, such as those that may occur in grasslands, along a freeway, or at an airport. For example, a pumper at an airport fire department carries equipment for the manufacture of large quantities of foam that may be needed either to extinguish a fire or to coat a runway to facilitate an aircraft's landing.

The primary function of ladder trucks is, of course, to provide accessibility for firefighters to fires in the upper reaches of a building. Ladder trucks are easy to recognize because they are often built in an articulated semitrailer style with a second driver seated at the back of the truck to assist in negotiating the truck around corners. Ladder trucks may carry up to sixty meters (200 feet) of ladder in hinged sections that can be brought into position by a hydraulic pump. *See also* FIREFIGHTING; HALL OF FLAME.

Further Reading: Conway, W. Fred. *Chemical Fire Engines*. New Albany, IN: Fire Buff House, 1987; DaCosta, Phil. *100 Years of America's Firefighting Apparatus*. New York: Bonanza Books, 1964; Daly, George Anne, and John J. Robrecht. *An Illustrated Handbook of Fire Apparatus*. Philadelphia: INA Corporation Archives Department, 1972; "Fire Service History," <http://www.fire.org.uk/>, 12/12/98; Halberstadt, Hans. *The American Fire Engine*. Osceola, WI: Motorbooks International, 1993; Mahoney, Gene. *Introduction to Fire Apparatus and Equipment*, 2nd edition. New York: Fire Engineering Book Service, 1985; Wood, Donald F. *American Volunteer Fire Trucks*. Osceola, WI: Motorbooks International, 1993.

FIRE EXTINGUISHER

Fire extinguishers are handheld or portable devices for putting out fires. They are used most effectively on small fires and/or on fires that have just begun to burn.

The first fire extinguisher was invented by George William Manby (1765–1854), a retired member of the British militia. Manby appreciated the importance of having such a device after watching firefighters in Edinburgh attempt to put out fires in the upper stories of a burning building. His invention consisted of a copper cylinder that held three gallons of water. The cylinder also contained air under pressure. When a valve was opened in the cylinder, the compressed air pushed water out of the cylinder in the form of a spray.

Types of Fire Extinguishers

One disadvantage of Manby's original invention was its bulkiness. It could not easily be carried around to use on small fires. The solution to this problem was the chemical fire extinguisher, first invented by a Russian, Alexander Laurent. In Laurent's device, a chemical reaction produced inside a metal container forced water and carbon dioxide out of the container and onto the fire. Laurent's invention became the model for many later fire extinguishers, one of the most famous of which was the soda-acid extinguisher.

The soda-acid extinguisher consists of a metal cylinder, much like the bright red fire extinguishers seen everywhere today. The cylinder contains a solution of sodium bicarbonate ($NaHCO_3$) and a small vial of sulfuric acid (H_2SO_4). When the cylinder is tipped upside down, the acid mixes with the sodium bicarbonate solution to produce carbon dioxide (CO_2):

$$H_2SO_4 + 2\ NaHCO_3 = Na_2SO_4 + 2\ H_2O + 2\ CO_2$$

The carbon dioxide gas formed in this reaction acts as a propellant, pushing water out of the cylinder and through a hose and nozzle onto the fire.

The soda-acid fire extinguisher is a simple, inexpensive firefighting device for many types of fires. However, it cannot be used on certain types of fires, such as those in which fats, grease, or oil are burning. It also cannot be used on electrical fires.

In the case of burning oils, water is not an effective agent because oils are less dense than water. Thus, even if water is sprayed on an oil fire, the oil floats on top of the water and continues to burn. In the case of an electrical fire, an electrical current decomposes water. The hydrogen gas formed in this reaction burns easily, so that adding water to an electrical fire makes the situation worse rather than better.

A number of alternative types of fire extinguishers have been invented to remedy the limitations of the soda-acid extinguisher. For example, some fire extinguishers contain carbon dioxide gas under high pressure in a steel cylinder. When a valve is released, the gas escapes from the cylinder, forming an inert covering that prevents oxygen from reaching the fire.

Another type of fire extinguisher is a foam device. A foam device is one in which a chemical reaction releases a thick, bubbly material

Fire extinguisher used for putting out small fires. *Courtesy of Ansul Incorporated.*

that covers a fire and cuts off its oxygen supply. For example, in one type of foam extinguisher, aluminum sulfate ($Al_2[SO_4]_3$) reacts with sodium bicarbonate ($NaHCO_3$) to produce carbon dioxide. The carbon dioxide bubbles out of the extinguisher in a frothy form that looks something like soap suds. The excess of carbon dioxide bubbles cuts off the oxygen supply and helps to cool the temperature of the fire.

For many years, a very popular type of fire extinguisher contained a chemical known as Halon. Halon is the registered trademark for a class of compounds known as polytetrafluoroethylene. These compounds all contain carbon, bromine, fluorine, and, sometimes, chlorine. The members of the Halon family are distinguished from each other according to their molecular structures; that is, the relative amount of each element they contain. For example, the compound known as Halon-1211 has the chemical formula $CBrClF_2$. Halon-1301 has the formula $CBrF_3$, and Halon-2402 has the formula CBr_2F_2.

Halons have been very popular as fire extinguishers for two reasons. First, they are not combustible (they do not burn) themselves. So, like carbon dioxide, they can put out fires by smothering them. Second, Halons actually react chemically with materials that are burning, interrupting the process by which the materials react with oxygen in the air.

Questions, however, have arisen regarding the use of Halons in fire extinguishers (and other applications). Scientists have found that these compounds escape into the atmosphere and break down. When they do so, they release bromine atoms into the atmosphere. These bromine atoms react with ozone molecules in the stratosphere. This reaction has potentially serious consequences for life on Earth since ozone protects organisms from the harmful effects of ultraviolet radiation in sunlight.

Because of the environmental concerns about Halons, these chemicals are now being replaced by other substances as fire extinguishers. For example, chemicals known as hydrochlorofluorocarbons (HCFCs) have many properties similar to those of Halons. However, they are less likely to break down in the atmosphere and to damage ozone molecules. Some types of fire extinguishers, therefore, now contain HCFCs rather than Halons as the extinguishing agent.

Principles of Fire Extinguishing

In order to extinguish a fire, at least one of three steps must occur. First, the burning material must be removed. Second, the burning material must be prevented from having access to oxygen. Third, the temperature of the burning material must be reduced. In actual practice, most fire extinguishers operate on one or both of the last two steps.

For example, suppose a pile of leaves catches fire. The easiest way to put out the fire may be to pour water on the fire. The water lowers the temperature of the leaves below the ignition point, the temperature at which the leaves will burn. Steam formed by the water may also prevent oxygen from reaching the fire.

Carbon dioxide fire extinguishers operate primarily on the principle of preventing oxygen from reaching the fire. Carbon dioxide is more dense than oxygen. When sprayed on a fire, therefore, the carbon dioxide spreads out over the fire, pushing oxygen up and out of the way. Since carbon dioxide does not burn and does not support combustion (help other materials burn), it prevents the material from burning.

Classes of Fires

The National Fire Protection Association has developed a system for classifying fires in terms of the most effective method by which they can be extinguished. That system includes four classes, as follows:

Class A fires are the simplest type of fire, often involving wood, cloth, and paper. They can be extinguished by water or water-based chemicals or with certain dry chemicals. These materials lower the temperature of the fire and prevent oxygen from reaching the burning material. A bucket of water or a soda-acid fire extinguisher can be used with Class A fires.

Class B fires are more serious fires that usually involve a flammable liquid. They must be extinguished by cutting off the supply of oxygen, by preventing the burning material from giving off combustible vapors, or by interfering with the burning process itself (as was described above for the Halon reaction). Water or water-based materials cannot be used on Class B fires. Instead, certain dry chemicals, carbon dioxide, or Halon fire extinguishers (or environmentally-acceptable alternatives) can be used.

Class C fires are those that involve electrical equipment. The extinguisher used cannot be a conductor of electrical current (as water is). Dry chemicals, carbon dioxide, and Halon extinguishers can be used with Class C fires.

Class D fires are those that involve burning metals, such as burning magnesium. These fires generate very large quantities of heat and require special fire extinguishers that lower the temperature of the fire and prevent oxygen from reaching the burning material. Carbon dioxide and Halon extinguishers are designed for Class D fires. A special powder known as Met-L-X has been developed for use with Class D fires.

Further Reading: Bryan, John L. *Fire Suppression and Detection Systems* (Facsimile of original). New York: MacMillan Publishing Company, 1993; *Fire Extinguishers, Rating and Fire Test, Ul 711*. Northbrook, IL: Underwriters Laboratories, 1995; *NFPA 10 Standard for Portable Fire Extinguishers: 1998 Edition*. Quincy, MA: National Fire Protection Association, 1998; Rossotti, Hazel. *Fire*. Oxford: Oxford University Press, 1993, Chapter 20.

FIREFIGHTER

A firefighter is a man or woman whose primary job it is to put out fires. Firefighters may also have many other responsibilities in connection with this job, including fire education, fire prevention, and emergency medical services (EMS).

Census of Firefighters

The total number of men and women employed as firefighters in 1998 was about 314,000. This number includes both full- and part-time employees. More than 90 percent of these firefighters were employed by municipal or county fire departments. Such departments range in size from a small handful to many thousands of firefighters. A relatively small fraction of firefighters is employed by other agencies, such as federal or state agencies, airports, and private firefighting companies.

About three-quarters of all firefighters are volunteers, men and women who serve their community as unpaid workers. As of 1999, there were about 815,500 volunteer firefighters in the United States. The majority of volunteer firefighters work in small communities. In 1997, for example, the last year for which data are available, there were 410,500 volunteer firefighters and only 6,800 paid firefighters in communities with a population less than 2,500. Overall, there were 754,200 volunteers compared to 69,150 in all U.S. communities with populations of less than 25,000. By comparison, there are relatively few volunteer firefighters in larger communities. In 1997, as an example, there were 35,000 paid versus 150 unpaid firefighters in communities with populations greater than one million. The National Association of State Foresters has estimated that volunteer firefighters save U.S. communities about $36.8 billion annually.

The average annual salary earned by a firefighter depends on experience and title. In 1998, the median annual salary for a firefighter in the United States was $31,170. Firefighters of higher rank had higher earnings: $36,000 for an engineer; $38,100 for a fire lieutenant; $41,100 for a fire captain, $50,650 for a battalion chief; and $59,350 for a fire chief. Many firefighters augment income from these base salaries, however, by putting in substantial amounts of overtime work.

Firefighters come from virtually every part of society, although the vast majority are white males. In 1998, there were about 5,200 full-time women firefighters in the United States, about 60 of whom were in upper-level positions, such as district chief, battalion chief, or chief. Eleven fire departments, ranging in size from Cobb County, Georgia (550 members) to Tiburon, California (21 members), were headed by women. In addition to paid firefighters, there are about 30,000–40,000

Firefighters extinguishing hotspots, making sure that the fire is totally out. *Courtesy of the Bureau of Land Management at the National Interagency Fire Center.*

women volunteer firefighters in the United States. The professional organization representing women firefighters in the United States is Women in the Fire Service, Inc., a nonprofit, membership-based organization founded in 1982.

Fire departments outside the United States tend to have fewer women members than those in the United States. There are, for example, currently about 200 full-time and 200 volunteer women firefighters in Great Britain. Other nations with women firefighters include Australia, Brazil, Canada, Chile, Costa Rica, France, Germany, the Netherlands, New Zealand, Panama, South Africa, Trinidad, and Tobago.

Black men and women make up about 10 percent of all firefighters. The professional organization for Black firefighters is the International Association of Black Professional Fire Fighters (IABPFF), a nonprofit organization founded in 1969 to encourage recruitment of Black men and women into the fire service, improve relations between firefighters and communities, and better relationships

between Blacks and non-Blacks in fire departments. IABPFF maintains "Black Women in Fire Service" committees in many parts of the United States. The goal of these committees is to address the special issues faced by Black women in the firefighting profession.

Job Description

At one time, a career in firefighting meant that one dealt almost exclusively with fire suppression. Today, firefighting has taken on a much broader meaning. For example, the majority of calls to which firefighters respond do not involve fires, but medical emergencies. Firefighters are called on to treat injuries, accidents, emergency health problems, and other medical crises. About one-half of all fire departments now provide ambulance or other ambulatory EMS services along with their firefighting equipment.

Fire prevention has also become a major part of a firefighter's responsibility. Firefighters may be called upon to visit schools and give talks, lead tours through a fire house, give advice on proposed fire legislation, and provide other forms of education about fire prevention.

Other areas in which firefighters may specialize include: arson investigation; hazardous materials treatment; building inspections; sprinkler and alarm systems; OSHA (Occupational Safety & Health Administration) regulations; hydraulics and water distribution systems; design, purchase, and maintenance of equipment; supervision and administration; and fire protection planning.

Requirements and Training

All firefighters are expected to meet certain minimal standards dictated by the nature of the job. These standards include physical strength and stamina, sound medical condition, and ability to pass certain written examinations. Prospective firefighters must also pass a drug screening. Applicants must normally be at least eighteen years of age and have completed high school or a high-school equivalency examination.

Fire departments typically require that applicants complete some specific training program. In larger departments, that program may be operated by the department itself at facilities built for that purpose. In smaller departments, applicants may be accepted on an apprenticeship basis during which they learn the business of firefighting on a day-to-day basis.

Public and private institutions that provide training in firefighting are also available. Hundreds of community and four-year colleges offer programs leading to a degree in fire science. Increasingly, the number of people who go into firefighting have completed at least one year of post–high school education before applying for a job.

As an example, the Massachusetts Fire Academy, a division of the Massachusetts Department of Fire Services, annually offers nearly 200 different courses in fire sciences. These courses range in length from three to ninety hours and cover topics such as aerial ladders, basic ice rescue, cardiac emergencies, elevator rescue, fire control, firefighter in court, ground ladders, hazardous material recognition and identification, large diameter

hose, legal issues, positive pressure ventilation, rural water supply, and school violence preparedness. By completing various combinations of these courses, an individual may receive various types of certifications, such as firefighter I or II, fire inspector I or II, hazardous materials technician, or fire officer I or II.

There is now a growing trend in the United States to achieve some level of standardization in the training of firefighters. The lead organization in this trend is the National Board on Fire Service Professional Qualifications (NBFSPQ). The board was created in 1972 by a group of firefighting professionals who were concerned about the uneven level—and, in some cases, the inadequate level—of training being provided to firefighters in various parts of the country. The board created a National Professional Qualifications System that is now used to accredit the training programs of various fire-training entities. Such entities submit their training programs for review to the NBFSPQ, which passes on the adequacy of the program. Programs that pass the NBFSPQ review are provided a Certificate of National Certification, listed in the NBFSPQ national registry, and encouraged to develop reciprocity agreements with other register programs throughout the nation.

Programs for training in wildfire fighting are generally similar to those described above, although they often include specialized topics, such as fire watch, interface fires, and controlled burn techniques. Currently, about 270 community and four-year schools offer some form of wildfire training instruction in the United States.

Volunteer firefighters are expected to master many of the same skills as those required of their paid colleagues. Programs for volunteers are generally shorter and less intense than those for professional firefighters, however. For example, prospective volunteer firefighters may be expected to attend class one or two nights a week and spend part of their weekends developing firefighting skills. Such programs may last no more than a few months before the individual is certified as a volunteer firefighter. Refresher courses and training sessions are almost always required of volunteer firefighters also.

Issues

As is true in many occupations, firefighters must deal with a number of social, economic, and political issues. Members of minorities, including women and racial minorities, tend to be confronted with such issues most frequently. For example, the IABPFF was organized at least partly because "many union locals and city governments failed to institute an affirmative action plan in the testing, recruiting and promotional process areas" (IABPFF Main Page). One of its announced goals is "to compile information concerning the injustices that exist in the working conditions in the fire service, and implement action to correct them" (IABPFF Main Page).

Women firefighters also deal with a number of issues in the workplace. Women in the fire service have identified issues that fall into four primary categories: resistance from some elements of the workforce; institutional barriers; effects of the male firefighting tradition, and of societal beliefs about women and men; and obstacles that are not gender-specific—that all firefighters face. As an example of the first class of issues, many members of the fire service question whether women have the physical, mental, and emotional skills necessary to serve as firefighters. Others are attached to the all-male environment characteristic of most firehouses.

Some examples of institutional barriers include the fact that most fire stations have sleeping, bathing, bathroom, and other facilities for only one gender, men, with little provision for women's needs. Protective gear has generally been designed for men only and may not always be suitable for women. The male firefighter tradition has also contributed to the creation of self-doubt among women with regard to their own ability to serve as firefighters. In addition, many women might want to serve as firefighters, but don't necessarily want to take on the additional role of being the first of their kind in a department.

Finally, firefighting is a physically challenging and dangerous occupation for anyone, no matter one's gender. Stress, lack of sleep, and long work hours are difficult for both men and women.

Further Reading: Basic Information on Women in Firefighting," <http://www.wfsi.org/WFS.basicinfo.html>; "Certification Frequently Asked Questions," <http://www.state.ma.us/dfs/mfa/cert_faq.htm>; Chetkovich, Carol. *Real Heat: Gender and Race in the Urban Fire Service.* Rutgers, NJ: Rutgers University Press, 1997; Dunn, Vincent. *Command and Control of Fires and Emergencies.* Saddle Brook, NJ: Fire Engineering Book Department, 2000; Ertel, Mike, and Gregory C. Berk. *Firefighting: Basic Skills and Techniques.* Tinley Park, IL: Goodheart-Wilcox, 1997; "Fire and EMS Resource Center," <http://www.fire-ems.net/resources/schools.html>; *Firefighter's Handbook: Essentials of Firefighting and Emergency Response.* Albany, NY: Delmar Publishing, 2000; "IABPFF Main Page," <wysiwyg://171/http://www.iabpff.org/index2.htm>; "NPQS (National Board on Fire Service Professional Qualifications)," <http://www.npqs.win.net/>; "Schools and Education," <http://www.wildfiretrainingnet.com/links/cat13.shtml>; Yoder, Curt, and Karen Yoder. *The Heart behind the Hero.* Trabuco Canyon, CA: Stoney Creek Press, 2000.

FIREFIGHTING

In its strictest sense, the term *firefighting* refers to the process used to put out fires. In the modern world, however, firefighting is a much larger profession that involves research on fire, training, and fire prevention efforts as well as the actual practice of extinguishing fires. This entry deals primarily with the fighting of fires in urban areas, while the entry on Fire Management focuses primarily on the fighting of wildland fires.

History

For most of human history, only very primitive methods of fighting fires were available. By far the most common of these methods was the bucket brigade. A bucket brigade consists of a line of men and women who pass buckets full of water from a water source to a burning building. This method is inherently inefficient because it provides only small

Firefighter keeping flames inside the established fire line. *Courtesy of the Bureau of Land Management at the National Interagency Fire Center.*

amounts of water at a time. Also, individuals at the end of the line are generally not able to get very close to the fire itself. For these reasons, fires have traditionally been one of the most feared—because of the most uncontrollable—of all human disasters.

Inventors as far back as the fourth century B.C. searched for mechanical devices that would replace or augment the bucket brigade. For example, the Alexandrian Greek Ctesibius developed a simple hand pump that could be used to propel water at a fire from a short distance away. His invention, eventually called the *siphona,* seems to have become popular in many urban areas. But it represented only a modest improvement over the bucket brigade.

The first organized system for fighting fires was apparently developed by the Romans. In about 32 B.C., Caesar Augustus formed fire brigades known as the *vigeles.* These primitive "fire departments" consisted of 100–1,000 men each and were headed by a *praefectus vigilum,* or "fire chief." When a fire was discovered, signals were sent out by

means of trumpet-like horns to the nearest fire brigade. The fire brigade was responsible for organizing the bucket brigade as well as such other activities as might be possible. For example, they sometimes found it necessary to pull down burning buildings to keep a fire from spreading.

Early British Efforts. Firefighting techniques did not advance very far over the next 1,600 years. In most instances, efforts were made to *prevent* fires since methods for putting them out or keeping them from spreading were largely ineffective. For example, William the Conqueror issued an edict in the eleventh century calling for a nightly firewatch curfew. Officers walked through the streets warning people to put out their candles and extinguish their fires. Shortly thereafter, King Richard I ordered the construction of high, thick walls between urban neighborhoods to prevent the spread of fire from one region to another. These walls were the earliest form of the modern fire wall.

Particularly serious conflagrations were often the cause of new concerns for the development of better firefighting methods. The Great Fire of London in 1666, for example, destroyed a quarter of the city. City officials realized that more efficient systems for preventing and extinguishing fires would be necessary to prevent a repetition of this catastrophe. Fortunately, this concern coincided with the birth of modern technology, and a number of new firefighting devices were soon to appear. For example, the Dutch engineer Jan van de Hieden invented a new type of fire hose in the early 1670s. His hose was made of leather and was sturdier than cloth hoses previously used by firefighters. Van de Hieden's hoses could be joined together to make very long systems that allowed firefighters to remain at some distance from the fire itself.

At the same time, more advanced firefighting vehicles were being developed. For example, steam power was being used to pump water from large cisterns with enough force to spray dozens of feet from the vehicle. Longer hoses, ladders, and more powerful sprays meant that firefighters could attack a fire with much more efficiency and much less danger to themselves than had been possible with more primitive equipment.

A powerful force in the development of well-organized groups of firefighters was the introduction of fire insurance in Great Britain in the late 1600s. Fire insurance companies were organized to collect premiums from owners of buildings who wanted financial protection in case their property was destroyed by fires. To protect their policyholders, fire insurance companies organized companies of paid firefighters who responded to calls at buildings covered by their companies. These buildings were generally marked with a metal tag so that firefighters could find them in case of a fire. The system developed by fire insurance companies was not very effective, however. In addition to the technical problems that remained with firefighting equipment, political rivalries developed among different companies. A fire brigade might respond to a call to a burning building covered by a rival company and refuse to put out the fire because it was not one of "their" buildings. Indeed, rival companies sometimes spent more time fighting each other (to convince prospective customers of choosing their service) than they did fighting fires!

The First Fire Departments. The first full-time, formally organized group of paid

Denver firefighters, c. 1910, on a Seagrave Arial Ladder fire engine. *Courtesy of The Denver Firefighters Museum.*

firefighters in Great Britain was created in 1824. The city of Edinburgh, Scotland, arranged for the consolidation of a number of fire brigades operated by insurance companies. The new group was called the Edinburgh Fire Engine Establishment and was headed by James Braidwood. Braidwood soon became famous for his rigorous training of firefighters and introduction of new and improved firefighting technology. He was so successful in Edinburgh that he was hired away by the city of London in 1832 to form a comparable firefighting department in that city.

Braidwood insisted on a vigorous training program for his firefighters, drilling them at any time of the night or day. He insisted that they become skilled in the correct placement of ladders on roof tops so that they could get as close as possible to a fire. He also trained them in the ability to move their fire equipment quickly through the narrow streets of Edinburgh and required that they actually be involved in fighting a fire within a minute of their arrival at the burning building. Braidwood also emphasized the importance of moving fire hoses quickly and efficiently through narrow passageways and around street corners to get the stream of water they produced as close to the fire as possible.

Firefighting in Colonial America was handled by a combination of private and public organizations. By the mid-1600s, most large cities had adopted some sort of publicly mandated and paid-for system to watch for, prevent, and fight fires. In 1648, for example, New York governor Peter Stuyvesant appointed four men to act as fire wardens. Their job was to inspect all chimneys and report violators of regulations. In 1631, Massachusetts governor John Winthrop outlawed wooden chimneys and thatched roofs in the colony.

The first formal fire department in the United States was organized in 1678 in Boston. The department consisted of twelve men and one captain who were hired originally to care for a new "state of the art" fire engine that had just been purchased from England.

The fire engine consisted of a wooden tub three feet long and eighteen inches wide with carrying handles, a pump, and a small hose.

In spite of these early efforts, many historians credit Benjamin Franklin with establishing the first full-service fire department in Philadelphia in 1740. Franklin's department was a privately operated, subscription organization, similar to earlier firefighting associations in England.

Elements of Modern Firefighting

A firefighting episode begins with notification to a fire department of a fire. That notification may come in a variety of ways, such as by telephone call. Probably the most common system of reporting a fire, however, is the fire alarm call box. The earliest fire alarm call box was invented in 1830 by Ithiel Richardson. Today, call boxes are sophisticated electronic devices that relay to firefighters the location from which a fire is being reported.

Response to an alarm by the nearest fire station is typically very fast, usually taking less than a minute. Appropriate firefighting equipment is dispatched to the fire, including one or more pumpers, ladder trucks, rescue trucks, and other specialized equipment.

The act of bringing a fire under control usually consists of five major steps. The first of these steps involves pinpointing the exact location of a fire in a building. Next, firefighters attempt to determine whether anyone is in the burning building and, if so, where they are located. Removing endangered occupants then becomes the next step in the firefighters' agenda.

The next step in firefighting may be to limit the fire to some given area. For example, if a fire has already reached significant proportions, it may be necessary to pull down adjacent endangered structures or to cover them with a wall of water to prevent them from also igniting. Only at this point can firefighters turn their attention to the main fire itself. Water and other fire extinguishing chemicals can then be used to put out the main fire.

When the fire appears to have been extinguished, "mopping up" operations are necessary as the final step. Primary in importance among these operations is the search for any small remaining flames or smoldering materials that may be rekindled at a later time. *See also* FIRE ENGINE; FIREFIGHTER; FIRE INSURANCE; FIRE MANAGEMENT.

Further Reading: Cannon, Donald J., general ed. *Heritage of Flame—The Illustrated Encyclopedia of Early American Firefighting.* Garden City, NY: Doubleday, 1977; Ditzel, Paul. *Fireboats, A Complete History of the Development of Fireboats in America.* New Albany, IN: Conway Enterprises, 1989; Ditzel, Paul. *Fire Engines, Firefighters.* New York: Crown Publishers, 1976; Ditzel, Paul. *Firefighting, A New Look in the Old Firehouse.* New York: Van Nostrand Reinhold, 1969; Goudsblom, Johan. *Fire and Civilization.* London: Allen Lane, 1992, pp. 115; 143–150; Green-Hughes, E. *A History of Firefighting.* Paris: Klincksieck, 1965; Lyons, John W. *Fire.* New York: Scientific American Books, 1985, Chapter 6; "Fire Service History," <http://www.users.globalnet.co.uk/~pekszy/>, 12/12/98; Martin, Stanley B., and A. Murty Kanury, "Fire Technology," in *McGraw-Hill Encyclopedia of Science & Technology*, 8th edition. New York: McGraw-Hill Book Company, 1997. Volume 7, pp. 138–142; O'Brien, Donald M. "Fire Fighting and Prevention." In *Encyclopedia Americana.* Danbury, CT: Grolier, 1999. Volume 11, pp. 244–253; Rossotti, Hazel. *Fire.* Oxford: Oxford University Press, 1993, Chapter 21; Souter, Gerry, and Janet Souter. *The American Fire Station.* Osceola, WI: Motorbooks, International, 1998; Wallington, Neil. *Images of Fire: 150 Years of Firefighting.* Newton Abbott, England: David & Charles, 1989.

FIRE HYDRANT

A fire hydrant is a pipe through which water can be released from an underground reservoir, such as a municipal water system, for the purpose of fighting fires. Fire hydrants come in many designs, but have certain common characteristics. They usually have a stubby barrel shape, about twenty-four inches in height and eight inches in diameter. Traditionally, fire hydrants have been painted red, to make them easy to find in an emergency. Some communities are now painting their fire hydrants yellow since that color may be even easier to see.

At one time, the water needed to fight fires typically came from one of two sources: an underground cistern or a tank mounted on a cart or truck. In either case, water was pumped from the cistern or tank into fire hoses.

As communities grew, they began to develop more sophisticated methods for obtaining water. One of the earliest of these systems was developed by the city of Philadelphia. In 1801, the city had opened one of the largest and most modern public water systems in the United States. Water was transported through wooden pipes buried beneath city streets. At certain critical points throughout the city, T-shaped metal pipes were inserted into the water pipes. These pipes had two openings, one that could be used for drinking water, and the other for use in case of fires.

In some communities, less sophisticated methods were used for obtaining water from public water supplies. For example, firefighters might simply cut a hole into the wooden pipes of such systems and attach their hoses to the opening. When the fire had been put out, firefighters sealed the hole by forcing in a wooden plug. The term *fire plug*, sometimes used to describe a fire hydrant, arose from this practice.

The fundamental design of a fire hydrant is relatively simple. A valve at the bottom of the hydrant can be opened or closed to allow water to enter the hydrant from the water supply. Screw caps on two or more sides at the top of the hydrant allow hoses to be attached to the pipe. Most fire hydrants today are of the "dry-barrel" design. Such hydrants get their name from the fact that water is released from the hydrant when it is not in use. Dry-barrel hydrants have the advantage of preventing water inside the pipe from freezing in cold weather, and they also reduce the opportunity for children to open hydrant valves and release water from the system.

Further Reading: "FireHydrant.Org," <http://www.firehydrant.org>; *Installation, Field Testing, and Maintenance of Fire Hydrants* AWWA Manual M17, 3rd edition. Denver: American Water Works Association, 1989; *The Fire Protection Handbook*, 18th edition. Quincy, MA: National Fire Protection Association, 1999.

FIRE INSURANCE

Fire insurance is protection a person or company can purchase against financial losses resulting from a fire. As with other forms of insurance, the insured person or company makes regular payments (premiums) based on the relative amount of risk and the value of the property insured. In case of an actual fire, the insurer pays the insured some agreed-upon sum of money, the face value of the policy.

History

Probably the first mention of fire insurance known is that found in the Code of Hammurabi from the seventeenth century B.C. Named after an early Babylonian ruler, the code provided that anyone who lost property because of a fire could appeal to the state for financial reimbursement. Such provisions were rare, however, until the late 1600s. Then, disastrous fires sweeping through large cities made people realize the value of banding together to create fire insurance programs. The Great Fire of London in 1666, for example, left more than 200,000 people homeless and caused millions of dollars in damage to property.

Soon after the London fire, groups of individuals and companies began to join together to form fire insurance companies. These companies were paid for by the premiums paid by members and were composed of paid and unpaid firefighters who were called to fires at properties owned by members. Local and national governments were generally not involved in the operation of these companies in any way.

Modern firefighting companies arose largely as a function of fire insurance companies. These companies had a stake in reducing losses due to fires at member properties. On the other hand, they had no stake at all in fires that broke out at properties covered by other fire insurance companies. Confusion and conflicts over firefighting at properties covered by rival insurance companies were not uncommon during the late seventeenth and eighteenth centuries.

Early Fire Insurance

The first fire insurance company in the United States was the Philadelphia Contributionship. It was established on 13 April 1752 and continues to exist today. Over the next half century, fire insurance companies grew up in other major cities. These included the Mutual Assurance Company, in 1784; the Knickerbocker Fire Insurance Company, in 1787; the Baltimore Equitable Society, in 1794; the Insurance Company of North America, in 1794; and the Hartford Insurance Company, in 1803.

Many early companies had one common financial problem. They charged a relatively low premium for membership. As a result, they never developed large monetary reserves sufficient to pay off the costs resulting from a very large fire. This problem became obvious after a huge fire that swept through New York City on 16 December 1835. Nearly 700 buildings were destroyed with a financial loss far greater than insurance companies could cover. As a result, many of those companies were forced to declare bankruptcy.

One result of the 1835 fire in New York was that governments became more closely involved in the control of fire insurance companies. In 1837, for example, Massachusetts passed a law requiring that companies be sufficiently capitalized to cover major disasters like the New York fire. Other states soon followed suit in setting standards for fire insurance companies.

Another important step was the creation in 1866 of the National Board of Fire Underwriters (NBFU). The board was created by major fire insurance companies to establish industry-wide standards for determining risk and premium rates, to create standardized forms, and to develop programs for fire prevention. The board soon moved forward in a number of directions, including the creation of fire maps, indicating where fire risks were especially high, and the establishment of uniform insurance forms and rates. Formation of the NBFU was soon followed by other national and regional associations, including the Western Union in 1879, and the

Underwriters Association of the Middle Department in 1881.

One consequence of the formation of these associations was the development of industry-wide schedules. In earlier years, an insurance company decided on coverage, premium, and other factors based on an individual assessment of the property to be covered. In theory, two very similar properties might receive very different assessments depending on the person and company making the decisions.

By the early twentieth century, most companies had begun to make decisions based on industry-wide schedules. These schedules were based on standard features of a property, such as the method of construction, its location, type of business, risk factors, and other elements involved in determining the likelihood of a fire and potential financial loss. Greater attention was also paid to the collection and analysis of statistical data about the causes, locations, costs, and other aspects of fires. These data provided an even more scientific and rational basis for setting rates in the industry.

Fire Insurance Today

About 3,000 insurance companies today write fire insurance policies. A third of these companies account for the vast majority of those policies. Fire insurance accounts for about 10 percent of all insurance dollars spent in the United States, and fire insurance companies employ about 600,000 people in one or another aspect of the business (Hall 256).

Fire insurance is now available as a separate policy that covers fire risks only, or, much more commonly, it is included in a comprehensive policy that covers a variety of risks. For example, the homeowner's policy that most people purchase today includes, as one of its provisions, protection against losses from fires. Companies or individuals with unusual fire risks may choose to purchase separate policies that cover fire losses exclusively.

Whether part of comprehensive coverage or a separate policy, the elements of fire insurance coverage are essentially the same no matter what company one deals with or the region of the country in which one lives. National associations have largely standardized application forms, risk determination, claims forms and procedures, and other issues involved in fire insurance.

Further Reading: Goudsblom, Johan. *Fire and Civilization*. London: Allen Lane, 1992, pp. 150–152; Hall, John W. "Fire Insurance," *The Encyclopedia Americana*. Danbury, CT: Grolier, 1999. Volume 11, pp. 254–256.

FIRE LOOKOUT. *See* LOOKOUT TOWERS

FIRE MANAGEMENT

Fire management refers to all those policies and practices that humans use to deal with fire. This entry deals with the control of more rural fires, such as forest and wilderness fires, while the entry on Firefighting deals with the control of urban fires.

Fire management includes a range of policies and practices that differ widely across time and in various parts of the world. In general, it can include at least one or more of the following fundamental philosophies:

1. "Let 'em burn," the philosophy that fire is a natural phenomenon and that it should be allowed to have essentially free rein. For all its environmental logic, the "let 'em burn" philosophy obviously has some inherent limitations as, for example, when human lives and property are threatened.

2. "Stop 'em dead," a policy and practice based on the philosophy that fire in natural settings is usually bad or, at the least, a disadvantage for human societies at least partly because it deprives them of valuable natural resources, such as timber and wildlife.

3. Prescribed burns, or controlled burns, a set of policies and practices based on the assumption that fire provides many benefits to plants, animals, the physical en-

vironment, and humans, but, in a world so largely dominated by humans, it must be used judiciously in certain places at certain times under certain circumstances.

In this text, it is not possible to provide a complete survey of fire management practices in various cultures throughout history. It is possible, however, to review some of the kinds of decisions that various human societies have made about their relationship with fire.

Primitive Cultures

As a rule, primitive cultures appear to have understood the critical role that fire places in the natural world and learned how to use it to their benefit while maintaining its place in the environment. Early humans used fire not only for cooking, light, heating, and the manufacture of tools and other implements, but also in such basic practices as hunting and agriculture. These practices were especially suitable for nomadic people who moved regularly to find new hunting grounds or new land for their crops. The relatively few societies that still use slash-and-burn practices today are modern-day representatives of this attitude toward fire.

Fire Prevention and Suppression

In most parts of Europe, a very different philosophy of fire management developed over the centuries. As people began to settle down in cities and towns or on permanent farms, fire became an enemy. It deprived people of the wood they needed to build houses, to heat their homes and cook their meals, and to use for the production of charcoal and other useful products. Fire also tended to drive animals away from an area, making hunting more difficult. As a consequence, the fire management practice of choice became one of prevention and suppression. Every effort was made to keep wildfires from getting started and, if begun, to extinguish them.

Policies of fire prevention and suppression exact a severe price, however. For one thing, natural fires can almost certainly never be completely eliminated. As wildfires are brought under control, more and more fuel (unburned trees and other vegetation) begins to build up. At some point, a stroke of lightning or a carelessly dropped cigarette butt is almost certain to ignite a new fire. Over time, developing the equipment and personnel to manage fires may have become prohibitive.

Perhaps more important, however, is the ecological damage caused by fire suppression. The role of fire in the reproduction and growth of many kinds of plants and animals was only poorly understood until a relatively short time ago. Societies that adopted policies of fire control learned much too late, and sometimes to their great dismay, the damage they had brought to their biotic and abiotic environments.

Stephen Pyne's *Cycle of Fire* series of books provides a superb account of instance after instance in which people have struggled mightily to wipe out fire, only to see their "successes" in this regard followed by the loss of irreplaceable plant and animal life and totally unexpected disruptions of their economies and social structures. To choose a single example, Pyne describes the effect of fire suppression on the *taiga biome* found in much of Russia and Siberia.

The biome depends absolutely on fire for its health. Fire makes possible a faster rate of decomposition of organic matter than would occur without it, a critical factor in the short growing season available in the taiga. "In this model of nature's economy," he points out, "fire becomes an invisible hand, seeking out available fuels, transferring critical materials along pathways to where they are most needed, balancing supply and demand. Interrupt that idealized, laissez-faire flow and the system slows and its geography becomes lumpy with maldistributions. Eventually it may simply collapse, the ecological equivalent of an economy in which national wealth is buried in socks or stuffed in mattresses" (Pyne 1997, 132).

Policies in the United States

In the United States, a variety of fire management policies have been adopted. The early pioneers, for example, found themselves on a continent with seemingly endless natural resources. The usual practice was for settlers to burn down forests to obtain land for farms. When these lands could no longer support crops, they were abandoned, and new farmland was created by burning down more trees.

For nearly two centuries after the Pilgrims first arrived, few people worried about this practice. Everyone knew that there were more virgin lands to the west. Hardly anyone was concerned that people would ever destroy—or even have a serious impact—on the continent's enormous forest reserves. Some pioneers did learn about and practice the technique of prescribed burns used by Native Americans. But, for the most part, the concept of preserving or conserving any natural resource was a foreign one.

Thus, for much of American history, fire was a force to be used freely and without concern, not a phenomenon that required any particular attention or planning. Thus, one observer wrote about conditions in northern California in 1904, observing that:

> the people of the region regard forest fires with careless indifference. . . . The white man has come to think that fire is a part of the forest, and a beneficial part at that. All classes share in this view, and all set fires, sheepmen and cattlemen on the open range, miners, lumbermen, ranchmen, sportsmen, and campers. Only when other property is likely to be endangered does the resident of or the visitor to the mountains become careful about fires, and seldom even then. (As quoted in Pyne 184)

But this laissez-faire attitude could not survive the passing of the frontier. As settlers reached the West Coast, a new consciousness about the natural environment slowly began to develop. For the first time in American history, some men and women began to work for the conservation or preservation of natural resources. These efforts eventually came to fruition with the creation of Yellowstone National Park in 1872, followed a decade later by the establishment of Sequoia and Yosemite National Parks.

Pyne argues that the origin of a national U.S. policy about fire management can be traced to the last decade of the nineteenth century, a period when vast amounts of forested lands were transferred from the Department of the Interior to the control of the National Park Service and the U.S. Forest Service (USFS). That process culminated in the Transfer Act of 1905, which assigned to the USFS the responsibility for managing fires on huge stretches of forests. Pyne then outlines five periods, each about twenty years in duration, in which one or another fire management philosophy held sway in the USFS and, later, in other agencies responsible for fire management on federal lands.

1910–1930. The first of these periods, from about 1910 to 1930, opened with a vigorous debate between those who favored an aggressive program of fire suppression and those who argued for controlled burns. In some ways, the debate was weighted in favor of the suppressionists since many of them were trained in Europe. Upon their return to the United States, they brought with them the traditional distrust of fire that had become prevalent in most parts of Europe.

For this and other reasons, the pendulum gradually swung towards fire suppression during this period as the "official" national fire management policy. One important element in deciding that debate was passage of the McSweeny-McNay Act of 1928, which assigned responsibility for federal fire research to the USFS. Already strongly biased toward suppression as a technique of fire management, the USFS quickly used the McSweeny-McNay Act to create the Shasta Experimental Fire Forest. A major function of the Shasta unit was to test and demonstrate the effectiveness of various fire suppression techniques. Prescribed burns did not completely disappear from the landscape of fire management, but they became a trivial part of the federal government's overall program of fire control.

1930–1950. By the 1930s, the policy of fire suppression had essentially carried the day with the USFS and most other agencies responsible for fire control on public lands. As the Great Depression came to an end, funds began to pour into the USFS budget, as well as the budgets of other federal agencies, for the purpose of streamlining, upgrading, and improving fire suppression techniques. One of the most obvious elements of this thrust was the creation of the Civilian Conservation Corps (CCC), one of whose main tasks was the development of fire prevention and suppression systems. Fire suppression entities found they had more than enough money with which to build new fire control roads and trails, modern telephone communication systems, lookout towers, smoke-jumping teams, chemical fire suppressant systems, air tankers, and well-drilled teams of highly trained firefighters. The key USFS fire control policy during this period reflected the confidence with which fire suppression efforts were imbued. That policy was the "10 A.M." policy, whose goal it was to extinguish any fire by 10 A.M. of the day following its discovery (or, if that were impossible, by 10 A.M. of the next day, and so on).

Still, serious weaknesses in an aggressive policy of fire control were already becoming obvious towards the end of this period. For one thing, the USFS had begun to spend huge amounts of money to protect lands of questionable economic or aesthetic value. In addition, fires that were intentionally set on private lands constantly posed a threat to adjoining public forests and wilderness areas. Finally, of course, the more successful fire control programs were, the larger the amount of fuel that accumulated and, hence, the greater the risk of new fires. The philosophy that had eliminated to a large degree natural forest fires in Europe was much less likely to succeed on the enormous American continent with its vast expanses of protected natural forests and wilderness areas.

1950–1970. World War II had its own impact on fire management policies. Professional foresters and the general populace were constantly being reminded during the war of the horrible consequences of uncontrolled fire. The German air attacks on Great Britain in 1940, the Allied air attacks on German cities later in the war, the dropping of the first atomic bombs, and the widespread use of incendiary and other fire-based weapons kept the horrors of uncontrolled fire continuously in the public and professional eye.

By the end of World War II, fire suppression policies had received a shot in the arm from this kind of negative publicity. At the same time, and partly in response to the fire horrors of war, some aggressive public education programs were developed to further convince Americans that fire was a dangerous and destructive phenomenon and that everyone had an obligation to help control fire. Smokey Bear and the Keep America Green programs were only two of these widely promoted educational programs.

Just at its peak of popularity as a fire management scheme, fire prevention and suppression began to fall under increasing scrutiny in the 1960s. For one thing, the wisdom of the policy itself came more into question as more and more resources were being called upon to fight too many relatively small fires on remote lands. Perhaps more important, however, was the growth of the environmental movement in the United States (and the rest of the world). More and more people began to understand the importance of the natural environment and of the need for human efforts to protect that environment from despoliation. For example, the Wilderness Act of 1964 specified that certain public lands were to be maintained in their native and natural condition. In such cases, that requirement often meant that fire—certainly a natural part of the environment—had to be returned to these previously protected areas. The age of prescribed burns seemed to have returned.

Accepting the importance of and need for prescribed burns was not, unfortunately, the same as making that policy work in practice. By 1970, most foresters still knew too little as to when and how prescribed burns should be used. As a consequence, far too many prescribed burns got out of control and became fires that, themselves, needed to be suppressed. Pyne points out that "for more than ten damning years the country's most disastrous fires were the results of breakdowns in the management of prescribed fires" (Pyne 206).

1970–1990. During the decade of the 1980s, fire management policy in the United States was somewhat in a state of confusion. The general public, most legislators, and many fire experts still believed that forest and wilderness fires were inherently bad and needed to be prevented, contained, and/or suppressed. Probably the majority of experts had become convinced, however, that some form of prescribed burn was essential to the health and productivity of natural lands. The ability to carry out such prescribed burns, however, had not as yet been perfected.

Finally, some fire authorities still clung to some form of the "let 'em burn" philosophy. They argued that scientific studies of forest ecology and fire had shown beyond doubt that fire was an essential part of the natural environment and humans simply had to stay out of its way until and unless human lives and property were threatened. That philosophy was sorely tested in the mammoth Yellowstone fires of 1988 when park officials decided to let fires burn out of control throughout the park (except for areas that were inhabited or contained buildings) for most of the summer.

1990 and Beyond. The decade of the 1990s appears to have indicated that uncertainty and controversy over fire management techniques are far from resolved. Prescribed burns have become a fundamental tool of fire management in almost every part of the country, and fire suppression and prevention are both still essential tools in the control of fire. But deciding what the correct mix of these forms of fire management is has not yet been determined.

In fact, some of the fundamental arguments about fire management techniques extending back more than a century are still going on, with relatively little resolution having been reached. For example, experts are still debating whether suppression or prescribed burning is the better fire management technique. As late as June of 1999, an important study on this question with regard to brushland fires was published in the journal, *Science*. The study reports on research con-

ducted by Jon E. Keeley of the Ecological Research Center of Sequoia-Kings Canyon Field Station, C. J. Fotheringham, of the Center for Environmental Analysis at the University of California at Los Angeles, and Marco Morais of the Santa Monica Mountains National Recreation Area. The Keeley-Fotheringham-Morais study attempted to measure the effect of fire suppression on the number and intensity of brushland fires in California. The researchers came to the conclusion that fire suppression had *not* ultimately led to larger and more intense fires. They also suggested that prescribed burns designed to reduce the number and intensity of fires had been largely unsuccessful. Response to this report became intense almost immediately, and a website was created (http://www.fs.fed.us/fire/operations/ecology/keeley) to deal with the exchange of views about the research and its implications. It seems likely that the debate over fire management policies and practices in the United States and other parts of the world is likely to continue into the foreseeable future. *See also* FIRE ECOLOGY.

Further Reading: Biswell, Harold H. *Prescribed Burning in California Wildlands Vegetation*. Berkeley: University of California Press, 1989; Brown, Arthur A., and Kenneth P. Davis. *Forest Fire: Control and Use*, 2nd edition. New York: McGraw-Hill, 1973; Hendee, John C., George H. Stankey, and Robert C. Lucas. *Wilderness Management*. Golden, CO: North American Press, 1990, Chapter 12; Keeley, Jon E., C.J. Fotheringham, and Marco Morais, "Reexamining Fire Suppression Impacts on Brushland Fire Regimes," *Science*, 11 June 1999, 1829–1832; Long, James N., ed. *Fire Management: The Challenge of Protection and Use*. Logan: Utah State University, 1985; Parfit, Michael, "The Essential Element of Fire," *National Geographic*, September 1996, 116–139; Pyne, Stephen J. *World Fire: The Culture of Fire on Earth*. Seattle: University of Washington Press, 1997.

FIRE MODELING

Fire modeling is the process by which scientists attempt to predict the course and consequences of a fire by using physical models or mathematical equations. Fire modeling is used both for structural fires, such as those that occur in homes, commercial buildings, and road or railway tunnels, and for fires in nature, such as forest or grassland fires.

Modeling has become a very important technique in fire science because it provides one of the few controlled ways in which scientists can learn about the details as to what takes place during a fire. It is possible, of course, to make observations and measurements about the changes that occur during the course of a fire or after the fire has run its course. But a fire scene itself is hardly the best environment in which to conduct a scientific experiment, and observing the results of a fire can give only limited information as to what happened *during* the fire. Modeling has therefore become a useful tool for fire scientists, not only because it gives a better step-by-step view of the fire process, but also because it can provide clues as to how future fires can be prevented.

Physical Models

A physical model of a fire is a scaled-down version of some setting in which a fire might be expected to occur. It could be a room, a building, or a natural setting (forest or grassland, for example) of limited dimensions. A fire can actually be set using the scaled-down version and observations made as to the changes that take place during the event.

Researchers at the U.S. Fire Administration, for example, have built a one-eighth scale model of an office and apartment complex that can be set on fire to use for both research and training purposes. The structure has been nicknamed "National Towers" by researchers who work with it. The model is four-feet tall and contains three stories. It has a retail area on the first floor, offices on the second floor, and residential apartments on the third floor. The model is made of fireproof clear plexiglass and has working smoke detectors, fire doors, and elevator shaft, and a complex ventilation system. Smoke is generated by a device similar to that used in model railroad engines, and heat is produced by a special lighting device developed especially for the model. Fires can be simulated at any one of eleven locations in the model,

and the path of flames and smoke can then be tracked throughout the building.

Physical models of forest fires, grassland fires, and other fires in nature are more difficult to build. Yet, such models are now being built and used. For example, U.S. and Canadian fire researchers have been working together to develop a physical model for crown forest fires. A crown forest fire is one that sweeps through the upper branches of trees in a forest. Such fires are by far the most important single type of forest fire in Canada and in some parts of the United States.

To model crown forest fires, the U.S./Canadian research team chose a forested region about 50 kilometers (30 miles) north of Fort Providence, Northwest Territory. The area is covered with jack pine about 80 years old and surrounded by marshlands. The marshlands provide a boundary across which fire normally is not able to escape. The test area has been divided up into individual "burning plots" on which experiments can take place. Each burning plot ranges in size from about 75 by 75 meters (250 by 250 feet) to 150 by 150 meters (500 by 500 feet). Firelines—strips of bare land 50 meters wide—were constructed around each plot to further contain the test fires.

As fires are lit on each test plot, a number of measuring devices and techniques are used to track the development of the fire. Studies are made of the kind and amount of fuel present on the test plot; measurements are made of weather conditions; ground, helicopter, and tower-based video cameras make visual records of the fire; heat and temperature measurements of soil, air, and fuel are made; evaluations are made of natural and human-made fire shelters; and many other observations are made. As of January 2000, eleven crown fires had been set in the three-year period from 1997 to 1999. Data from those fires are still being evaluated.

Mathematical Models

As is often the case, mathematical models have a number of advantages over physical models. They do not require, as in the case of forest fire modeling, very large physical facilities that could be damaged or destroyed by an out-of-control fire. The progress of mathematical models is often easier to follow than is the course of an actual fire. And mathematical models are usually much "cleaner," in that they provide clear-cut options and outcomes.

But mathematical models have their disadvantages also. Perhaps most important, it is usually difficult to reduce the reality of physical phenomena to numbers and equations. An actual flame may be difficult to control and observe in nature, but it may be even more difficult to represent in a mathematical model.

Still, scientists in many fields have begun to represent a host of physical phenomena by means of mathematical equations. The tool that has made it possible for them to do so is the computer, the most sophisticated of which can often perform trillions of calculations per second. Thus, solving large sets of complex simultaneous equations, which would once have been an impossible task, is now a challenge that can be handled reasonably well.

One of the earliest models of fire behavior was a program called BEHAVE, developed in the early 1970s by Richard Rothermel, then of the Northern Fire Laboratory, and now at the U.S. Forest Service's Rocky Mountain Research Station. BEHAVE was a program designed to help firefighters and fire administrators understand the direction a forest fire was likely to take. It also suggested certain options for prevention and suppression that would work best for different kinds of fires. BEHAVE has recently been updated and revised in a form known as Beyond-BEHAVE.

Many programs for modeling forest fires are now available. Some examples include EMBYR, produced by the Oak Ridge National Laboratory in Oak Ridge, Tennessee to study fires similar to those that struck Yellowstone National Park in 1988; Fire Storm Pro, developed by Fire Safety Services at Blackwood, South Australia; IGNITE, a project of the Charles Sturt University in New South Wales, Australia; PEGASUS, a program being written at the U.S. Forest Ser-

vice Pacific Southwest Research Station in Riverside, California; SPARKS, developed for use in Swiss National Parks; and WiFE, currently being developed at the National Center for Atmospheric Research at Boulder, Colorado. All of these programs are similar in attempting to convert physical factors involved in wildfires into mathematical symbols and equations that can then be manipulated by computers. Most programs, however, include some features that make them suitable for specific topographic or geographic conditions.

Researchers have been trying to model structural fires for many years also. The earliest efforts were, of course, relatively primitive. The approach was to model the course of a fire in an idealized cubic room of some given dimensions. If a fire were to break out in such a room, certain physical phenomena would occur that could be described, at least in principle, by well-known chemical and physical laws and equations. For example, air in the room would become heated near the point at which the fire first appeared. As that air was heated, it would expand and move upward by means of convection. The moving column of air would, in turn, heat other air in its vicinity. This phenomenon could be described, at least in general, by equations involving the heat capacity, transference of heat, temperature change, expansion of gases, and other familiar phenomena. Mathematical equations could be written that would describe the changes taking place during the process of heating by convection.

Other changes could also be described mathematically. For example, the loss of heat through the walls of the room would depend on the conductivity of the material and, hence, on the type of material used in the walls. Heat changes would also depend on factors such as the number and size of doors and windows, the source of the original flame, and the size and shape of the room.

Many such factors would determine how the fire would spread through the room, but they could all be expressed in mathematical terms. A mathematical model of the fire spreading through the room would show how these factors interacted with each other during each second (or other time interval) during which the fire was spreading. Models dealing with this problem for the whole room all at once could not possibly be very accurate or precise. But they were the only approach researchers could take at first because even such a relatively "simple" problem involved a huge number of calculations for each second the fire was in process.

Mathematical modeling became a much more efficient and more successful tool when it was possible to break the room down into smaller subunits. In later models, a typical room was broken down into more than 100,000 such units, or "cells." Calculations that had been performed for the room as a whole could then be repeated for each individual cell, and the effect of changes in one cell on adjacent cells could be determined. These changes could be calculated for much smaller time units, such as thousandths of a second, than was possible with the first models.

By the 1990s, fire scientists had developed models that could predict the course of a fire through a room with relatively good accuracy. The test for such models was to select a set of criteria (such as a room of certain size with certain characteristics in which a particular type of fire breaks out), predict the course of fire with the mathematical model, and then conduct an experiment with a room having exactly those characteristics. In tests of this kind, mathematical modeling has been found to be a powerful tool in predicting the course of a fire.

Fire models can be used in various ways by various type of fire management and firefighting organizations. For example, modeling a room fire doesn't have much value in actually fighting that fire. But it can have a great deal of value in suggesting ways of building and furnishing a room in order to make it more fireproof.

In the case of forest fires, grassland fires, and the like, models have some value in warning firefighters the way a conflagration is likely to move across the landscape. It can also sug- gest steps to take to reduce the likelihood that the fire will break out in the first place. *See also* BEHAVE.

Further Reading: "About the International Crown Fire Modeling Experiment," <http://www.nofc.forestry.ca/fire/fmn/nwt/about.html>, 08/11/99; Agee, James K. *Fire Ecology of Pacific Northwest Forests.* Washington, DC: Island Press, 1993, pp. 47–52, 144–150; Lane, Earl, "Crisis Forecasting," *Newsday,* 30 December 1997, sec. C04; Lyons, John W. *Fire.* New York: Scientific American Books, 1985, pp. 95–101; "Simulating Fire Patterns in Heterogeneous Landscapes," <http://www.esd.ornl.gov/ern/embyr/embyr.html>, 08/11/99.

FIRE MUSEUMS. *See* HALL OF FLAME

FIRENET: THE INTERNATIONAL FIRE INFORMATION NETWORK

FireNet is an internet-based international information retrieval and exchange system designed for anyone interested in rural and landscape fire management. The system was established in 1993 and is operated by the Department of Forestry at Australian National University (ANU). Subscription to FireNet is free. The latest statistics available, from 1995, indicated that more than 700 individuals are members of FireNet.

FireNet provides a wide variety of services, including an e-mail list, information about education and training programs, fire-related software, information on emergency services and management, a virtual library, and a current event bulletin board.

The e-mail function makes it possible for subscribers to FireNet to correspond with other members around the world almost immediately. As soon as a new message arrives at the host site, it is transmitted to all other e-mail subscribers. Thus, information about fast-breaking fire stories can be made known worldwide within a matter of minutes. E-mail messages are also used to provide information on job opportunities, ongoing research, curriculum development, new courses in fire science and management, and to request information about specific research topics.

The educational and training function of FireNet provides extensive information on ANU courses and in-service workshops, as well as comparable information from other universities and governmental agencies. Software that has been developed specifically for fire management and control is also available on FireNet. Some examples include IGNITE, an interactive fire model; CHARADE, a decision support system designed for firefighting in Italian forests by Italian, French, and Spanish scientists; and a prototype for simulation of forest fires written by Tiago Oliveria, a Portuguese fire scientist.

Access to emergency services, such as the Global Emergency Management Information Network, the U.S. Federal Emergency Management Agency, and Emergency Services in Finland is also available through the website.

Bibliographic sources that can be accessed through FireNet include a bibliography on fire ecology in Australia, *Wildfire Magazine*, the Canadian Forest Service Fire Management Network bibliography, and the U.S. National Aerial Firefighting Safety and Efficiency Project.

Further Reading: "FireNet Information Network," <http:online/anu.au/Forestry/fire/firenet.html>, 07/31/99; Green, D.G., A.M. Gill, and A.C.F. Trevitt, "FIRENET—An International Network for Landscape Fire Information." *Quarterly Journal of the International Association for Wildland Fire*, Spring 1993, 22–30; Trevitt, Chris, "FireNet—An Information Exchange System for Forest and Rural Fire Managers, Researchers, and Educators." *Institute of Foresters Newsletter*, October 1994. Also at <http://online/anu.edu.au/Forestry/fire/ACFT/IFA-News-Oct94.html>, 07/31/99.

FIREPLACE

A fireplace is an opening in the wall of a room, usually built of stone or bricks, designed to hold a fire. Historians believe that the fireplace evolved over the eons from what began as a notch in the wall of a cave that had been provided with a vertical opening—the chimney—through which smoke could escape. Fireplaces have traditionally had one or both of two major functions: as a way of heating a room and/or as a place for cooking.

In all but the simplest of fireplaces, the fuel (usually wood, but sometimes coal) is placed on a grate, an iron platform that rests about six inches or so above the floor of the fireplace. This arrangement permits the air needed for combustion to pass upward through the fuel as it burns.

Fireplace Heating

Fireplaces today are used primarily for their romantic or aesthetic appeal. They are not very efficient systems for heating a room. The primary reason for this inefficiency is that heat produced by burning fuel in a fireplace tends to pass upward through the chimney and out of the room. More heat is lost by conduction through the back and sides of the fireplace, especially if they are on the outside wall of a room or house, as they often are.

Almost the only source of heat from a fireplace to a room comes by way of radiation. A person standing in front of the fireplace is warmed by radiation, while his or her back remains cold. This phenomenon was immortalized in a comment by Benjamin Franklin that, "A Man is scorch'd before, while he's froze behind" when standing near a fireplace.

The efficiency of a fireplace depends significantly on the outside weather. When the outside air is still and clear, room air flows easily into the fireplace and up the chimney. This flow of air causes the fire to burn with a hot flame. Smoke travels easily up the chimney. Under these conditions, the fireplace "draw" (movement of air through the system) reaches a maximum. However, the fireplace actually tends to produce a cooling effect in such cases because warm air is removed from the room more rapidly than the fire generates heat.

The presence of winds, haze, clouds, or other atmospheric conditions reduce the efficiency with which a fireplace operates. Under these conditions, outside air may flow downward through the chimney, forcing smoke back into a room and choking off the fire.

Fireplace Adaptations

Inventors have struggled to find ways of improving the efficiency of a fireplace. One of the first improvements came in the early seventeenth century with the invention of the smoke shelf and damper at the back of the fireplace. A smoke shelf is a nearly horizontal projection on the back of the fireplace wall at the base of the chimney. The smoke shelf prevents smoke from backing up into a room. When smoke flows downward in a chimney, it is deflected by the smoke shelf and carried back up the chimney by the hot flow of gases coming from the fire.

The damper is a sheet of metal that projects outward from the smoke shelf. Depending on its position, it can exaggerate the effect of the shelf and increase or decrease the movement of air up the chimney. At its most extreme "closed" position, the damper can prevent air from leaving the fireplace entirely (as when it is not being used). Changing the shape of the fireplace space can also affect its efficiency. Since nearly all of the heating produced by the fireplace comes from radiation, the goal is to find a shape that maximizes radiation from the burning fuel while, if at all possible, obtaining some heat transfer by means of convection and/or conduction. One design that achieves this objective is to tilt the back and side walls inward, producing a "cupping" effect around the fire.

The effectiveness of a fireplace can also be increased by placing furnishings that will absorb and reradiate heat in the room where the fireplace is located. For example, brick walls and stone floors will absorb and reradiate heat more efficiently than will plaster walls and wood floors. Many homeowners choose not to take these factors into consideration, however, since the aesthetics of room design are more important than the modest function that fireplaces have in providing heat to a room.

Another mechanism for increasing the efficiency of a fireplace is the installation of glass doors on the front of the fireplace. These doors reduce significantly the amount of room air that can flow through the fireplace and up the chimney. The doors can be adjusted to allow just enough air to enter to keep the fire burning, but prevent excess air not needed for combustion from leaving the room. The material selected for the glass doors is of critical importance. Heat does not radiate through regular window glass and pyrex very efficiently. Fused quartz is far more transparent to heat than either of these two materials and is used, therefore, in the best fireplace doors.

The flow of air through the fireplace can also be controlled by installing an "air-flow box" below, behind, and above the area in which combustion occurs. This "jacket" allows cold air to enter below the fire itself, become warmed, flow upward through the jacket behind the fire, and leave the jacket through the space above the fire, returning to the room. The flow of air through the jacket can be made more efficient by installing a fan that moves the air through the system. *See also* Heat; Thompson, Benjamin (Count Rumford).

Further Reading: *Fireplaces and Wood Stoves*. Alexandria, VA: Time-Life Books, 1997; Lyons, John W. *Fire*. New York: Scientific American Books, 1985, pp. 42–55; Orton, Vrest. *Observations on the Forgotten Art of Building a Good Fireplace: The Story of Sir Benjamin Thompson, Count Rumford, an American Genius and His Principle*. Dublin, NH: Yankee Press, 1974; Rossotti, Hazel. *Fire*. Oxford: Oxford University Press, 1993, Chapter 4; Tecton, Mike. *Traditional Fireplaces*. McLean, VA: Mike Tecton Publishing, 1989.

FIRE PLUGS. *See* Fire Hydrants

FIREPROOFING. *See* Flame Retardants

FIRE PROTECTION HANDBOOK

The *Fire Protection Handbook* is the "bible" for individuals and companies involved in fire prevention, protection, and fighting. It was first published in 1935, and its 18th edition was released in 1999. It is revised and expanded, as needed, on a regular basis. The 18th edition consists of 190 chapters written by more than 200 professionals in the field and cover about 2,500 pages. Examples of some of the topics covered include fire hazards of alternative fuels; sprinkler system design and installation; venting systems; planning for fire station locations; disaster education programs; legal aspects of fire protection; and fire protection for special facilities, such as libraries and museums. The book is published by the National Fire Protection Association. The cost of the handbook is $122.75 for nonmembers, and the book can be ordered from NFPA at 1 Batterymarch Park, Quincy, MA 02269–9101; telephone (617) 770–3000; fax (617) 770–0700. *See also* NATIONAL FIRE PROTECTION ASSOCIATION.

FIRE REGIME

Fire regime is a term used to describe the frequency, extent, intensity, effects, and times of occurrence of fires in any particular ecosystem. A number of different fire regime systems have been developed for various ecosystems and have been used with greater or lesser value. One of the early systems of a fire regime was suggested by M.L. Heinselman in 1973. Heinselman proposed a system in which fire regimes were rated from 0 to 6. Fire regime number 0 represented a situation with very little or no natural fire and, consequently, essentially no damage to an ecosystem from natural fires. Fire regime number 3, by contrast, was used for an ecosystem in which infrequent but severe surface fires dominate, with an interval between fires of at least twenty-five years. Fire regime number 6 was taken to represent severe surface fires combined with crown fires with very long intervals of at least 300 years between them.

Another system of defining fire regimes is based not on the types of fires that strike but on the kinds of vegetation in an area. Of course, the two systems are related in some ways in that the nature of vegetation in an area is an important factor in the types of fires that occur there. In the regime system based on vegetation, one category includes dry lumber pine; another warm, dry ponderosa pine; a third, cool, dry Douglas fir; and a fourth, moist, lower, subalpine vegetation. A total of eleven such regimes are included in this system of organization. *See also* FIRE ECOLOGY.

Further Reading: Agee, James K. *Fire Ecology of Pacific Northwest Forests*. Washington, DC: Island Press, 1993, Chapter 1; Heinselman, M.L. *Fire Intensity and Frequency as Factors in the Distribution and Structure of Northern Ecosystems*. In H.A. Mooney et al., eds., *Fire Regimes and Ecosystem Properties*. USDA Forest Service General Technical Report WO-26, 1981; Wright, Henry A., and Arthur W. Bailey. *Fire Ecology: United States and Southern Canada*. New York: Wiley-Interscience, 1982, passim.

FIRE RESISTANCE

Fire resistance is any practice in which a person handles flames, fire, and hot objects without apparent harm. Some examples of fire resistance include fire eating, fire handling, fire breathing, firewalking, and surviving exposure to very high temperatures.

History of Fire Resistance

Fire resistance practices go back at least 2,500 years, and perhaps longer. Such practices usually occurred within one of two settings: as part of religious ceremonies, and as part of conjuring or magical acts. For example, the use of trial-by-fire to determine the guilt or innocence of accused people was a common practice in the early Christian church. There is evidence that some leaders of the church had learned techniques for handling fire without causing damage to themselves. In one of the most complete references on fire resistance, Harry Houdini's, *Miracle Mongers and Their Methods*, reference is made to a case involving Simplicius, Bishop of Autun. Simplicius had been married before being promoted to his current status. To prove that his act was approved by God, he offered to have burning coals placed on his body. He survived this trial by fire without being harmed.

The ability to withstand the effects of fire was often regarded as an act of grace among early Christians. An individual frequently mentioned in this regard was Francis of Paola, who died in 1507. This monk was said to be entirely comfortable with the handling of fire. On one occasion, he was visited by church officials who doubted the fire-handling skill attributed to him. When Francis demonstrated his abilities to them, the officials prostrated themselves before him. The monk was later canonized in 1519.

The handling of fire gradually disappeared from the Christian tradition, perhaps because of the growing role of eternal fire and damnation in Christian theology. Still, remnants of the practice remain today. Small Christian sects in the southeastern United States, for example, continue to handle hot coals in their worship ceremonies. These sects are among those that also handle poisonous snakes during these ceremonies. They believe that their faith in God is sufficient to protect them from any harm that may come from hot coals or snakes.

Houdini also mentions many historical examples in which magicians and conjurers have demonstrated unusual talents for handling fire without being harmed. One of the earliest individuals about whom there is extensive information is an Englishman by the name of Richardson. A 1667 article in the *Journal des Savants* claims that Richardson was able to "devour brimstone on glowing coals before us, chewing and swallowing them; he melted a beere-glass and eate it quite up; . . . Then he melted pitch with wax and sulphur, which he drank down as it flamed; . . . he also took up a thick piece of iron, such as laundresses use to put in their smoothing boxes, when it was fiery hot, held it between his teeth, then in his hand, and then threw it about like a stone . . ." (Houdini, p. 18).

Another famous fire resister was Ivan Ivanitz Chabert (1792–1859), who billed himself as the "only really incombustible phenomenon." In addition to eating and handling flames and very hot objects, Chabert was famous for his ability to survive inside a very hot oven. During his act, Chabert entered an iron cabinet similar to those used by cooks and bakers. He carried with him a leg of mutton and remained within the cabinet until the meat was thoroughly cooked. He then emerged with the meat without any apparent damage to his own body.

Explanations of Fire Resistance

In his account of fire resisters, Houdini comments that he is not so surprised at the acts magicians and priests have performed but at the fact that they have continued to amaze the general public for so long. Explanations for these acts have been around, he points out, as long as have the acts themselves. He observes that "it is much easier to become fireproof than to become exposure proof" (Houdini 1920, 98).

In general, a person can develop the ability to carry out fire resistance acts in one of two ways. First, one can build up an immunity to flame and heat over time. The secret to Richardson's success, for example, was revealed late in his life by one of his servants. According to that explanation, Richardson developed an immunity to fire and heat by "rubbing the hands and thoroughly washing the mouth, lips, tongue, teeth, and other parts which were to touch the fire, with pure spirits of sulphur. This burns and cauterizes the epidermis or upper skin, till it becomes as hard and thick as leather, and each time the experiment is tried it becomes still easier" (Houdini 1920, 102).

A second method for increasing one's resistance to fire is by using some type of covering on the skin or other body parts that are to come into contact with flames and heat. One of the earliest recipes in existence is found in a book, *De Secretis Mulierum*, by Albertus Magnus (1193–1280). Albertus recommended using a mixture of marshmallow juice, egg white, flea-bane seeds, lime, and radish juice as a wash for any part of the body to be exposed to heat. According to Magnus, once this mixture dries, it provides a thin, heat-resistant layer that will allow a person to handle very hot objects. Over the centuries, many other ointments, washes, lotions, and salves have been developed to provide heat resistance for magicians and priests.

In the case of Chabert's oven trick, some straightforward deception was involved also. After entering the hot cabinet, Chabert apparently was able to don an asbestos cloak and lie quietly at the bottom of the oven for the time required to permit the leg of mutton to cook.

Fire Breathing

Fire breathing is an act by which one seems to exhale flames from his or her mouth. This act is fundamentally different from the various forms of fire resistance described above in that, under normal circumstances, no fire, flame or heated object is actually taken into the body or comes into contact with the body.

Again, fire breathing is a very old art. An 1817 book on the *History of Inventions* tells of a Jewish leader, Rabbi Bar-Kochba, who practiced fire breathing during the reign of the Roman Emperor Hadrian (ruled 117–138 A.D.). The purpose of the performance was to prove to Roman Jews that Bar-Kochba was, indeed, the messiah for whom they had been waiting.

Fire breathing is still a popular art among magicians, conjurers, and other performers. Advertisements by those who perform the act can be found on the Internet, and an excellent outline of the practice is available at the "Fire Eating and Fire Breathing FAQ" listed in the Further Reading section below.

A variety of techniques have been used by fire breathers. In general, the principle is to hold in one's mouth some flammable liquid, which is then expelled and ignited as it leaves the mouth. One of the oldest procedures is to make two small holes at the opposite ends of a small nut shell and fill the shell with some flammable liquid. The nut is then held in one's mouth and the liquid is expelled by breathing through the nut. As the liquid leaves the mouth, it is ignited.

Another method is to soak a small rag, string, or piece of cloth with a flammable liquid, which is then hidden inside the performer's mouth. The liquid must be not only flammable, but also easily vaporized so that it can be blown out of the mouth before being ignited. Specially built pipes and tubes can also be made to hold in the mouth and through which a flammable liquid can be expelled.

A key warning to be observed in performing fire-breathing acts is to use liquids that are not toxic, carcinogenic, or otherwise dangerous in and of themselves. For example, gasoline has been used as the fuel in fire-breathing acts, but it is not recommended because of its potential harmful effects if inhaled, even if not burning. Fire breathers also are cautious to point their mouths upward to avoid having the flammable liquid or flames escape upward into the nasal passages. Readers should under no cirumstances attempt fire breathing or any other types of fire resistance.

Further Reading: "Fire Eating and Fire Breathing FAQ," <http://www.juggling.org/help/circus-arts/fire-eat/fire-eat.html>. See also related pages on equipment and safety cautions on this Web site; Garth, Benjamin. *Fire Eating: A Manual of Instruction*. New York: Brian Dube, 1993; Griffen, Charles Elridge. *Fantastic Feats with Fire: Original Explanation of Fire Breathing Phenomena*. Suffern, NY: Chas. E. Griffen, Publisher, 1897; Houdini, Harry. *Miracle Mongers and Their Methods*. New York: E. P. Dutton and Company, 1920. This book is also available in its entirety on the Internet at http://www.conjuror.com/archives/houdini/miracle_mongers/; Musson, Clettis, V. *Fire Magic: Fire Eating Plus Many Fire Tricks*. Chicago: Magic, 1952.

FIRE SAFETY INSTITUTE

The Fire Safety Institute (FSI) is a not-for-profit institute whose purpose it is to find and recommend ways of reducing life and property loss from fire through rational fire-safety-making decisions. The Institute is located in Middlebury, Vermont, and was founded in 1981.

FSI emphasizes a threefold approach to its mission: information, research, and education. As part of its information function, the Institute collects information about fire safety practices from a variety of sources and compiles them into a series of educational publications intended primarily for fire safety specialists. It also provides editorial and review services for technical fire safety journals, such as the journal *Fire Technology*, published by the National Fire Protection Association.

The research function of the Institute includes designing methods by which fire safety decisions can be made, developing concepts of fire risk analysis and fire risk assessment, and assessing alternative approaches to fire safety evaluation.

FSI pursues its educational goals in a variety of ways, such as producing textbooks and tutorial papers on innovative and fundamental fire-protection engineering, presenting seminars and lectures on technical aspects of fire safety, and developing fire-protection engineering curricula for engineers, fire service personnel, and other professionals.

Fire Safety Institute
P.O. Box 674
Middlebury, VT 05753
(802) 462–2663
http://middlebury.net/firesafe/
e-mail: firesafe@middlebury.net

FIRESTORM

A firestorm is a weather phenomenon in which very high temperatures created by an intense fire causes the heating of a column of air. As the heated air rises, it produces strong winds that move into the area vacated by the hot air. Firestorms can be produced both by severe forest fires or by urban conflagrations.

Examples of Firestorms

One of the most famous firestorms caused by a natural phenomenon—a forest fire—swept the town of Peshtigo, Wisconsin, in October 1871. The fire swept across more than 2,000 square miles of forest, killed about 2,300 people in and around the town of Peshtigo, and essentially destroyed the entire city. A similar forest-fire-induced firestorm struck the region around Sundance, Idaho, in September 1967. Winds as high as 120 miles per hour were measured over a region at least ten miles in diameter. The fire was contained to some extent by the mountainous terrain around Sundance.

The mammoth air raids used during World War II sometimes resulted in firestorms. For example, British and United States air forces attacked the city of Dresden, Germany, on 13–14 February 1945, drop-ping nearly 3,500 tons of high explosive and incendiary bombs. A firestorm swept the city that may have killed more than 100,000 people. A similar attack had been launched on the German city of Hamburg on 27 February 1943. More than 20,000 people were killed, including many who were roasted to death while trying to hide in air raid shelters. The heat produced by the firestorm was so intense that steel girders used in buildings were melted.

Mechanics of a Firestorm

A firestorm begins when an intense fire is created in a relatively short period of time. For example, firestorms may begin when a brush or forest fire breaks out in a very dry area. If large amounts of vegetation burn very quickly, a firestorm may develop. When a bomb or group of bombs strikes the Earth, for example, the temperature at point of impact may rise to several thousands of degrees. Air at the impact point is heated and begins to rise as a result of convection currents. Cold air surrounding the impact point begins to flow inward to replace the warmer air being carried upward.

This phenomenon is similar to those events that accompany the formation of a tornado, cyclone, or hurricane. As with those phenomena, the path taken by rising winds is affected by a number of factors, one of which is the Earth's rotation. As warm air moves upward from the point of impact, it is diverted into a counterclockwise (in the Northern Hemisphere) circular pattern.

The movement of a firestorm is also affected by local topographic features. When the terrain around the impact point is essentially flat, there is little to stop the movement of the firestorm, and it may travel across the land like a tornado. If the terrain is hilly or mountainous, the firestorm may be confined to a smaller area and its force diminished.

The term firestorm has also come to be used in a more general fashion. For example, newspaper stories sometimes talk about terrible forest fires as "firestorms," although they do not have all or any of the characteristics listed above. *See also* FOREST FIRES; PESHTIGO FIRE.

Further Reading: "Hermann Kroger—World War Two," <http://www.valourandhorror.com/BC/Stories/Kroger.htm>, 08/10/99; "Holocaust at Dresden," <http://www.fen.baynet.de/~na1723/army/dresden.html>, 08/10/99; Lyons, John W. *Fire*. New York: Scientific American Books, 1985, pp. 117–118.

FIRE SUPPRESSION. *See* FIRE EXTINGUISHER

FIRE SWALLOWERS. *See* FIRE RESISTANCE

FIRE TOWER. See LOOKOUT TOWERS

FIRE TRUCK. See FIRE ENGINE

FIREWALKING

Firewalking is the practice of walking barefoot across very hot coals or rocks. The practice is part of many cultures throughout the world, from Asia and the South Pacific to various parts of Europe. It has also become popular in some New Age religions and self-discovery movements that have developed in the United States and other developed countries.

Elements of Firewalking

Some forms of firewalking make use of materials heated by natural processes. In the Hawaiian Islands, for example, firewalking has traditionally been practiced on very hot rocks only recently formed from molten lava.

In most cases, hot coals used in firewalking are produced by the burning of wood, an act that is sometimes part of the ceremony itself. The temperature of the coals produced by this process may reach 650°C (1,200°F). Firewalkers walk across these hot coals with apparently no damage to their feet. In some cases, the walk is directed to an idol or deity to whom honor is given.

Firewalkers may also pick up and handle the hot coals or rocks. In such cases, damage to the hands and other parts of the body is limited to simple blisters or, in most cases, is nonexistent.

Historical Traditions

The custom of firewalking goes back hundreds and thousands of years in many parts of the world. In each instance, folklore explains the origin and significance of the practice. The residents of Beqa Island in Fiji, for example, believe that they were the first firewalkers on Earth. According to their legends, the practice was introduced to them by a little man found in a river hole. In order to save his life, the little man promised the earliest members of the Beqa tribe that he would show them how to walk on fire, giving them the greatest of all powers over nature.

Firewalking traditions occur in many other cultures. The !Kung tribe of the Kalahari Desert in Africa is an example. The earliest Westerners to visit the !Kung remarked on their apparent lack of fear of fire. Firewalking to the !Kung, in fact, was an entirely natural act in which they had engaged as long as the tribe has existed.

Explanations

Scientists have long tried to find rational explanations for the effects of firewalking. Sufficient evidence appears to exist that people who walk on and/or handle hot coals do not, in fact, experience serious burns or other damage to their bodies. Two scientific explanations have been developed for this fact: the so-called Leidenfrost Effect and the Low Conductivity Theory.

The Leidenfrost Effect is named for the German physicist who first described the phenomenon. He explained that a drop of water placed on a hot surface evaporates from the bottom up. That is, water at the bottom of the drop takes heat from the hot surface, but the upper part of the drop maintains a thin cooling surface between the hot surface and the air above the drop. According to this explanation, moisture on the bottom of a firewalker's feet provides a very thin cooling layer that protects the feet from being burned.

The Low Conductivity Theory is based on the fundamental difference between heat and temperature. Heat depends on two factors: temperature and mass. Two objects can have the same temperature, but the one with the larger mass contains the greater amount of heat. According to this explanation, the burning coals on which a person walks may have a high temperature, but have so little mass that they contain little heat. They cannot, therefore, burn the person's foot.

For those who practice firewalking, scientific explanations of their experiences are neither satisfactory nor necessary for the protection they receive when walking on or handling fire. In the first place, they point out that many firewalkers are, indeed, burned. So the scientific explanations given above are insufficient to explain why some people experience burns and others do not.

In addition, those who practice firewalking tend to rely on religious, mystical, or other explanations. Firewalkers believe that the world is ordered by two very different sets of laws: those of the physical world and those of the spiritual world. Just because fire might harm the body in the physical world does not mean that it can or will do so in the spiritual world. Indeed, from this point of view, fire can be a positive, healing, and cleansing agent. It provides the means by which people may be cured of disease or purified of their sins and evil deeds.

Modern day firewalkers have explained their experiences by suggesting the presence of psychic energy given off by the human body. That energy is not unlike spiritual forces posited in Chinese medicine (the *ch'i*), in some religions of the Indian subcontinent (*prana*), and by the !Kung tribe (*n'um*). This energy is presumed to surround the human body like a halo and to protect it from influences in the natural environment. One is not burned by walking on hot coals, according to this explanation, because of the body aura that stands between one's feet and the hot coals.

Firewalking Today

Whatever its explanation, firewalking has become a popular activity among many ordinary people today. Tolly Burkan, Director of the Firewalking Institute for Research and Education, estimates that there are forty to fifty firewalking instructors in the United States today, and over 300,000 people have taken part in at least one firewalk.

For modern firewalkers, the experience is far more than a demonstration of the human body's ability to withstand the effects of fire. It is an opportunity for individuals to conquer their fears (of fire and other events in their lives) and to discover and learn how to draw on the inner strengths that all humans presumably have available to them.

As with fire breathing, readers are discouraged from firewalking. *See also* RITUAL FIRE.

Further Reading: Danforth, Loring M. *Firewalking and Religious Healing: The Anastenaria of Greece and the American Firewalking Movement.* Princeton, NJ: Princeton University Press, 1989; Doherty, Jim, "Hot Feat: Firewalkers of the World." *Science Digest*, August 1982, 369–384;

Firewalking has become a popular activity. Here a participant in the Firewalk Challenge, in Palm Beach County, FL, walks through hot burning coals. © *Greg Lovett/*The Palm Beach Post.

Dylan, Peggy, "Red Hot and Healing," <http://sundoor.com/fwessc.html>, 12/11/98; Frazer, J.G. *Balder the Beautiful: The Fire-Festivals of Europe and the Doctrine of the External Soul*, 2 vols. London: Macmillan and Company, 1913, Chapter 6; Margrave, Tom, "The Firewalk Experience," <http://heartfire.com/firewalk/pages/experience.html>, 12/11/98.

FIREWORKS

Fireworks are a form of pyrotechnics, substances that produce heat, light, noise, smoke, and motion when they are ignited. Other forms of pyrotechnics include flares, incendiaries, smoke bombs, and explosives. Fireworks are different from other forms of pyrotechnics in that they are used primarily for the purpose of entertainment.

History

The chemical principles behind the use of pyrotechnics date back well over a thousand years. The first pyrotechnic used was gunpowder, discovered in China before 1,000 A.D. Gunpowder is a mixture of potassium nitrate (KNO_3), charcoal (nearly pure carbon), and sulfur. The usual proportion for these three ingredients is 75 percent, 15 percent, and 10 percent. In this mixture, the potassium nitrate is the oxidizing agent and charcoal and sulfur are the fuels. When the mixture is lit, oxygen released by the potassium nitrate causes the charcoal and sulfur to burn very quickly, producing an explosive reaction.

Gunpowder was originally used by the Chinese primarily for ceremonial purposes, for blasting rock, and in warfare. Its function as a ceremonial device was to scare away evil spirits.

It was not until the mid-thirteenth century that gunpowder was introduced into Europe. Europeans found out about the substance from Islamic tribes with whom they battled. The first mention of a recipe for gunpowder in the English language is found in the writings of the English scientist Roger Bacon (1220–1292). In 1242, Bacon wrote down a set of instructions for making the explosives. Unlike the Chinese, early Europeans used gunpowder at first almost entirely for utilitarian purposes, primarily in construction operations and for warfare.

Eventually, Europeans were introduced to the use of gunpowder for ceremonial and entertainment purposes. Fireworks first became popular in Italy in about 1500, and then spread rapidly through the rest of Europe. Records indicate that Queen Elizabeth I was present at a fireworks display in Warwick in 1572. The person who may have used fireworks most lavishly was King Louis XIV of France, the Sun King, who ruled from 1643 to 1715. Louis was especially fond of providing fireworks displays at his magnificent Palace of Versailles.

Today, fireworks demonstrations are popular at holidays and other special events throughout the world. Most communities in the United States, for example, celebrate the Fourth of July with fireworks displays.

Composition

Probably the best known form of fireworks today is a rocket fired out of a cannon into the sky. At some high point in its ascent, the rocket explodes producing a very loud sound and a dazzling display of colored light.

The rocket used in such displays consists of three parts: the casing, color-producing compounds, and a pair of fuses and a lift charge. The rocket casing is usually made of cardboard or laminated paper in the shape of a cylinder. Some kinds of fireworks are enclosed in casings with other shapes. For example, the device known as a Catherine wheel consists of a cardboard cylinder wrapped in a spiral pattern around a central core. When the wheel is ignited, the chemicals inside burn upward through the cylinder producing a spinning shower of light.

Packed inside the casing are the compounds that give the rocket its distinctive colors. Fireworks designers can choose among a large variety of chemical compounds to obtain very distinctive colors. For example, compounds of calcium, lithium, and strontium give off reddish colors when they burn. Barium, copper, tellurium, thallium, and zinc compounds produce various shades of green, from a whitish-green for zinc to a bluish-green for thallium. A variety of blues are produced when compounds of arsenic, copper, lead, or selenium are used. Bluish-purple, reddish-purple, and violet are produced when compounds of cesium, potassium, and rubidium, respectively, are used. Finally, compounds of sodium are used to obtain yellow light. Magnesium powder in a rocket releases a brilliant white display when it is ignited.

The rocket also contains two fuses and a lift charge. The two fuses are the *time fuse* and the *side fuse*. The time fuse is attached to the top of the cylinder and the side fuse to its side. The lift charge is attached to the bottom of the cylinder. It consists of a small amount of gunpowder. The cylinder with its colored compounds and the two fuses and lift charge are then wrapped up inside another paper or cardboard cylinder. Finally, a *leader*, is connected to the time fuse at the top of the cylinder. The leader is a thin, wire-like fuse.

The Explosion

The fireworks display is initiated when a match is touched to the leader, causing it to start burning. The leader burns down to the top of the cylinder, igniting both the time fuse and the side fuse at almost the same time. The time fuse gets its name from the fact that it does not go off immediately, but burns for a certain period of time. The side fuse does ignite immediately, and sends a burst of fire down a wire to the lift charge.

When the fire reaches the lift charge, it ignites the gunpowder of which it is made. The gunpowder then explodes like a small bomb. That explosion propels the rocket into the sky. Meanwhile, the time fuse continues burning. At some point, it goes off and ignites the chemicals contained inside the cylinder. As the chemicals begin to burn they produce the characteristic colors and sounds for which they were chosen.

Safety Issues

Fireworks present a serious safety issue for both manufacturers and consumers. After all, one of the fundamental parts of a fireworks device is a small bomb. When fireworks explode, they always present a potential threat to humans who are in the area.

People who make and work with fireworks are, of course, especially familiar with the hazards posed by these devices. For example, some forms of clothing pick up static electricity easily. This static electricity is sufficient to ignite a fireworks device accidentally. To avoid this problem, those who work with fireworks wear nothing but cotton clothing since cotton is less likely to develop static electricity on itself.

Fireworks are also stored very carefully to prevent their accidental ignition. Some manufacturers keep their products in metal sheds separated by sand banks. In this way, if one box of fireworks should accidentally catch fire, others will be protected from the explosion. Still, accidents at fireworks factories do occur. When they do so, they can result in some of the most serious of all industrial disasters.

Consumers can be at risk from fireworks accidents also. People who are careless with fireworks devices run the risk of being injured by premature or unexpected explosions. Each year in the United States, hundreds of people are burned or hurt by fireworks over the Fourth of July weekend.

According to the most recent data available from the Consumer Products Safety Commission, about 8,500 people were treated in 1998 in hospital emergency rooms for injuries associated with fireworks. Just over half of those injuries were burns involving the hands, eyes, and face. About a third of these injuries occurred to boys and girls between the ages of five and fourteen years. Another third occurred among young men and women between the ages of fifteen and twenty-four. Overall, about seven out of every ten injuries occurs with boys and men. In response to the rise of fireworks injuries, some communities and states have banned the sale of fireworks.

Readers are encouraged to attend community fireworks displays and are strongly discouraged from using fireworks themselves.

Further Reading: Conkling, John A. *Chemistry of Pyrotechnics: Basic Principles and Theory*. New

York: Marcel Dekker, 1985; Grayson, Martin, ed. *Kirk-Othmer Encyclopedia of Chemical Technology*. New York: John Wiley, 3rd edition, 1978, volume 19, pp. 484–499; Rossotti, Hazel. *Fire*. Oxford: Oxford University Press, 1993, Chapter 13.

FIRING OF POTTERY

Firing of pottery is the process by which some type of ceramic product is made by the heating of clay. Firing involves both a physical and chemical change in the original material. As an example, consider the changes that take place in the mineral known as kaolinite, a common type of clay.

Chemically, kaolinite is primarily hydrated aluminum silicate with the chemical formula $Al_2Si_2O_5(OH)_4$. As it occurs in nature, kaolinite contains a certain amount of absorbed moisture. It may be wetter or drier depending on the location in which it occurs, but even the "driest" clay has some moisture within it. When kaolinite is heated to a temperature of more than 100°C, this absorbed moisture is driven off.

As kaolinite is heated to higher temperatures, a chemical reaction occurs. In this reaction, water held chemically within kaolinite molecules is expelled:

$$3Al_2Si_2O_5(OH)_4 \xrightarrow{heat} Al_6Si_2O_{13} + 4SiO_2 + 6H_2O$$

The water formed in this reaction is driven off as steam and the material that remains is modified aluminum silicate ($Al_6Si_2O_{13}$), known as mullite, and silicon dioxide (SiO_2), or silica. Crystals of mullite and silica interlock with each other as they are formed in the heating process, producing a new kind of material that will not reabsorb water. It is now a ceramic product that can be painted, glazed, and otherwise treated.

The conversion of kaolinite to mullite and silica takes place at temperatures above 800°C. Firing of other types of clays require different temperatures. The temperatures needed for the firing of pottery require special types of ovens, known as kilns. Firing in a kiln usually occurs over an extended period of time, sometimes as long as three days.

The kiln is first heated to relatively moderate temperatures to allow absorbed water to escape without forming bubbles and cracking the clay. After about a day, the temperature is raised to that which is needed to bring about the appropriate chemical reaction.

Further Reading: Birks, Tony. *Tony Birks Pottery: A Complete Guide to Pottery-Making Techniques*. Oviedo, FL: Gentle Breeze Publishing Company, 1998; Kenny, John B. *The Complete Book of Pottery Making*. Iola, WI: Krause Publications, 1976; Rossotti, Hazel. *Fire*. Oxford: Oxford University Press, 1993, pp. 85–91; Yanagida, Hiroaki, Kunihito Kawamoto, and Masaru Miyayama. *The Chemistry of Ceramics*. New York: John Wiley & Sons, 1996.

FLAG BURNING AMENDMENT

The flag burning amendment is a proposed amendment to the Constitution of the United States that would prevent the burning or other desecration of the American flag. Some form of the flag burning amendment has been submitted in every session of the U.S. Congress since 1989. There have been few reports that the burning of a national emblem, such as a flag, has been an ongoing, contentious, political issue in nations other than the United States. In its earlier forms, the proposed amendment consists of a single sentence, which reads:

> The Congress shall have the power to prohibit the physical desecration of the flag of the United States.

History of the Issue

The recent controversy over desecration of the American flag began in the summer of 1984 when Gregory Johnson participated in a demonstration during the Republican National Convention in Dallas, Texas. Johnson burned an American flag during the demonstration and was arrested, charged, tried, and convicted of violating a Texas law against desecration of the American flag. Johnson's appeals slowly worked their way through the courts and, on 21 June 1989, the U.S. Supreme Court overturned his conviction. The Court ruled on a five-to-four vote that Johnson had been deprived of his rights of

The Flag Burning Amendment would ban the desecration of the flag such as occurred at this anti–Vietnam War protest in Washington, DC, 1969. © *Terry Atlas/Archive Photos.*

proposed flag-burning amendments, which then failed to make it through the Senate.

The fundamental argument in favor of the flag-burning amendment is that the U.S. flag is an important national symbol and treasure that needs to be honored and revered. People who desecrate the flag, supporters say, also dishonor the memory of all the men and women who have died defending the flag and the nation it represents.

Opponents of the amendment believe that the flag-burning amendment is an unwarranted intrusion on the right of free speech, guaranteed by the First Amendment. In addition, they argue, the amendment would put the federal government in the business of regulating political protest, a step that has never been taken in the nation's history.

Further Reading: ACLU Action Alert: Say No to the Flag "Desecration" Amendment! <http://www.aclu.org/action/flag106.html>, 07 July 1999; "Citizens Flag Alliance: Who We Are." <http://www.cfa-inc.org/about.htm>, 07 July 1999; "Pro-Free Expression Groups." <http://www.FlagAmendment.org/pro-freedom.shtml>, 07 July 1999.

free speech under the First Amendment to the Constitution.

Within months, the U.S. Congress adopted the "Flag Protection Act of 1989" making it a crime to burn the American flag. President Bush declined to sign the law, indicating that he preferred a Constitutional amendment to achieve the same result. In January of 1990, Federal District Court judges in Seattle and the District of Columbia declared the Flag Protection Act illegal.

In response to this set-back, the House of Representatives voted in the summer of 1990 in favor of a proposed Constitutional amendment banning flag burning. The Senate failed to support the same legislation by a very narrow margin. Although the bill received a majority in the Senate, it fell three votes short of the two-thirds majority required for a proposed Constitutional amendment. In succeeding years, the House continued to adopt

FLAME

A flame is the visible, glowing evidence of a burning object. For most people, no formal definition of a flame would seem to be necessary. Flames are a familiar and common aspect of everyday life. But scientists now point out that flames are actually complex physical and chemical systems whose structure and behavior are still not entirely understood.

Two general types of flames are recognized: the *premixed flame* and the *diffusion flame*. A premixed flame is one in which the fuel and oxidizing agent are artificially combined in some predetermined way. Probably the best example of a premixed flame is the flame produced by a Bunsen burner or the burner in a kitchen gas stove. (For more information on such flames, see *Bunsen burner.*

Diffusion Flames

A diffusion flame is named because of the fact that the fuel is supplied at one part of the reaction, and the oxidizing agent needed for combustion (usually, oxygen) diffuses into the fuel from some outside source. By far the most common type of diffusion flame is that produced by a burning candle.

Candle flames are of great interest to scientists because they provide relatively simple models of the far more complex fire flames produced by larger, more complex burning objects. Thus, scientists interested in learning more about the processes that occur when a building burns down are stymied by the enormous complexity of the physical and chemical changes that take place during that process. But they can gain some basic understanding of the process by studying the dynamics of the combustion of a candle.

Mechanics of a Candle Flame

A candle is made of wax into which a rope-like wick has been inserted. The wax itself is a hydrocarbon, a compound of carbon and hydrogen, of high molecular weight. The chemical formula of the wax typically has values in the range from about $C_{20}H_{42}$ to $C_{35}H_{72}$.

When the wick of a candle is lit, the heat produced by combustion of the wick causes the upper surface of the candle wax to melt. The molten wax moves upward through the wick by capillary action until it comes into contact with the burning wick. At that point, the wax itself vaporizes. The wax vapor then catches fire.

Combustion occurs at this point because oxygen from outside the candle has diffused inward toward the burning wick. This oxygen reacts with the vaporized wax, producing a flame. The burning flame represents a condition of steady state, in which oxygen and wax vapor are provided to the flame at the same rate at which the products of combustion are removed from the flame.

The structure of a candle flame is surprisingly complex, containing a complex mixture of combustion reactants (materials needed for the combustion reaction to take place) and combustion products (materials formed as combustion occurs). For example, in the region directly around the wick, large hydrocarbons from the wax break down to form simpler hydrocarbons with molecular formulas in the range from $C_{10}H_{22}$ to $C_{20}H_{42}$.

A number of complex chemical changes take place at this point. For example, saturated hydrocarbons (hydrocarbons that contain single bonds only) are converted to unsaturated hydrocarbons (those that contain double bonds). Also, straight-chain hydrocarbons (those in which the carbons are arranged in a long chain) are converted to ring compounds (in which the carbons are arranged in a circular pattern).

The presence of these hydrocarbons is made apparent by the red glow of the candle flame. The glow is produced by unburned hydrocarbons and unburned particles of carbon that have been heated to a high enough temperature to cause them to give off light. But there is insufficient oxygen present at these locations for the particles to actually catch fire.

Oxygen continues to diffuse inward toward the wick, however. The presence of this oxygen causes these unburned hydrocarbons (as well as the original larger hydrocarbons and unburned particles of carbon) to begin burning, forming combustion products, such as carbon dioxide (CO_2), carbon monoxide (CO), and water vapor. Carbon dioxide and water vapor cannot be oxidized further, and they escape from the flame as final oxidation products. Carbon monoxide can, however, be further oxidized, and it is converted to carbon dioxide in other parts of the flame. Under normal circumstances, then, the final products of the combustion of candle wax are carbon dioxide and water vapor.

The combustion of carbon and hydrocarbons is an exothermic process. That is, heat is released as these substances are changed into carbon dioxide and water vapor. The release of this heat makes possible the continuation of the sequence of reactions (melting of wax and combustion of wax vapor) originally made possible by a lighted match. The reaction continues until the fuel is used up (the candle wax is gone), the temperature is reduced, or the oxygen supply to the candle is cut off. *See also* HEAT.

Further Reading: Lyons, John W. *Fire*. New York: Scientific American Books, 1985, Chapter 2; Martin, Stanley B., and A. Murty Kanury, "Fire Technology." In *McGraw-Hill Encyclopedia of Science & Technology*, 8th edition. New York: McGraw-Hill Book Company, 1997, Volume 7, pp. 150–153; Walker, Jearl, "The Amateur Scientist: The Physics and Chemistry Underlying the Infinite Charm of a Candle Flame." *Scientific American*, April 1978, 154–162; Wilbraham, Antony C. et al. *Addison-Wesley Chemistry*. Menlo Park, CA: Addison-Wesley Publishing Company, 1990, pp. 159–160.

FLAME BALLS

A flame ball is a spherical flame that forms in very-low-gravity atmospheres. Flame balls were originally predicted in 1944 by the Russian physicist Yakov B. Zeld'ovich. Zeld'ovich said that, in the absence of gravity, the flame produced by a burning object would take a spherical shape rather than the typical tear-shape with which we are generally familiar.

For about fifty years, there were no simple methods available for testing Zeld'ovich's theory. Low- or zero-gravity environments were very difficult to produce, so no tests of his theory could be conducted. The nearest approximation to such tests involved the use of devices known as *drop towers*. A drop tower is a tall structure from which various objects can be released. Cameras inside the walls of the drop tower are able to record changes that take place during an object's fall. Experiments on the shape of a flame in drop towers are, however, difficult to perform with much accuracy.

Another approach to the study of Zeld'ovich's hypothesis has been to use aircraft that travel in a parabolic arc through the atmosphere. At the top of this trajectory, the aircraft experiences a nearly gravity-free environment in which flame studies can be conducted. Again, these conditions are far from ideal in testing Zeld'ovich's hypothesis.

SOFBALL Experiments

The recent availability of Earth-orbiting spacecraft has finally made it possible to test Zeld'ovich's hypothesis with some degree of accuracy. The gravitational force inside an orbiting spacecraft such as the U.S. Space Shuttle is only about one-millionth of that on Earth. These conditions are ideal for carrying out experiments on the structure of flames in the absence of gravity.

In the mid-1990s, a series of experiments designed to test Zeld'ovich's hypothesis was designed and given the name Structure of Flame Balls at Low Lewis-Number, or *SOFBALL*. The first of these experiments was planned for the flight of STS-83 on 4 April 1997. Only two of the planned fifteen experiments were actually carried out, however, because fuel problems forced the space orbiter to shorten its mission. A second series of experiments was planned for the flight of STS-94 on 1 July 1997. In this case, all thirteen experiments were completed as planned.

The design of these experiments called for the ignition of a mixture of 4.9 percent hydrogen and 9.8 percent oxygen, diluted with 85.3 percent carbon dioxide. Once the mixture was ignited, the flame was allowed to burn until it went out of its own accord or the experiment was brought to an end after 500 seconds.

A number of important observations were collected during these experiments. First, at least one spherical flame ball of the kind hypothesized by Zeld'ovich was formed in each of the experiments. In some cases, more than one flame ball was produced during ignition, up to a maximum of nine balls in one case. Second, the balls were generally spherical, as predicted, by Zeld'ovich, and they remained suspended in space for anywhere from 150 to the maximum 500 seconds. Third, the flame balls were the weakest flames ever produced or observed by humans. They released an average of about one watt, which compares to about 50 watts for the average birthday candle.

The SOFBALL experiments have some obvious applications to the problem of accidental fires on spacecraft. Such fires would behave very differently from those that occur in an Earth-gravity condition with which scientists are familiar. Data from the SOFBALL experiment can be used to develop methods for protecting against and fighting such space-based fires.

The SOFBALL experiments have also provided information on fuel mixtures used only rarely on Earth—those with very low concentrations of fuel (such as hydrogen, in the SOFBALL experiments) and oxidizer (oxygen). This information may be useful in designing more efficient combustion systems for transportation and industry.

Further Reading: "Candle Flames in Microgravity— Mir/Priroda," <http://zeta.lerc.nasa.gov/expr/ cfm.htm>, 01/04/99; "Combustion in Microgravity," <http://www.microgravity.com/ combustion.html>, 01/04/99; Newton, David E. *Chemistry: Oryx Frontiers of Science*. Phoenix: Oryx Press, 1999, pp. 106–110; "Structure of Flame Balls at Low Lewis Number (SOFBALL): Preliminary Results from the STS-83 and STS-94 Space Flight Experiments," <http:// cpl.usc.edu/sofball/sofball.htm>.

FLAME RETARDANTS

Flame retardants are chemicals added to materials to prevent them from catching fire when subjected to a low-energy flame, such as that produced by a burning match. When so treated, such materials are sometimes said to be *nonflammable* or *fireproof*. These terms are somewhat misleading, however, as the materials may be more or less resistant to flames depending on the treatment to which they have been exposed and the intensity of the flame applied to them.

History

Humans appear to have been searching for methods to make materials fireproof for centuries. There is evidence that artisans as far back as the fourth century B.C. had learned how to coat wood with vinegar and clay to make it more resistant to fires.

By the nineteenth century, the study of flame retardants had become an active field of chemical research. The French chemist Joseph Louis Gay-Lussac (1778–1850) began a study of flame retardant chemicals in 1820. He discovered that ammonium sulfate, ammonium phosphate, ammonium chloride, and boric acid were all more or less effective in protecting a material against flames. An-

other breakthrough was made by the English chemist Sir William Henry Perkin (1838– 1907) in the late nineteenth century. Perkin found that metallic compounds of stannic and tungstic acid could also be used to make a material fire-resistant.

Principles

The principles that underlie the use of flame retardants are similar to those for other forms of fire prevention and firefighting. That is, to prevent or suppress a fire, one must take one of three actions: remove the fuel, cut off the supply of oxygen, or reduce the temperature of the fuel below the ignition point.

These objectives can be achieved by either physical or chemical means. For example, the surface of a material can be covered with a flame-resistant chemical that prevents oxygen from reaching the surface of the underlying combustible material. Or, a substance can be added to the material that releases a gas when heated. The gas may then reduce the temperature of the material and/or cut off the supply of oxygen.

An example of a chemical mechanism is one in which a substance is added to the combustible material that breaks down easily when heated to form elemental carbon. The carbon thus produced may then coat the surface of the combustible material, forming a protective layer that prevents oxygen from reaching the material. Another example of a chemical mechanism is the addition of a substance to a material that becomes liquid when heated. As the liquid forms, it may cause the combustible material to flow away from the zone of combustion, thus removing the fuel and suppressing the fire. In general, chemical mechanisms for providing flame protection tend to be more effective than physical methods.

Methods

The objectives of using flame retardants can be accomplished by one of three general methods:

1. The fire retardant material can simply be applied as a coating to the combustible material to be protected. This method is one of the easiest to use and one of the

oldest employed by humans. For example, the Chinese have for many centuries covered the thatched roofs of their homes with a layer of mud to protect them from heat and fire.

2. The combustible material can be soaked in a solution of the fire retardant, allowing the retardant to penetrate the interior space of the combustible material. Two substances commonly used for this purpose are ammonium carbonate $[(NH_4)_2CO_3]$ and ammonium sulfamate $(NH_4OSO_2NH_2)$. When heated, ammonium carbonate decomposes to form carbon dioxide (among other products) and ammonium sulfamate decomposes to form sulfur dioxide (and other products). Neither of these gases is flammable, so they protect the outer surface of the material from further oxidation. In addition, they tend to absorb and carry away heat, reducing the temperature of the burning material and contributing to its dying out. Another combination of additives is borax (sodium borate; $Na_2B_4O_7 \bullet H_2O$) and sodium silicate (Na_2SiO_3). When heated, these two substances react to form a glassy substance that protects the material from further oxidation.

3. The combustible material can be altered chemically. As an example, most combustible materials contain an abundance of hydrogen atoms in their chemical structure. These hydrogen atoms combine readily with oxygen (and other oxidizing agents), increasing the likelihood that combustion will occur. One way to reduce the flammability of such materials is to modify their chemical structure by replacing some hydrogen atoms with those of fluorine, chlorine, or bromine. These atoms do not react with oxygen, as does hydrogen, so the material becomes less combustible.

These three methods represent increasingly effective methods of fireproofing a material. With the first procedure, the flame retardant may wear off easily during wearing and/or cleaning. With the third procedure, however, the flame retardant becomes an integral part of the material and is not easily removed from it.

New Materials

Another approach to fireproofing is to design entirely new compounds from which nonflammable fabrics can be made. Many of the materials from which cloth is made are organic compounds rich in carbon and hydrogen atoms, both of which oxidize readily. Such materials are, therefore, inherently flammable. But chemists have now invented entirely new kinds of polymers with fewer hydrogen atoms in their chemical structure, making them inherently less flammable. At this point in time, those materials tend to be somewhat more expensive and less suitable for clothing and other uses than more traditional, flammable materials.

Further Reading: Ash, Michael, and Irene Ash. *The Index of Flame Retardants*. Brookfield, VT: Ashgate Publishing, 1997; Kroschwitz, Jacqueline I. "Flame Retardants." In Mary Howe-Grant, ed. *Kirk-Othmer Encyclopedia of Chemical Technology*, 4th edition, Volume 10; Lyons, John W. *The Chemistry and Uses of Fire Retardants*. New York: Wiley-Interscience, 1970; Lyons, John W. "Flame Retardants (An Overview)." In *Kirk-Othmer Encyclopedia of Chemical Technology*, 4th edition, Volume 10. New York: John Wiley & Sons, 1980, pp. 348–354; Troitzsch, Jürgen. *International Plastics Flammability Handbook*. Munich: Hanser Gardner Publications, 1990; *Also see* an excerpt from this book at <http://www.firesafety.org/FlameRetardants/FlameRetardants.htm>, 01/08/99.

FLAME TEST

A flame test is a laboratory procedure by which the presence of certain elements in a material can be identified. The test is possible because various elements produce flames of distinctive colors when they are heated in a hot flame.

Procedure

A flame test is typically conducted with a platinum or Nichrome wire that has been mounted in a glass rod. Nichrome is a nickel-chromium alloy. The wire is first cleaned by dipping it in dilute hydrochloric acid and then heating it in the flame of a Bunsen burner. Any color that appears in the flame, other than that of the burner flame itself, indicates that the wire is not clean. With such a result, the process of dipping in acid and heating is repeated until no colored flame is produced.

The clean wire is then dipped into the material to be tested and placed into the burner flame again. The presence of any color other than that of the burner flame itself indicates the presence of one of the elements listed below.

Flame Colors

The characteristic color produced by an element when heated depends, in part, on the chemical state in which the element occurs. Generally speaking, compounds of an element with chlorine or bromine tend to give the most distinctive colors. In some instances, the unknown material may first be moistened with or dissolved in hydrochloric acid to convert the elements present into their chlorides.

Color analysis may be complicated by the presence of more than one element in a material. For example, sodium compounds tend to give a very clear yellow color when heated. This color is so pronounced that it may mask the presence of other elements with which it commonly occurs, such as potassium.

The following table provides only a general guideline to the use of flame tests to identify elements. It is subject to the conditions mentioned above. The elements marked with an asterisk (*) may have to be identified with the use of a colored glass. For example, if both sodium and potassium are present in a material, the yellow color of the sodium flame will mask the violet color of the potassium flame. But if the flame is viewed through a cobalt glass (a bluish glass that contains a cobalt compound), the yellow flame is rendered invisible, and the violet flame becomes visible.

Flame Tests to Identify Elements	
Color of Flame	**Element Indicated**
BLUE	
Azure	copper, lead, selenium
Pale blue	arsenic, selenium
Greenish-blue	antimony, arsenic, copper, lead
GREEN	
Pale green	antimony, ammonium compounds
Whitish-green	zinc
Yellow-green	barium, molybdenum, boron
Emerald green	copper, thallium
Pure green	tellurium, thallium
Blue-green	boron, phosphorus compounds
RED	
Carmine	lithium*
Scarlet	calcium*
Crimson	strontium*
VIOLET	cesium, potassium, rubidium*
YELLOW	sodium*

FLAME THROWERS

A flame thrower is a device, usually a weapon, that shoots a burning stream of liquid out of a nozzle. The device is especially useful in some aspects of battle because the stream it emits is able to enter small openings and flow around corners in ways unmatched by other forms of military ordnance.

History

The concept of propelling flaming liquids at an enemy can be traced to at least 500 B.C. when Chinese soldiers are believed to have used this technology against enemies. During the Peloponnesian War, a primitive flame thrower was built of a hollow log fitted with a bellows to blow a mixture of flaming pitch, sulfur, and naptha against an enemy. Some time later, the practice of dumping burning oil on attackers from castle walls became a popular method of defense. In about 671 A.D.,

a type of burning liquid known as *Greek fire* was first used by the Byzantines in their defense of Constantinople against Muslim attackers.

The modern form of the flame thrower was first developed by German military engineers in the first decade of the twentieth century. The device was produced in two forms, one small enough to be carried by a single person and the other large enough that some form of mechanized transport was needed to move it about. In both cases, the weapon consisted of a tank containing the flammable liquid, a source of compressed air, and a long hose capped by a nozzle. A source of ignition was also provided at the nozzle. The fuel most commonly used was gasoline.

To use the flame thrower, a trigger was pressed to release compressed air into the tank, forcing the flammable liquid out of the tube and nozzle. As it reached the nozzle, the liquid was ignited. The stream of burning liq-

A U.S. Navy Patrol boat tests a ZIPPO flame thrower in Vietnam, 1969. *Courtesy of DOD Defense Visual Center.*

uid was able to travel about twenty meters for the smaller device and forty meters for the larger device. In either case, the burst of flame could be sustained for only a brief period, usually less than ten seconds.

Flame throwers were used sporadically by the German army against France during World War I. But they had very limited effect. The flames they produced had too short a range to be of much value, and the equipment was dangerous and unreliable. Both the British and the French eventually developed flame throwers also, but the United States Army did not.

World War II Applications

Interest in the use of flame throwers was rekindled in the late 1930s as World War II approached. The Italians used the weapons against their enemies in the conquest of Abyssinia (Ethiopia) in 1935–1936. While they were still not very effective weapons, they did have a powerful psychological effect on those who came into contact with them.

By 1940, the United States Army had decided to begin the development of its own flame-thrower weapons. Researchers developed a series of models, known as the E1, the E1R1, and the M1A1 before finding one that worked satisfactorily. That model was known as the M2-2. It was used as a portable weapon, capable of being carried and used by a single soldier, throughout World War II. Researchers later developed a larger model of the M2-2 which was mounted on tanks.

Modern flame throwers use a modified form of gasoline as their fuel. A thickening agent is added to the gasoline to form a fuel known as *napalm*. The advantage of napalm is that it burns hotter and longer than does gasoline. Also, it tends to stick on any surface on which it lands, setting fire to that surface also.

The most recent versions of the U.S. flame thrower weigh about 25 kilograms (55 pounds) and are filled with 16 liters (4 gallons) of fuel. The compressed air used to drive the fuel exerts a pressure of about 140 kilograms per square centimeter (2,000 pounds per square inch), providing the weapon with a range of about 50 meters (50 yards). It is typically fired in short bursts that last about six seconds each. The larger version of the device can fire its charge a distance of about 200 meters (200 yards).

Flame throwers still suffer from many of the limitations discovered in World War I. However, they also have some unique applications. For example, during World War II, Japanese forces in the Pacific often constructed nearly impenetrable fortifications. These bunkers were built with layers of wood and steel, and protected with sandbags or oil drums filled with sand. Ordinary ordnance was essentially useless against such fortifications. However, flame throwers had the ability to project their burning fuel into any opening that existed in the bunkers, such as through gun turrets or breathing holes. They were first used in combat on 15 January 1943 during a battle on Guadalcanal by the Second Marine Division. The flame throwers were entirely successful in penetrating a particularly stubborn Japanese bunker, killing all of the defenders.

The Guadalcanal experience convinced military strategists that flame throwers, even given all their limitations, still had important applications in military encounters.

Further Reading: Kleber, Brooks E., and Dale Birdsell. *The Chemical Warfare Service: Chemicals in Combat*. Washington, DC: Office of the Chief of Military History, United States Army, 1966, Chapter XIV; Mountcastle, John Wyndham. *Flame On!: U.S. Incendiary Weapons, 1918–1945*. Shippensburg, PA: White Mane Publishers, 1999.

FLARES

A flare is a device for producing an intense light, usually of short duration, for the purpose of signaling and illuminating an area at night, or producing some special effect. A flare is a type of pyrotechnic device; that is, one that achieves a dramatic visual display as the result of a combustion reaction. Flares are somewhat similar in their operation and effect to fireworks.

History

It seems likely that humans have used fire for the purpose of signaling and illumination for untold centuries. However, devices with the two most important characteristics of flares—intense light and brief duration of burn—were invented only quite recently. The first patent for a pyrotechnic signaling system was awarded to Martha Coston (b. 1826) in 1859. The system made use of a torpedo-shaped tube filled with an oxidizing agent and fuel that could be fired into the air. The light produced could then be used to send signals between two ships or between a ship and the shore.

Perhaps better known for his invention of signal flares was an American naval lieutenant, Edward Very. In the 1870s, Very invented a system for sending messages using red and green flares. Very designed a special type of pistol for use with this system. It had a steel barrel that was about nine inches long and was tapered at the muzzle. The barrel could be broken open in order to accept a paper cartridge containing the charge. The charge consisted of musket powder and a coloring agent that produced either red or green light. The red color was produced by using compounds of calcium or strontium, while the green color was produced with compounds of barium, copper, or zinc. Very's signaling system consisted of various combinations of red and green flares that had numerical equivalents which could be used as numbers or converted into alphabetic equivalents.

Modern Flares

Flares were first invented to provide a way for military units to contact each other over short distances. Coston's invention, for example, was credited with contributing to the success of Union naval units in sea battles against Confederate ships in the Civil War. This application of flares became less important as other means of communication, such as radio, were developed.

Today, signal flares are still used in military operations for specialized purposes, such as locating lost individuals and military units and for sending signals over short distances

when other means of communication have failed. One of the most important applications of flares today is as a component of survival kits. Boat owners, hunters, hikers, and others who may travel to remote or isolated areas usually carry flares with them in case they become lost or disabled. Flares are commonly used also to indicate the presence of roadside hazards, such as a disabled vehicle. Foresters sometimes use flares to set off controlled burns when the fire is planned for an isolated or remote area.

Composition

Flares have three primary components: a fuel, an oxidizing agent, and a material to provide color to the flame produced in the combustion reaction. The specific chemicals used for each of these components differs depending on the use to which the flare is to be put. For example, highway flares are usually made of potassium perchlorate (the oxidizing agent); a mixture of sawdust, wax, and sulfur (the fuel); and strontium nitrate (the coloring agent). This combination of materials is encased in a cardboard tube and ignited by means of a wick. The combination of sawdust and wax reduces the rate at which the flare burns.

Emergency flares like those used by boaters or hunters omit sawdust and wax and include a fast-burning fuel to increase the intensity of the light produced when the flare is ignited. These flares are available in many different sizes, shapes, and colors, but are usually small enough to be fired from a small pistol or held in one's hand. Some flares, for example, are small enough to be fired from a twenty-five-millimeter pistol. Others are designed to be fired from specially built miniature rocket tubes that direct the flare in a given direction. Still others come attached with small parachutes that allow them to remain in the sky for an extended period of time.

FLASH POINT

Flash point is the temperature at which a liquid or a volatile solid produces a sufficient amount of vapor to form an ignitable mixture with the air near the surface of the liquid or solid. A flammable liquid, such as gasoline, normally does not catch fire directly. Instead, some portion of the liquid first evaporates. The mixture of gasoline vapor and oxygen may then ignite.

The term *flash point* comes from the fact that the vapor may "flash," or burn briefly, at the flash point, but it will not continue to burn. The point at which the vapor will continue to burn after being ignited is referred to as the *fire point*.

Flash point and fire point differ from ignition (or kindling) temperature in two ways. First, the former two terms require that an outside source of combustion, such as a match, be provided for a fire to begin. Ignition temperature, by contrast, is the point at which a flammable liquid begins to burn on its own. Second, flash point and fire point are terms used only for those substances that produce vapors, such as liquids or volatile solids, while ignition temperature refers to any object that burns.

FOREST FIRES

A forest fire is any type of fire that affects forested lands. Forest fires are a normal and natural part of the cycle of life in wilderness areas. In one stage of that cycle, carbon dioxide and water are used up to produce cellulose and other plant materials by means of photosynthesis. In another stage of that cycle, those materials are broken down by decay, fire, or other processes to release the carbon dioxide and water from which they were originally formed.

Causes of Forest Fires

The vast majority of forest fires today are caused by human activity. That fact was not always true. At one time, most forest fires were caused by natural forces, such as lightning, volcanoes, and rock slides. The relative importance of natural versus anthropogenic sources of forest fires depends to a large extent on the development that has taken place in an area. As humans move into a region, build homes and other structures, and create cities, towns, and other populated regions, their own activities become more important in the ignition of a forest fire.

A forest wildfire in the early stages. *Courtesy of National Fire Academy, Emmitsburg, MD.*

For many centuries, those who have been responsible for the monitoring of forest reserves promoted a policy of fire prevention and suppression. This policy was based on a philosophy that forest reserves should be protected and harvested for human use. As the ecological folly of that policy became apparent in some parts of the world, humans began to adopt a practice of prescribed burns, in which they tried to duplicate on a controlled scale fires that had once been natural in an area. As a result, prescribed burns are now an important source of forest fires in many parts of the world, not only as a controlled source of fire, but also as a source of runaway fires that grow out of prescribed burns that grow beyond their intended boundaries.

By some estimates, humans are now responsible for about 90 percent of all forest fires in the United States. The most important single cause of such fires appears to be cigarette smoking, followed by the burning of residential wastes, sparks and "hot spots" created from machines and fireplaces, and campfires.

Factors Affecting the Course of a Forest Fire

Forest fires differ dramatically in their size and intensity. The primary factors affecting the nature of a fire include the amount and type of fuel available in a region, weather conditions, and topography. The term *fuel* when applied to a forest fire refers to any material, living or dead, that can burn. Most forests have a wide variety of fuels, including living trees of various ages, brush, grass, moss, dead leaves and branches, and other decaying organic matter.

One factor that affects significantly the ability of such materials to ignite and burn is their moisture content. For example, living plant material tends to have a high moisture content and may require a higher temperature to ignite and burn. Plant material that has been dead for an extended period of time has usually dried out sufficiently to make it highly flammable.

Weather can affect the course that a forest fire takes and may even determine whether the fire begins in the first place or continues to burn for very long. Hot, dry weather tends to increase the likelihood that fires will break out, and high winds tend to promote the spread of a fire. Conversely, fires can produce their own weather by creating convection currents that spiral upward producing tornado-like storms. Such storms may then make their own contribution to the spread of a fire.

Topographic factors can also determine the direction and intensity of a fire. In some parts of the United States, for example, the topographic features of a region almost ensure that serious fires will spread through the same area year after year. The fires created by the Santa Ana winds and the topography of the land near Santa Barbara, California, are an example of this phenomenon.

Types of Forest Fires

Forest fires are often classified into one of three major groups: surface, ground, and crown.

1. Surface fires run along the top of the ground, feeding off grass, shrubs, dead organic matter, and other plant material. These fires are often relatively easy to deal with because they stay well within the reach of firefighters.

2. Ground fires, by contrast, actually burn beneath the surface of the ground. They require special kinds of material and topography in order to keep burning. In a bog, for example, combustible material may be covered by a thin layer of water. Slow fires may develop and burn for long periods of time underground without breaking through to the surface. Ground fires also tend to be relatively easy to deal with, although they do present the threat of showing up at unexpected times in unexpected places.

3. Crown fires burn through the tops of trees. Crown fires are often the fastest moving, the most destructive types of forest fires. They spread quickly from one tree to the next and may throw off sparks and firebrands that are carried hundreds

of feet away to new trees. In the most severe cases, crown fires can create their own "fire weather" in which swirling convection currents produce tornado-like effects that spread flames great distances.

Detection and Warning

The detection of forest fires has long been a serious and difficult problem since a fire agency may have to monitor thousands of acres of forest looking for the outbreak of a fire. At one time, virtually the only system available for detection was the fire tower or lookout tower. In the United States, these towers became especially popular during the 1930s when the manpower available from the Civilian Conservation Corps was available for the construction of these structures.

In recent years, fire agencies have come to rely more on monitoring by aircraft and even satellite. Technology has developed to the point that an airplane flying at an altitude of nearly four kilometers (about 12,000 feet) can now detect fires as simple as those used by campers to cook their dinner. Systems have been developed by which fire warnings can be relayed from scout aircraft to central monitoring stations and then to fire suppression teams in a matter of minutes. For example, the National Oceanic and Atmospheric Administration's AVHRR (Advanced Very High Resolution Radiometer) satellite uses a high-resolution radiometer to detect the smoke produced by fire and fire scars. The Defense Meteorological Satellite Program Operational Linescan System uses two telescopes to detect fires. One telescope operates in the visible range of the electromagnetic spectrum, and the other, in the infrared range. The Total Ozone Mapping Spectrometer, designed primarily to study ozone depletion in the atmosphere, is also able to detect dust and smoke particles produced by forest fires. Information from sources such as these can be transmitted almost instantaneously to firefighting organizations responsible for the areas observed.

Suppression

Firefighting has become a highly sophisticated operation involving not only men and women with picks and shovels on the ground, but also air tankers, smoke jumpers, chemical suppressants, specialized firefighting tractors, and sophisticated communication systems. Fire suppression can be said to take one of two general forms: fuel removal and extinction of burning fuel. Fuel removal simply means depriving a burning fire of any additional fuel it may need to continue its progress.

In a city fire, fuel removal may take the form of the destruction of buildings that may be in the course of a moving fire. In a forest fire, fuel removal involves the construction of a firebreak, a strip of land from which trees, bushes, grass, and any other combustible material has been removed. Fuel removal is often effective for moderate fires, although, in the case of more serious conflagrations, sparks and firebrands may leap through the air over a firebreak and extend the blaze to new, untouched fuels. The latter objective can be achieved by dropping water or chemical suppressants on a fire. In both cases, the water of chemical suppressant tends to produce a layer (steam or a chemical vapor) that deprives the fire of oxygen and also lowers the temperature of the fire itself.

In the case of large forest fires, simply getting sufficient amounts of water or chemical suppressant to a fire is a major problem. Giant tanker aircraft and "bombs" filled with water or chemicals are used for this purpose, but may be effective only on relatively moderate-sized fires.

Prevention

For most of the twentieth century, the primary goal of fire management programs in the United States was one of preventing fires. Programs such as Smokey Bear and Keep America Green were efforts by federal agencies to educate Americans about the dangers of forest fires and to enlist their help in preventing such fires.

Another approach to the prevention of forest fires is fuel removal. Fires are much less likely to begin if fuels such as dead trees and slash are removed from an area. Slash is the debris left after logging in a forested area. It consists of leaves, branches, brush, and other plant matter that is often highly flammable.

The problem with fuel removal is that it is often difficult and expensive to carry out such actions deep within a forested area. As a result, fires begun because of a carelessly dropped match or a lightning strike often have a rich source of fuel on which to feed.

The fire-management community in the United States and many other parts of the world now believe that prescribed burns are an important and essential component of forest fire prevention. By carrying out carefully controlled burns of a forest area, the fires that will inevitably strike that area eventually will have less fuel on which to operate and will, therefore, be less severe.

Fire Management Policies

Human attitudes about the control (or absence of control) of forest fires have changed dramatically over the centuries and from nation to nation. In those places and at those times when trees are valued primarily for their use as lumber or fuel for human consumption, fire prevention and suppression are likely to become national policies. Such policies almost inevitably lead to one of two results, however. On the one hand, the nation's forests may become so small and carefully controlled in size that the original objective is achieved. Such tends to be the case in many European nations. On the other hand, fuel on the forest floor eventually accumulates to a point that devastating and uncontrollable fires break out. Such has been the pattern in the United States over the past few decades.

Destruction of the Tropical Rain Forests

The issue of fire in tropical rain forests has become a topic of significant concern to many people interested in the global environment. One factor that led to the growing interest in this problem has been the availability of satellite photographs that show fires in the Amazon jungle, the forests of Indonesia, and other rain forest regions. These photographs have given a first-hand reality to the problem of burning forests that once came only from relatively dry statistical data. Those concerned

with rain forest fires estimate that, on average, one hectare (2.5 acres) of forest burn up every second somewhere in the world. In Brazil alone, one of the nations most affected by this problem, about two million hectares (five million acres) are destroyed every year (Rainforest Information: Fact Sheet 4B).

One of the fundamental problems in dealing with fire in the tropical rain forest has been to identify a culprit that can be blamed for the problem. On the one hand, some observers point to the practice of shifting cultivation ("slash-and-burn cultivation") practice by many nomadic people throughout the world. On the other hand, this practice has been in use for millennia and hardly seems to be a serious enough force to cause the devastation now being observed.

The problem may, in fact, not be the practice of shifting cultivation itself, but its adoption by large farms and industrial operations. Instead of setting fire to an acre or two of land, these entities may actually burn off thousands of acres at a time. When dry weather conditions appear during the use of such practices, as they did in the El Niño year of 1998, such fires may get out of control and create smoke and combustion problems that cover a whole region of the globe. *See also* BEHAVE; CLIMATE CHANGE; FIRE ECOLOGY; FIRE MODELING; FIRESTORM; INTERMIX FIRE; PRESCRIBED BURN.

Further Reading: "Fire Monitoring by Satellite," <http://modarch.gsfc.nasa.gov/fire_atlast/fires.html>, 08/24/99; "Forest Fires," <http://www.enviroliteracy.org/forest_fires.html>, 08/24/99; Fuller, Margaret. *Forest Fires*. New York: John Wiley & Sons, 1991; Pyne, Stephen J. *America's Fires: Management on Wildlands and Forests*. Durham, NC: Forest History Society, 1997; Pyne, Stephen J. *World Fire: The Culture of Fire on Earth*. Seattle: University of Washington Press, 1997; "Rainforest Information," <http://www.ran.org/info_center/>, 06/29/98; "Series of Emergency Net News Reports on the Indonesia Fire/Environmental Disaster," <http://www.emergency.com/indofire.htm>, 12/21/98.

FORGES AND BLACKSMITHING

A forge is an open fireplace or hearth, to which a bellows is attached, used for the heating of metals, especially iron. The word *forge* comes from Latin, Middle English, and Old French terms that refer to the person who works at the forge, the *blacksmith*, or "smithy."

History

The earliest form of the forge was probably developed during the Iron Age (about 1,000 B.C.) when humans first learned how to work with iron metal. The key to ironworking was finding a method for heating iron and its ores to temperatures of at least 500°C (about 900°F). A simple and efficient way to accomplish this task is to place the iron or ore in a bowl-shaped depression in the ground; fill the depression with coal, wood, or charcoal; ignite the fuel; and force air over the resulting fire.

References to blacksmithing and the use of forges goes back thousands of years. One of the Olympian gods honored by the ancient Greeks was Hephaestus, a blacksmith. One of Hephaestus' responsibilities was the forging of thunderbolts for use by Zeus in controlling the behavior of men and the Titans. Legend has it that Hephaestus' forge was located at the base of a volcano on the island of Lemnos. In Roman mythology, Hephaestus was known as Vulcan, and his counterpart in Norse mythology was Thor.

The art and craft of blacksmithing reached its peak during the Middle Ages when superb examples of kitchenware, weaponry, armor, and decorative objects were produced. Many blacksmithing tasks were taken over during the Industrial Revolution when all types of iron and steel products were produced by machines. Still, blacksmithing remained a crucial occupation in many parts of the world as late as the early twentieth century. Forges were needed to make an almost endless list of products, including hammers, knives, files, spears, swords, wagon wheels, ship anchors, nails, hinges, spikes, gates, fences, kettles, pots, horseshoes, and children's toys. Today, there are very few iron and steel products that cannot be made by machines. Still, blacksmithing remains popular among hobbyists and workers who specialize in handmade iron products.

Forge Construction and Operation

The basic design of the forge has changed remarkably little in three thousand years. Forges come in all sizes and shapes, but the most common form is made of brick or stone about 75 centimeters (30 inches) high and 60 to 100 centimeters (25 to 40 inches) square. The interior of the forge is filled with fire clay shaped to form a bowl with a hole in the bottom. The fuel to be used in the forge, coal or charcoal in most cases, is placed into the bowl and then ignited by means of kindling.

An essential adjunct to the forge itself is the bellows, a device that forces air through the fire, making it burn more hotly. Some type of bellows-like action was necessary in even the most primitive of forges, since a coal or charcoal fire alone does not burn hotly enough to permit the working of iron. Archaeologists have uncovered a number of mechanisms for increasing the flow of air in a forge. Perhaps the most primitive technique may have been nothing other than vigorous fanning of the flames by a servant working with the blacksmith. A more advanced technique was the use of blowpipes through which air could be forced over the fire.

The modern-day bellows was probably first invented in Rome around the fourth century A.D. It operates on a simple, but ingenious, principle. When the bellows is opened, air rushes in through an opening at the back of the device. When the bellows is closed, a valve closes this opening, and the trapped air is forced out the front of the device.

In a forge, the bellows is attached to the forge by means of a long pipe known as a *tuyere* (also known as the *tweer* or *tue iron*). The tuyere runs from the mouth of the bellows underneath and up into the hole in the base of the forge. Air rushing out of the bellows, through the tuyere, and up through the fire raises the temperature of the fire to the extent necessary to permit working with iron.

Further Reading: Bealer, Alex. *The Art of Blacksmithing.* New York: Funk and Wagnalls, 1969; Butt, Daniel E. "The Folk Art of Blacksmithing: Teach Yourself Blacksmithing," <http://www.win.bright.net/~wbforge/edu.html>; Heath, Robert M. "History of

Blacksmithing: Techniques and Materials," *Anvil Magazine*, April 1997; "Learning Blacksmithing," <http://www.anvilfire.com/FAQs/getstart/how-to.htm>; Postman, Richard A. *Anvils in America*. Berrien Springs, MD: Postman Publishing, 1998.

FUELS

The term *fuel* usually applies to any substance that reacts with oxygen to produce heat. The materials used in fission reactors, such as the isotopes of uranium and plutonium, are also referred to as *nuclear fuels*, but since no combustion or fire-related process is involved in the nuclear generation of energy, that topic is not discussed here.

Fuels are one of the three components essential for a fire. The other two are an oxidizing agent that can set the fuel on fire, such as oxygen, and sufficient heat to raise the fuel to its kindling temperature (the lowest temperature at which a material will begin to burn).

By far the most common type of fuel is a *chemical fuel*. Most chemical fuels contain carbon and hydrogen, which react with oxygen during the process of combustion, forming carbon monoxide, carbon dioxide, water, heat, and other products. The heat produced by the combustion of chemical fuels can be used to make steam, to drive machinery, to warm a house or building, to facilitate some chemical process, or for many other purposes. A great variety of chemical-related fuels exist, including wood, coal, gasoline, heating oil, diesel oil, natural gas, synthetic gases of various types, charcoal, and coke. Each type of fuel has its own properties and uses.

Wood as a Fuel

For the greatest part of human history, the most important fuel by far was wood. Trees provided the wood which, when dried, could be burned to make fires for cooking, heating, frightening away animals, waging wars, conducting religious ceremonies, and other applications. In certain locations, peat was also used as a fuel. Peat is a type of immature coal which forms when vegetable matter decays in boggy conditions.

Neither wood nor peat is a particularly efficient fuel. The efficiency of fuels is sometimes expressed in comparison to that of a standard, anthracite, or "hard", coal. Wood has a coal efficiency rating of about 0.500 and peat, about 0.429. Both fuels, then, are about half as efficient as a comparable amount of hard coal.

Humans learned early on, however, how to convert wood to a more efficient fuel, charcoal. Charcoal is formed when wood is heated in the absence of air. The heating process drives off the volatile components of wood, primarily water and hydrocarbons, leaving behind nearly pure carbon. When heated, charcoal burns more cleanly and efficiently than does wood.

Fossil Fuels

Over the past two centuries, the most important fuels by far have been *fossil fuels*: coal, petroleum, and natural gas. These fuels are so-called because of their origin. They are all believed to have formed within the Earth's crust as the products of the decay of plant and animal life.

Formation of Fossil Fuels. When living organisms die, they usually undergo a process of slow oxidation known as *decay*. In this process, the organic materials of which their bodies are composed are gradually converted to carbon dioxide, water, and other by-products:

$$(C_6H_{12}O_6)_n + 6O_2 = 6CO_2 + 6H_2O$$

This process requires, of course, the presence of oxygen, which is abundant in the atmosphere.

However, circumstances exist in which the slow oxidation of plant and animal life cannot occur. For example, a tree may fall into a swamp, a bog, or a shallow lake. The water and mud surrounding the dead tree prevent oxygen from reaching and reacting with the dead material. Instead of oxidation, then, the compounds of which the dead tree consists undergo a slow process of *decomposition*. Hydrogen and oxygen in the tree are released in the form of water vapor, leaving behind carbon of various degrees of purity:

$$(C_6H_{12}O_6)_n = 6C + 6H_2O$$

This process takes place very slowly and conversion of organic compounds within a dead body to even relatively pure carbon generally takes millions of years.

The conditions under which decomposition occurs will determine whether some form of coal, petroleum, or natural gas will be formed. For example, woody material buried in swamps and bogs often undergoes a slow transformation, first to humus, then to peat, then to lignite, then to soft (bituminous) coal, and finally to hard (anthracite) coal. The various forms of decomposed organic matter differ in the amount of pure carbon present, the amount of other volatile organic compounds, and the fuel efficiency. The table below shows some relevant properties for four forms of coal and coal-type materials.

that the products of decomposition are liquid (petroleum) and gaseous (natural gas). In many cases, petroleum and natural gas occur together underground (often, deep underground). The petroleum tends to be distributed within a permeable layer of rock where it fills the space between individual rock grains. The natural gas may form a cap or dome above the saturated rock.

Petroleum consists of a very complex mixture of dozens or hundreds of liquid hydrocarbons, in which are dissolved other solid and gaseous hydrocarbons. Commonly, a variety of impurities containing primarily sulfur, nitrogen, and oxygen are found dissolved in the petroleum.

Petroleum as it comes from the ground, or *crude oil*, has virtually no commercial value. It must first pass through a refinery process known as *fractional distillation*, in which various components of the mixture are separated from each other. These compo-

Properties of Coal Materials				
	Woody Material	*Lignite*	*Bituminous Coal*	*Anthracite Coal*
Carbon	45–50%	60–75%	75–90%	89–95%
Hydrogen	6%	6%	4–6%	2–4%
Volatile material	45–50%	20–30%	15–25%	3–5%
Approximate heat content (in Btu/lb)	4,500	6,500	13,000	13,500

Coal can also be converted to a more efficient fuel by heating it in the absence of air. In this case, when soft coal is heated in such conditions, it is converted into *coke*. Coke is nearly pure carbon and is one of the cleanest and most efficient of all fuels. It is used widely in the extraction and preparation of metals, especially iron and steel. It is also used as a source of various synthetic gases used as fuels, especially synthesis gas (a mixture consisting primarily of carbon monoxide and hydrogen).

Petroleum and natural gas are formed by mechanisms similar to that for coal, except

nents are divided into "fractions," which yield many of the useful products obtained from petroleum, such as gasoline, kerosene, diesel oil, various types of heating oil, lubricating oil, paraffin wax, and asphalt. Other materials are obtained by exposing petroleum fractions to heat and catalysts, changing ("cracking" or "reforming") them into new and useful substances.

Natural gas consists primarily of methane (70 to 90 percent) and ethane (0 to 20 percent), with small amounts of other gaseous hydrocarbons, carbon dioxide, oxygen, nitrogen, and hydrogen sulfide. Natural gas

To obtain useful products such as gasoline or kerosene, crude oil must be processed at a refinery such as the one shown here. © F. Gordon/H.A. Roberts.

For about 200 years, humans used fossil fuels as if they were a renewable resource, one that would be available essentially forever. By the 1950s, however, experts began to warn that coal, oil, and natural gas were, to a great extent, resources that were formed at one time in the far distant past and that our reserves of those fuels is limited. There would, in fact, come a time when these reserves could be depleted.

For nearly four decades, most people have understood this fact, but it has had little or no effect on the way the fuels are used. Nor has it led to any serious consideration of what can and should be done when fossil fuels are exhausted. Yet, evidence suggests that the turning point in the Fossil Fuel Age has already occurred. Production and consumption of coal, oil, and natural gas all peaked in the 1970s and have since begun to decline. Enormous undiscovered reserves of all three fuels probably still exist to be exploited. But the fact remains that those reserves are not infinite and, at some time in the future, the world will find it necessary to develop new sources of energy to replace those supplies.

is also fractionated to remove unwanted materials, such as higher hydrocarbons, water, and hydrogen sulfide. The final product is often cooled and condensed to form liquefied natural gas (LNG), a form in which the substance can be stored and transported more easily.

Role in Society. It would be difficult to overestimate the role of fossil fuels in human society. Indeed, the period from about 1750 to the present day is sometimes referred to as the Fossil Fuel Age (in comparison to the Bronze and Iron Ages, for example) because of the dominant influence of this fuel on our culture. Fossil fuels are used to power nearly every form of popular transportation, from trucks and cars to trains and airplanes; to heat homes, office buildings, factories, and other facilities; to drive many chemical operations; to generate electricity; and to perform endless other tasks that require the expenditure of energy.

Environmental Effects. Awareness of the limitations of fossil fuel resources occurred at almost the same time that humans began to appreciate the severe environmental costs of using such fuels. Those costs begin in the fields, where land may be scarred by open-pit mining and other procedures used to extract coal, oil, and natural gas from the earth. They become even more severe, however, once the fuels are burned and their waste products are released to the atmosphere.

For example, most fossil fuels contain sulfur and nitrogen as impurities. When coal, oil, and natural gas are burned, the combustion products of sulfur and nitrogen are also produced. These include sulfur dioxide (SO_2), sulfur trioxide (SO_3), and a mixture of five nitrogen oxides. In addition, the combustion of the fuels themselves yield products that are harmful to plants, animals, humans, and physical structures. These products include finely divided carbon (soot or particulates), carbon monoxide, unburned hydrocarbons (volatile organic compounds), and ozone. Even carbon dioxide, an otherwise natural and harmless component of the atmosphere, has become a threat to the environment because of the huge quantities in which it is formed.

Other Fossil Fuels. Fossil fuels may occur in other forms also, most commonly as tar sands and oil shale. Tar sands are a rocky material consisting of sandstone heavily saturated with bitumen. Bitumen is a heavy, asphalt-like product similar to heavy oils. Experts estimate that tar sands hold very large amounts of petroleum-like oils that could significantly supplement our existing supplies of petroleum. The problem is that mining tar sands is an energy-intensive activity that is economically viable only as the price of petroleum itself begins to increase. A plant in Alberta, Canada, has been extracting oil from the Athabasca tar sands deposits since 1969, but it is essentially the only site at which that resource is currently being exploited.

A similar resource is oil shale, another form of sandstone that contains high concentrations of kerogen. Kerogen is similar to bitumen in that it is an oily solid consisting primarily of carbon (75–80 percent), hydrogen (10 percent), nitrogen (2.5 percent), sulfur (1 percent), and oxygen. Like tar sands, oil shales tend to occur along or just below the Earth's surface, making them easy to mine. Extracting a useful fuel from the material can be costly, however, and no large commercial plant for refining the resource has yet been developed. Experts predict, however, that the oil shale in just one region, the Green River formation at the junction of Colorado, Utah, and Idaho, could supply the U.S. fossil fuel energy needs for more than 300 years if an economically feasible way could be found to extract and refine the material.

Biomass Fuels

Biomass fuels are fuels obtained from a living plant. Wood is the oldest, best known, and still most widely exploited of all biomass fuels. Biomass fuels are very attractive for a variety of reasons. First, they are renewable since replacement plants can always be planted to replace those harvested for fuel. Second, they are environmentally neutral fuels since their major products of combustion—carbon dioxide and water—are the substances from which new plants develop. Hence, there is no net change to the atmosphere when biomass plants are grown, harvested, and burned.

Wood is by no means an antiquated or outmoded fuel. It is still the primary source of energy in most developing nations of the world. Even in the United States, it is used for special situations. There are currently more than 1,000 wood-fired plants in this country, most used to generate energy for the timber, paper, and pulp industries.

Biomass is used as a source of fuels in other ways too. For example, agricultural wastes can be converted to liquid and gaseous fuels by fermentation or distillation. These fuels can then be used to substitute for gasoline, heating oil, diesel oil, and other products that tend to be more expensive. In some cases, crops can be grown for the specific purpose of conversion to a liquid or gaseous fuel. For example, methanol (methyl alcohol; wood alcohol) is produced in large amounts in Brazil from sugar cane for mixing with gasoline to produce the synthetic fuel gasohol.

Derivative Fuels

Humans have been making and using a variety of synthetic fuels for many years. These fuels are produced by converting simple raw materials—often fossil fuels themselves—to gaseous or liquid products that burn more efficiently than the original material. One example is coal gas. Coal gas is a by-product of the production of coke by the destructive distillation of soft coal. It consists of a complex mixture of gaseous hydrocarbons. Coal gas is also known as bench gas and coke-oven gas.

Producer gas is formed when bituminous coal or coke is burned in a limited supply of air. The gas that is formed consists primarily of carbon monoxide and nitrogen. Carbon monoxide burns very well, and nitrogen, not at all. The fuel is, therefore, not very efficient, but it can be made inexpensively.

Water gas is made by a process similar to that by which producer gas is generated, except that hot steam is passed over very hot coke. The two major products of this reaction are carbon monoxide and hydrogen, both very good fuels. Water gas is an excellent fuel, but more expensive to make than producer gas.

The term *synthesis gas* is used for a wide variety of gases formed when steam or steam and oxygen are passed over carbon-rich substances. The products of such reactions are primarily carbon monoxide and hydrogen. As with water gas, synthesis gas tends to be a good fuel, although somewhat expensive. *See also* CARBON.

Further Reading: Berkowitz, Norbert. *Fossil Hydrocarbons: Chemistry and Technology*. New York: Academic Press, 1997; Borowitz, Sidney. *Farewell Fossil Fuels: Renewing America's Energy Policy*. New York: Plenum Press, 1999; "Fuels," <http://www.cas.org/vocabulary/12132.html>, 07/31/99; Klass, Donald L. *Biomass for Renewable Energy, Fuels, and Chemicals*. New York: Academic Press, 1998; Valone, Thomas, ed. *Future Energy: Proceedings of the First International Conference on Future Energy*. Washington, DC: Integrity Research Institute, 1999; Weber, R. David. *Energy Information Guide: Fossil Fuels*. San Carlos, CA: Energy Information Press, 1984.

FUNERAL PYRE

A funeral pyre is a stack of combustible material, usually wood, on which a body is cremated. In the vast majority of cases, the body is that of a dead person although in certain ceremonies, such as that of *sati*, the person may still be alive. Sati is an ancient Hindu ceremony in which a wife throws herself on the funeral pyre of her dead husband.

History

Throughout history, various human cultures have held different views as to the proper method by which a dead person's body should be disposed of. Among followers of the Greek philosopher Thales (624–546 B.C.), for ex-

ample, water was thought to be the fundamental material of which all things are made. It seemed appropriate, therefore, for the body to be buried, where moisture in the Earth could aid its return to its original elements.

By contrast, other philosophers held that fire is the fundamental element of which all things are made. To those who held this belief, the proper way to dispose of a dead body was by burning. In that way, the body's fundamental elements could return to the source from which they came. Discussions of these philosophies, along with descriptions of both burying and burning of the dead, occur extensively in the literature of ancient Greece and Rome.

An association between gods and goddesses and funeral pyres is found in the mythology of many cultures. For example, the Hindu god Agni was thought to be present in all kinds of fires, especially hearth fires and funeral pyre fires. In Greek mythology, Hercules met an untimely end when he threw himself on his own funeral pyre in the hope of ascending to the heavens. The gods were so impressed with this act that they made Hercules one of their own, the only man ever to be transformed from mortal to immortal in Greek mythology.

The funeral pyre played an important role in the myth of the phoenix also. This mythical bird was said to live for 500 years, at the end of which time it built its own funeral pyre and was cremated. Out of the ashes of the funeral pyre, a new phoenix arose, to live the next 500 years.

The funeral pyre was a common theme in Nordic mythology also. At one point, for example, Brunhild (also, Brunnehilde), leader of the Valkyries, has her husband Gunther murdered, and then commits suicide by throwing herself on his funeral pyre.

Funeral pyres have long been a widely used method of cremation in Asian cultures also. For example, the Lord Buddha was said to have been cremated on a large funeral pyre, after which his ashes were distributed among eight holy centers. Even today in Thailand, tradition calls for the cremation of a dead king, queen, or royal person on a funeral pyre made from wood gilded with gold leaf. The size of the structure, as is generally the case, reflects the importance of the person who is being cremated.

The use of funeral pyres for cremation was never condoned by the Christian church, and the practice generally is not found in Western Europe, the United States, or other nations with a Judeo-Christian heritage. However, funeral pyres are still widely used in certain cultures, most commonly in the Indian subcontinent. For example, Hindus generally construct funeral pyres on which to cremate the remains of their dead, often in especially holy places, such as along the great rivers of India.

Sati

One of the best known uses of funeral pyres is in the Hindu ceremony known as sati (pronounced *su-tee*). In this ceremony, a widow joins her husband's dead body on the funeral pyre. She is thus able to join him immediately in their journey to the afterlife.

Sati is mentioned in Hindu writings going back thousands of years. However, there is limited historical evidence that the practice was ever very common. In any case, the British government banned the practice during its occupation of the Indian subcontinent (1790–1947), and the national government that replaced the occupation has kept that ban in place. According to the best estimates, no more than about one sati per year now takes place in India.

Over the last two decades, however, there has been a move to rescind that ban and permit the practice of sati once more. The primary force behind that move has been a group of Hindu nationalists who are attempting to restore a number of traditional Hindu practices, sati among them. That movement has developed at least in part because of the neutral position taken by the Indian constitution and government about religion, permitting no one denomination to be preferred over others.

The push to restore sati has met with relatively little support among women, intellectuals, and many other groups within Indian society. They argue that almost no woman would voluntarily join her husband on a funeral pyre in modern times, and only the most severe religious, social, and psychological pressure could cause women even to consider such an act. *See also* RELIGIOUS ALLUSIONS TO FIRE.

Further Reading: "An Account of Sati from Vikrama's Adventures," <http://www.humanities. ccny.cuny.edu/history/reader/sati.htm>; Matsunami, Kodo. *International Handbook of Funeral Customs.* Westport, CT: Greenwood Publishing Group, 1998; "The Tradition of Sati in India," <http://www. kamat.com/kalranga/ hindu/sati.htm>; "Sati—Burning of the Hindu Widow," <wysiwyg://25/http://adaniel.tripod. com/sati.htm>.

FURNACES

A furnace is an enclosed structure in which heat is produced for the purpose of bringing about some physical or chemical change in a material. Furnaces are closely related to ovens, from which they get their name. The word for "oven" in Latin is *fornus*.

Originally, all furnaces operated on the heat produced by the combustion of wood or fossil fuels (coal, oil, or natural gas). Over time, other types of furnaces have been developed in which heat is produced by an electric current (electric arc, induction, or resistance furnaces), a nuclear reaction, or solar energy. This essay discusses only traditional furnaces in which heat is generated by means of combustion.

History

Simple furnaces, similar to modern-day kilns, were probably perfected by the early Sumerians and Egyptians by about 3000 B.C. The production of pottery requires careful attention to the way in which wet clay is heated. If the heating process occurs too quickly or at too low a temperature, the product formed is not durable and may crack easily.

Early craftspeople probably learned how to modify kilns for a second important function, the production of metals from their ores. When the oxide of a metal is heated with charcoal or coke in a furnace, a chemical reaction occurs in which the oxide is reduced to yield the pure metal. For example, with iron ore, the process is as follows:

$$2Fe_2O_3 + 3C = 4Fe + 3CO_2$$

Types of Furnaces

Today, furnaces are used for a great many purposes, including the heating of water, the manufacture of glass and other types of ceramic materials, the smelting of metal ores, the refining of petroleum, and the production of many types of chemicals.

Furnaces that Heat a Fluid. One way of classifying furnaces is according to the type of change they are designed to produce. For example, some furnaces are designed to do no more than simply heat a fluid, usually water. Many homes, offices, and other buildings were once heated by hot water systems in which the heat from a furnace was used to raise the temperature of a body of circulating water. As water was warmed in the furnace, it was passed upward by means of convection through a series of pipes into radiators throughout the building being heated. The hot water then gave off its heat from the radiators by means of conduction and radiation. Cold water was then returned to the furnace, where it was reheated.

Steam Boilers. Another type of furnace is designed to produce a change in the physical state of a liquid. A steam boiler is an example of such a system. In a steam boiler, water is heated until it begins to boil and change into steam. The steam is then carried away to some place where it can do useful work.

One example of the application of a steam boiler is in a steam heating system used in homes and other structures. A steam heating system is similar to a hot water system except that heating is accomplished when steam, rather than hot water, passes through a radiator. Another application of a steam heating system is in the production of electricity. Steam produced in a boiler is used to drive the turbines that operate an electrical generator.

Process Furnaces. A third type of furnace is designed to bring about some type of chemical change in a material. Process furnaces are generally of two types: batch furnaces or continuous furnaces. In a batch furnace, the mass of materials to be heated is placed into the furnace and heated. The product that forms during the heating process is then removed before a new batch of raw materials is placed into the furnace. In a continuous furnace, raw materials are fed into the furnace on a conveyor belt or some other moving system. They undergo a change as they pass through the heat of the furnace and leave the furnace in their transformed state at the opposite end.

One example of a process furnace is the blast furnace used to convert iron ore to iron or one of the furnaces used to convert iron to steel. Three kinds of furnaces, the open hearth, Bessemer convertor, and basic oxygen process furnace, are all used for the latter purpose.

The refining of petroleum is another example of a process furnace. Heat is produced at the bottom of a tall cylindrical building, called the fractionating or refining tower, in which refining occurs. Crude petroleum is introduced into the tower and is broken down into two primary components, one gaseous and one liquid. This change occurs because petroleum consists of hundreds of different chemical compounds, each with its own distinctive boiling point. Those components that remain liquid at the temperature of the furnace make up the liquid fraction of the product, while those that boil at the furnace temperature make up the gaseous component.

The gaseous component of the refining process is drawn off at the top of the refining tower and piped away for future use. The liquid component is then further separated into various fractions based on their different boiling points. For example, the liquid portion that boils off at less than about 200°C, the "gasoline" fraction, is drawn off near the top of the tower. Another fraction, the "kerosene" fraction, includes compounds that boil between about 200°C and 300°C and is drawn off at a somewhat lower level. A third fraction, the "heating oil" fraction, consists of compounds with boiling points between about 300°C and 350°C and is drawn off at a still lower level of the tower. The fractions produced can be further refined by arranging exit pipes at levels that correspond to even more limited boiling point ranges.

The structures used in making glass are another important type of furnace. Glass furnaces are usually continuous furnaces in which sand, soda ash (sodium carbonate; Na_2CO_3), lime (calcium oxide; CaO), and other raw materials are fed into one end of a long furnace with an arched ceiling. Very hot air is forced up one side of the furnace, over the top of the mixture, and then down the opposite side of the furnace. The air is heated by burning natural gas or some other fuel outside the furnace itself. Cool air leaving the furnace is returned to the heating chamber, where it is reheated and passed again into the furnace. As the mixture of raw materials passes through the furnace, it melts, and a chemical reaction takes place among the constituents, forming liquid glass. The glass is heated both by the hot air and by radiant energy from the hot ceiling of the furnace. The liquid glass leaving the furnace is then recovered for further treatment, annealing, and shaping.

Another important use for process furnaces is in the smelting of metal ores. Smelting is the process by which a compound of a metal, usually a sulfide or oxide, is separated from impurities with which it occurs in nature. It is often the first step in preparing an ore for the production of a pure metal.

A smelting furnace used in the separation of copper sulfide from its ores looks somewhat similar in general plan to that of a glassmaking furnace. The mixture of copper sulfide and impurities that has been mined from the earth is fed into one end of the furnace. It is carried on a conveyor belt through the furnace under a series of burners attached to the ceiling of the furnace. Heat from the burners and radiant energy from the top of the furnace melts the ore mixture. Upon melting, the mixture separates into two parts. One

part consists of a moderately pure copper sulfide called a copper *matte*. The other part consists of the earthy materials with which the copper sulfide is mixed in the earth. This part of the mixture is called a *slag* and floats on top of the matte. At the end of the furnace, the slag is poured off into a container and disposed of, while the matte is captured for further refining into copper metal or some other form.

Fuels

The earliest furnaces were heated with wood or charcoal. Wood was readily available in most locations and could be converted rather easily to charcoal. Charcoal burns more efficiently than wood, giving off more heat per unit of weight and releasing fewer pollutants into the air.

The beginning of the Industrial Revolution saw a change from the use of wood and charcoal to the use of coal and coke as fuels in furnaces. Coke is produced from coal by destructive distillation; that is, by heating in the absence of air. Until the mid-twentieth century, the vast majority of furnaces were heated with either coal or coke.

The pattern changed, however, as individuals, companies, and governmental agencies became more concerned about the environmental effects of burning coal and coke. Carbon monoxide, soot (particulates), and other harmful products are released during the combustion of coal, especially the forms of coal most commonly used in furnaces (bituminous, or soft, coal). Most coal-based furnace systems were gradually converted to either oil or gas.

The task of finding exactly the right fuel to use in any particular furnace is a challenging one. Engineers have to take into consideration a number of factors, such as the temperature needed in the furnace, the cost of the fuel, the way in which it must be mixed with air, the possibility of recycling partially burned gases, and harmful by-products that may be formed during combustion. Today, a whole range of fuels, including natural gas, oil, and many different types of synthetic gas, are used in different types of furnaces.

Furnace Construction

The design of a furnace is also an important factor in maximizing the efficiency with which it performs its specific task. For example, some materials are more efficiently heated with very hot gases passed over or through them, while other operations proceed more efficiently if flames are forced directly on the materials.

The materials of which the furnace are made also influence the nature of the reactions that take place within the furnace. The linings of most furnaces, for example, must be made of materials capable of withstanding very high temperatures and preventing heat from escaping from the furnace. Such materials are known as *refractories*. Some examples of refractories are fire clay, silica (a form of sand), magnesite, dolomite, alumina, and silicon carbide. Heat produced in, or introduced into, such furnaces, reflects off the lining of the furnace onto the materials being heated. In such cases, the furnace may be known as a *reverbatory furnace*.

Furnace linings may sometimes take part in the chemical reactions that occur within a furnace. Reaction mixtures that are acidic in character, for example, may require furnace linings that have a basic character, and vice versa. *See also* FIRE BRICK; FIRECLAY.

Further Reading: Derry, T.K., and Trevor I. Williams. *A Short History of Technology from the Earliest Times to A.D. 1900*. New York: Dover Publications, 1993, passim; "Furnace." In *The New Illustrated Science and Invention Encyclopedia*. Westport, CT: H.S. Suttman, 1989, pp. 1066-1069; "Furnace Construction." In *McGraw-Hill Encyclopedia of Science & Technology*, 8th edition. New York: McGraw-Hill Book Company, 1997, Volume 7, pp. 571–575; Lyons, John W. *Fire*. New York: Scientific American Books, 1985, Chapter 4; Reed, Richard J. *North American Combustion Handbook: A Basic Reference on the Art and Science of Industrial Heating with Gaseous and Liquid Fuels*. Cleveland: North American Manufacturing Company, 1993.

G

GLASSMAKING

Glassmaking is one of the oldest human crafts. The earliest examples of human-made glass can be traced to at least 4000 B.C. One reason for the antiquity of this process is that glassmaking also occurs naturally. For example, the glass-like mineral, obsidian, is formed during volcanic eruptions when magma (lava) is ejected from the Earth's interior and then cooled rapidly. Early humans probably made glass accidentally when the heat from their campfires melted mixtures of sand and other minerals.

History

The first glass factory was built in Egypt in about 1400 B.C. By that time, artisans had mastered many of the basic techniques needed to make glass and form useful objects. For example, molten glass was poured around or into sand casts to make mugs, bowls, glasses, jewelry, and other objects.

By the first century B.C., the blowpipe was also in wide use. The blowpipe is a long metallic tube to which a glob of molten glass is attached at one end. A worker blowing into the opposite end of the blowpipe pushes out the molten glass into a spherical, bubble-like mass. The shape of the mass can then be changed by swinging, twirling, or manipulating the blowpipe in other ways.

The blowpipe can also be used to make flat glass. In the *crown method*, the worker first blows a large bubble of molten glass. The blowpipe is then spun rapidly, causing the bubble to flatten into a thin disk. The disk can then be cut into any desired shape.

In the *broad glass method* of making flat glass, the worker also begins with a bubble of molten glass. The bubble is then shaped into a cylinder by swinging it around the inside of a sand pit. The two ends of the cylinder are then cut off, and the cylinder is cut into two halves lengthwise. When each half is reheated and softened, it can be flattened and shaped.

Modern Methods of Glassmaking

Today, glass is made by one of two techniques, the *pot method* or the *tank method*. In the pot method, raw materials are placed into a large clay container that holds an average of 600–700 kilograms (1,320–1,540 pounds) of glass. The three basic ingredients for any type of glass include silica (silicon dioxide; sand; SiO_2), soda (sodium carbonate; Na_2CO_3, or one of its variant forms), and lime (calcium oxide; CaO, or one of its variant forms). Other ingredients may also be added to this mixture to make glass with special color, hardness, thermal resistance, optical qualities, or other properties.

The inside walls of a pot furnace are lined with a refractory material that can withstand very high temperatures and that does not react with molten glass. The heat needed to melt the raw materials is provided by a gas, oil, or (less commonly) electric-fired furnace located beneath the pot. The raw materials begin to melt when temperatures inside the pot reach about 1,400°C (2,500°F). That temperature is maintained for a period of time while the gas is "refined"; that is, while bubbles and other impurities in the molten product escape from the mixture. The temperature is then allowed to fall to about 1,200°C (2,200°F), at which point the viscosity of the mixture is suitable for forming, shaping, and working. The pot is then emptied and refilled with another batch of raw materials. A single pot can be reused in this way about thirty times before it must be repaired or replaced.

A tank furnace is a continuous flow system in which raw materials are injected at one end and finished sheets of glass leave at the opposite end. The interior of a tank furnace, like that of a pot furnace, is lined with a refractory material capable of withstanding temperatures of up to 1,500°C (2,700°F) and resistant to chemical attack by molten glass. The interior surface of the tank roof contains burners fueled by either natural gas or fuel oil. When ignited, these burners produce flames that play directly downward onto the raw material. In the first section of the tank, raw materials melt and are converted to molten glass. The molten glass is then carried along through successively cooler sections of the tank, where it is finished, polished, and annealed.

Both pot and tank furnaces achieve fuel economy by recycling waste heat from the melting and glassforming processes. Air to be used in the combustion chamber is first passed through tubes or pipes surrounding the pot or tank. This air takes up heat from the reaction chamber and is then fed into the furnace. This preheating process reduces significantly the amount of fuel needed in the furnace.

Annealing is the final stage in the manufacture of glass. Molten glass that is cooled too quickly may develop areas of stress that can cause the final product to become brittle and shatter easily. To prevent this problem, the temperature of molten glass is usually lowered very slowly, over a period of many hours. In a tank furnace, the annealing oven is generally placed at the end of the production line. The material leaving the annealing oven is then ready for cutting, reheating, shaping, and other operations used to produce dozens of glass products.

Further Reading: Bray, Charles. *Dictionary of Glass: Materials and Techniques*. Philadelphia: University of Pennsylvania Press, 1996; Doremus, Robert H. *Glass Science*, 2nd edition. New York: John Wiley & Sons, 1994; "Glass Encyclopedia," <wysiwyg://48/http://www.glass.co.nz/encyclopedia/index.html>; "Making Glass," <http://www.britglass.co.uk/publications/mglass/making1.html>.

GLOBAL FIRE MONITORING SYSTEMS

Global fire monitoring systems are programs for detecting, observing, analyzing, and reporting on forest, grassland, and other types of wilderness fires around the world.

Interest in global fire patterns is motivated primarily by two concerns. First, extensive wildland fires release large amounts of carbon dioxide to the atmosphere. Scientists now know that increasing levels of carbon dioxide in the atmosphere may have a significant effect on global climate patterns. Second, such fires also destroy large areas of natural vegetation needed by humans for food, fuel, and shelter. In addition, since green plants are the primary avenue by which carbon dioxide is removed from the atmosphere, the loss of trees may itself contribute to changes in global climate patterns.

Historically, the study of wildland fires around the world has been difficult because such fires often occur in remote areas, such as the upper Amazon basin, where ground-based observations are virtually impossible to make. Today, this problem can be solved to a large extent by using Earth-orbiting satellites that can detect and monitor fires of virtually any size and type.

An important impetus to the growth of global monitoring systems was the massive wildfires that occurred in Southeast Asia in 1997. Many nations lack the scientific personnel, equipment, and experience to detect, monitor, and make informed decisions about such fires.

Global Fire Monitoring Center

Currently, the main clearing house for global fire monitoring systems is the Global Fire Monitoring Center (GFMC), located at the Fire Ecology and Biomass Burning Research Group of the Max Planck Institute of Chemistry in Freiburg, Germany. The center was established in 1998 as part of the United Nations International Decade of Natural Disaster Reduction (IDNDR). Its purpose is to provide real-time or near real-time information on the status of natural wildfires throughout the world. GFMC collects, collates, distributes, and maintains archives on such fires.

Cosponsors of the organization include IDNDR, the United Nations Educational, Scientific, and Cultural Organization (UNESCO), the World Bank and Disaster Management Facility, the International Union of Forestry Research Organizations (IUFRO), the International Boreal Forest Research Association (IBFRA), the International Geosphere-Biosphere Programme (IGBP), and the U.S. Bureau of Land Management (BLM).

Objectives of the Program

One goal of global fire monitoring systems is to collect reliable data about the extent of natural fires. Scientists attempt to determine the number of acres of forest and other lands being burned each year. They also attempt to measure the type and amount of emissions being produced by these fires. The data are then made available to local environmental organizations and used to develop more accurate computer models to predict climate change, worldwide air pollution, and emissions produced by the combustion of biomass.

Some of the kinds of data to be collected include the susceptibility of various areas for combustion, characteristics of different types of fires, monitoring of active fires, extent of burned areas, and nature of smoke and trace gases produced during burns. The data are collected by about ten satellites owned and operated by the European Union, France, Japan, and the United States. These satellites travel in a variety of orbits and have different types of monitoring instruments that provide many different kinds of data.

For example, the Advanced Very High Resolution Radiometer (AVHRR) of the National Oceanic and Atmospheric Administration (NOAA) was originally designed to collect meteorological data. However, it has the capability of measuring cloud cover, changes in vegetation, smoke plumes, and scars left by burns. The Defense Meteorological Satellite Program Operational Linescan System (DMSP OLS), by contrast, is able to detect small pinpoints of light associated with burning biomass at night. The French satellite SPOT (Systeme pour L'Observation de la Terre) was intended to monitor vegetation cover, but has proved to be valuable in detecting burn scars.

An important breakthrough in worldwide fire monitoring occurred in 1999 with the launch of NASA's first satellite in the new Earth Observing System, EOS AM-1 later renamed Terra. This satellite has five sensors that will greatly improve the accuracy and efficiency with which burns can be detected and followed. The satellite travels in a near-polar orbit that takes it over the equator in the morning. Its sister satellite, EOS PM-1 later called Aqua, will be launched in 2002 and will follow a similar trajectory, except that it will pass over the equator in the afternoon. Data from these two satellites will make it possible to follow the development of fires in near-real time and with great accuracy.

Global Fire Monitoring Center
Fire Ecology Research Group
Freiburg University
P.O. Box D-79100
Freiburg, Germany
Telephone: ++49–761=808011
e-mail: fire@uni-freiburg.de

Further Reading: "Fire Globe: The Global Fire Monitoring Center (GFMC)," <http://www.uni-freiburg.de/fireglobe/>.

GLOBAL WARMING. *See* CLIMATE CHANGE

GODS AND GODDESSES OF FIRE

Virtually every human culture known has one or more figures who is associated with fire. In many instances, the god or goddess of fire plays a fundamental role in explaining how the universe, the Earth, and the culture itself was created. For example, Brighid (also known as Bridget and Bride) is the goddess of fire and the sun in Irish mythology. She symbolizes creative inspiration and fertility and is often regarded as the primary goddess of the culture. She has a number of different functions, including protecting the Irish people, acting as the healer of physical injuries, and protecting children.

A full description of even a small fraction of the gods and goddesses of fire is beyond the scope of this book. The following provides a brief overview of a few of the figures who have been most intensively studied.

Agni, God of Fire, is one of the primary gods of the Hindu religion. This image of him is from a seventeenth century temple in Madras. *Courtesy of Victoria & Albert Museum, London/Art Resource, NY.*

Agni

Agni is one of the three primary gods, along with Indra and Surya, of the Hindu religion. He represents three forms of fire: sun, lightning, and the ceremonial fire known as Yagya. Fire is regarded as having major importance in the religion both for its practical applications (cooking, heating, and lighting) as well as its religious and mystical significance (driving away bad spirits). Fire is a part of all important ceremonies in Hinduism. A major role of fire is to carry prayers and messages from humans to the gods. Fire offerings are usually made in the name of Agni's wife, Svaha.

Loki

Loki is the Norse god of fire, the son of Farbauti and Laufey. He is a very complex character, sometimes acting as an impish troublemaker and sometimes as a truly evil demon out to cause havoc among other gods and among humans. Some of his most memorable exploits are remembered in the great Norse poem *Poetica Edda*. In this poem, Loki's role in the death of Balder is retold.

Balder was the son of Odin and Frigg, reputed to be kind, handsome, and wise. He was immune to harm by any agent whatsoever except the mistletoe. Disguised as an old woman, Loki discovers that Balder can be killed only by this agent, and convinces the blind god Hod to throw a mistletoe stem at Balder, killing him. The gods are told that Loki's life will be restored if they will all mourn and weep for him. All the gods do so with the exception of one, a giantess known as Thokk (Loki in disguise). Since Thokk will not mourn Balder, he is doomed to death and is cremated on his funeral boat. Legend holds that he will come back to life again at the end of the world, the Ragnarok.

Moloch

Moloch is a god of fire who is mentioned many times in the Old Testament of the Bible. He was apparently known as far back as the third millennium B.C. in ancient Assyria and other adjacent cultures. He is mentioned in the Bible as requiring the sacrifice of young children, an act which was thought to bring good fortune to an individual or his family. In the Book of Jeremiah, for example, God speaking through the Prophet warns that the Canaanites built palaces dedicated to the god Baal in the valley of Hinnom, and they "cause[d] their sons and their daughters to pass through the fire unto Molech; which I commanded them not, neither came it into my mind, that they should do this abomination, to cause Judah to sin" (Jeremiah 32:35). Moloch may also be the same god as, or identified with, Molech and Milcom.

Pele

Pele is the Hawaiian goddess of fire, who is believed to be responsible for volcanic activity on the Hawaiian Islands. Legends about Pele extend back into history and can be found in Polynesian tales that the earliest settlers of Hawaii brought with them to the islands between about 300–600 A.D. Pele is generally regarded as a stormy goddess who is easily angered. When she becomes upset, she stamps her feet and tears the ground apart, causing earthquakes and volcanoes. She is reputed to have first settled in the western-most island of Kauai, and then worked her way down the chain until she finally reached the Big Island of Hawaii. She finally settled into the crater of Kilauea, where some Hawaiians believe that she continues to live today. The tradition is that she can be appeased by offerings of gin or sacred ohelo berries.

Xiuhtecutli

Xiuhtecutli, also known as Huehueteotl, was one of the two primary gods of the Aztecs. Along with his female counterpart, Chantico, he was regarded as the creator of the universe, Ometecuhtli. Xiuhtecutli was also known colloquially as "the old god," or even "the old, old god," indicating his enormous antiquity and his role in the creation of the world.

Xiuhtecutli was represented in a variety of ways in Aztec drawings, sometimes as a serpent made of fire and sometimes as a very old man carrying a brazier of fire on his back. His role in the pantheon of Aztec gods reflects the importance of fire to that culture.

One of the most important priestly duties was maintaining the sacred fire in temples dedicated to Xiuhtecutli's honor. The belief was that at the end of a calenderic cycle of fifty-two years, the god might decide to withdraw his gift of fire and return the world to cold and darkness. To prevent this from happening, live victims were sacrificed to the god on hot coals immediately after having had their hearts torn from their chests.

Svarazic

In some myths and religions, gods and goddesses are so highly revered that their names are never spoken out loud. Such appears to have been the case with the god of fire worshipped by early Slavic peoples in Eastern Europe and Russia. This god was the son of the Sky and Earth gods, Svarog and Makosh. In some literature, the god is given the name of Svarazic. He is sometimes pictured as wearing a helmet and carrying a sword. On his chest is emblazoned the image of a black bison's head. Some authorities believe that human sacrifices were made to the god.

Svarazic became the center of the primitive Slavic religion, partly because the force he exerted was so powerful that it drove off all evil creatures. People often made their pledges "in the name of" the god to indicate the depth of their sincerity. For example, he was called to witness the marriage of a man and woman. Part of the marriage ceremony might involve the couple's jumping through a fire together with the understanding that this act would guarantee that their love would last forever.

An interesting source of further references to gods and goddesses of fire is the U.S. Geological Service's Web site: <http://wwwflag.wr.usgs.gov/USGSFlag/Space/nomen/jupiter/iopate.html>. The Web site lists the official names of all patera (craters) on Io, Jupiter's largest moon. Many of the patera have been named after gods and goddesses from many cultures.

A few examples of the patera names include the following:

Agni: Hindu god of fire

Asha: Persian spirit of fire

Catha: Etruscan sun god

Carancho: Legendary hero from Bolivia who received fire from an owl

Fo: Chinese god of fire and the sun

Gish Bar: Babylonian sun god

Hatchawa: Slavic god who, in the form of a boy, gave fire to humankind

Laki-oi: Hero from Borneo who invented fire

Mafuike: Goddess from Hawaii whose fingers were filled with fire

Mbali: Pygmy word for fire

Nyambe: Zambezi sun god

Podja: Tungu spirit who is keeper of fire

Sui Jen: Chinese hero who discovered fire

Taw: Monguor word for fire

Tohil: Central American god who gave fire to humans

Vahagn: Armenian fire god.

See also HEPHAESTUS; PROMETHEUS; RELIGIOUS ALLUSIONS TO FIRE; VESTA.

Further Reading: Andrews, Tamra. *Legends of the Earth, Sea, and Sky.* Santa Barbara, CA: ABC-CLIO, 1998, pp. 80–82, passim; "Encyclopedia Mythica," <http://www.pantheon.org/mythica/articles/>, 07/13/99; "Fire Gods: Elemental, Domestic." In Marjorie Leach. *Guide to the Gods.* Santa Barbara, CA: ABC-CLIO, 1992, pp. 295–303; Goudsblom, Johan. *Fire and Civilization.* London: Allen Lane, 1992, pp. 97, 110, 120; "Greece: Myth and Logic." In Pierre Grimal, ed. *Larousse World Mythology.* New York: G.P. Putnam's Sons, 1965, pp. 96–175; Rossotti, Hazel. *Fire.* Oxford: Oxford University Press, 1993, Chapter 23; Staal, Frits. *Agni: The Vedic Ritual of the Fire Altar,* 2 vols. Berkeley: University of California Press, 1983; "The Wiccans Home—Traditions & Deities—Deity Information," <http://home.rmci.net/idahopyro/>, 07/21/99.

THE GREAT CHICAGO FIRE OF 1871

The Great Chicago Fire of 1871 is one of the most famous urban fires ever to occur in the United States. The fire broke out on the evening of 8 October 1871 in a barn belonging to Patrick and Katherine O'Leary. The story is told that Mrs. O'Leary's cow knocked over a lantern to start the fire. No one will ever know whether or not that story is accurate. The fire burned for thirty-six hours destroying nearly every building in a three-square-mile portion of the city center. Three hundred people were killed, 90,000 were left homeless, and more than $200 million in property damage was sustained.

Reasons for the Fire

Conditions in October 1871 were ripe for a major fire in Chicago. The summer and fall had been unusually hot and dry. No more than an inch of rain had fallen on the city between July and October. Leaves had started to fall from the trees as early as mid-summer. The city had been experiencing an average of two fires a day throughout the year. More than thirty fires had broken out during the first week of October.

By the eighth of the month, some citizens had become accustomed to hearing fire sirens and seeing fire trucks racing to blazes. To some observers, the first sign of flames at the O'Leary home at 137 De Koven Street must have seemed almost routine.

Course of the Fire

In fact, the first fire company did not arrive at the O'Leary property until more than a half

Ruins of the courthouse after the Great Chicago Fire of 1871. *Courtesy of Chicago Historical Society.*

hour after the fire was reported at 8:45 P.M. Before long, firefighters recognized the seriousness of the blaze, and a general alarm was sent out to all companies. Weather conditions and other factors were such, however, that the fire was soon out of control. Winds blew steadily from the southwest with a force of twenty to thirty miles per hour. They spread the fire rapidly through communities dominated by simple houses built of highly flammable materials, such as very dry wood.

At first, many residents gathered to watch the fire. They had seen a similar fire sweep through the same area only a day earlier, consuming a two-square-block area. That fire was quickly extinguished. However the fire of 8 October soon took an entirely different course. It swept up to and then leaped over the Chicago River. Buildings north of the river that had first been thought to be fireproof or sufficiently remote from the fire broke into flames. Hot cinders from buildings were blown across wide spaces, setting fire to structures in every direction. Observers were continually forced backward as their "safe" viewpoints were threatened and then burned, one by one.

The editor-in-chief of the *Chicago Tribune* later wrote a chronicle of his experiences during the fire. He described the early hours of the fire in these terms:

> The dogs of hell were upon the housetops of La Salle and Wells streets, just south of Adams, bounding from one to another. The fire was moving northward like ocean surf on a sand beach. It had already traveled an eighth of a mile and was far beyond control. A column of fire would shoot up from a burning building, catch the force of the wind, and strike the next one, which in turn would perform the same direful office for its neighbor. It was simply indescribable in its terrible grandeur. (White)

By early morning on 9 October, a number of major structures had been consumed by the fire, including the Court House; Field, Leiter & Company, the city's largest department store; the First National Bank; the Illi-nois Central Railroad land office; and the central depot. More than a million dollars in bills caught fire and burned in the vaults beneath the Post Office and Custom House. Planners had surrounded the vaults with three inches of boiler plate and brick wall, but even that was not enough to protect the paper money stored within.

Probably the most disastrous loss was that of the city water works. Fire tore through the roof of the building, causing pieces of wood to fall into and disable the pumping machinery. In spite of the fact that it sat on top of an enormous underground source of pure water, the city suddenly had no water for firefighting, for drinking, or for any other purpose.

Citizens were soon grabbing whatever property they could and heading for safe ground. Amidst the melee that occurred, however, is was not clear where "safe ground" was. Most people headed for the shores of Lake Michigan, although some were already asking what would happen if the whole city chose the same escape route.

Aftermath

By the morning of 10 October, the city was able to start counting up its losses. In addition to deaths, injuries, and property loss, the city was faced with finding structures in which to lodge people, safe water to drink, supplies of food, and medical care for the injured. Survivors tell of landlords doubling, tripling, and quadrupling rents for any rooms that survived the fire. Six people to a small bedroom was not an unusual housing arrangement, even at outrageous prices.

Remarkably, the city's recovery occurred more quickly than most people would have predicted. By 1875, the city had been completely rebuilt and virtually no evidence of the terrible fire remained. Today, the last remaining symbol of that great fire is the city Watertower at the center of the downtown area. The tower was one of the few structures to survive the fire and has become the symbol of the bravery and tenacity of the city's residents. *See also* Peshtigo Fire.

Further Reading: "The Great Chicago Fire," <http://taiga.geog.niu.edu/nwslot/fire.html>, 12/07/98; "The Great Conflagration," <http://www.chicagohs.org/fire/conflag/>, 12/07/98; White, Horace. "The Great Chicago Fire," <http://www.project21.org/ChicagoFire.html>, 12/07/98.

THE GREAT LONDON FIRE OF 1666

The Great London Fire of 1666 broke out on Sunday morning, 2 September. The fire started in the house of Thomas Farrinor, baker to King Charles II of England. According to diarist Samuel Pepys, Farrinor had apparently forgotten to turn off his oven, and smoldering embers from the oven set fire to some nearby firewood at one o'clock in the morning.

At first, response to the fire was limited and halfhearted. People whose homes were adjacent to the bakery were soon grabbing hold of whatever possessions they could carry and escaping the area. But, given that Sunday was a day of rest, there was little or no general effort made to fight the fire.

In the absence of any such action, the fire spread very rapidly. Most houses of the time were built of wood and held many other combustible materials, such as tar, pitch, and alcoholic beverages. The houses were also built close to each other, with the roof of one building often hanging over onto the top of an adjacent structure. The weather conspired to intensify the fire also as winds blew through the city, carrying flames and sparks from one neighborhood to another.

When organized efforts were made to fight the fire, those efforts were of minimal effectiveness. Firefighting at the time involved primarily the use of bucket brigades operated by volunteers willing to stay in the neighborhood and fight the blaze.

No real progress was made against the fire until King Charles ordered that houses be torn down to create a firebreak against the flames. This traditional approach to urban fires proved unsuccessful as winds carried the fire over the firebreaks and into areas thought to be protected from the flames. Finally, the king ordered that the construction of firebreaks be accelerated by the use of gunpowder to blow up whole blocks of houses. Pepys himself rushed to the waterfront to find sailors who were willing to take on the risky job of setting off the explosives.

The fire continued to rage for nearly a week before it was brought under control. In some parts of the city, buildings and refuse from the conflagration continued to smolder for another six months. At its conclusion, the fire had destroyed about four-fifths of the city, an area of more than 430 acres. About 13,200 houses, ninety parish churches, and fifty company halls were burned down. The disaster was all the worse since the city had just begun to recover from the terrible plague epidemic of 1664–1665. Surprisingly, only sixteen people died as a result of the fire.

As with many urban fires, the Great London Fire of 1666 had its positive results also. For one thing, most of the rats that carried the bacillus that causes plague were killed, resulting in a dramatic decrease in the number of cases of the disease. In addition, city planners were presented with an essentially virgin territory in which to start building a new city, safer and better designed than the one destroyed by the flames. A major role in the redesign and reconstruction of the city was played by Sir Christopher Wren, who designed forty-nine new churches as well as the new St. Paul's Cathedral.

Further Reading: Extracts from Pepys' diary concerning the fire are available on the Internet at <http://edweb.camenty.gov.uk/hinchingbrooke/diaries/fire.html>, 09/28/99; A report about the fire can be found on the web in a reproduction of *The London Gazette* 3–9 September, 1666, at http://members.aa.net/~davidco/History/fire1.htm>, 09/28/99; *See also* Leonard W. Cowie. *Plague and Fire: London, 1665–1666.* East Sussex: Wayland Publishers, 1970.

THE GREAT SAN FRANCISCO EARTHQUAKE AND FIRE OF 1906

The city of San Francisco, California, was struck by a terrible earthquake and fire on the morning of 18 April 1906. Some experts have called the event the worst natural disaster in the history of the United States. More than 3,000 people were killed, more than 225,000 people were injured, and property damage was estimated at nearly a half billion dollars.

The disaster began at about 5:12 A.M. on 18 April. A break opened up along the San Andreas Fault that runs from the southeast to the northwest along the California coast. San Francisco had experienced many earthquakes along the fault during its history. But the 1906 event was by far the most serious of all. When the quake first struck, the ground in San Francisco was oscillating up and down with a height of more than one-half inch and a frequency of about 240 vibrations per minute. Ground to the east of the fault moved as much as twenty-one feet southward compared to ground to the west of the fault.

Fires broke out almost immediately after the initial shocks occurred. Pipelines below city streets were ruptured by the quake and began releasing gas into the air. Any source of ignition was sufficient to set fire to the gas. One specific fire at the corner of Hayes and Gough streets has come to be known as the "Ham and Egg" fire because it was started when a woman began to prepare breakfast. Flames from her stove set fire to gas seeping from broken pipes near her home.

Ironically, San Francisco was as well prepared for fires as almost any city in the country. It had experienced countless fires in the past, and had developed an efficient firefighting system. Many of the city's buildings had been built to be fireproof and were thought to be safe in even the worst conflagration.

Such was not to be the case, however. The worst problem occurred because the earthquake had broken water mains as well as gas pipes. When firefighters attached their hoses to fire hydrants, they found nothing more than a trickle of muddy water. Fires spread through the city much faster than alternative sources of water could be found.

Even fireproof buildings fell to the blaze. The Palace Hotel, for example, had long been regarded as the model of fireproof buildings in the city. It had been built with fireproof materials and had a large reserve of its own water for firefighting purposes. As buildings around the Palace caught fire, however, the hotel's own supply of water soon gave out and it, too, succumbed to the flames.

Firefighters were reduced to the construction of firebreaks as their only hope of controlling the blaze. They used dynamite to clear a path between the fire and untouched buildings. This effort was also unsuccessful, however, partly because of firefighters' inexperience with the use of dynamite. In some cases, they actually started new fires during the process of dynamiting untouched buildings.

By morning of 19 April, all hope of stopping the fire had been given up. It burned uncontrolled until there was no fuel left on which it could feed. It was not until three days after the original earthquake, on 21 April, that the fire was declared under control.

For many citizens of San Francisco, however, the disaster had just begun. The city's infrastructure was essentially destroyed. More than 200,000 people had been left homeless. Many were provided with temporary housing in tents set up in Golden Gate Park. Many others simply left the area to find new homes in other parts of the state. Finding food and water and providing sanitation and health care for the sick and homeless became a task of gigantic proportions.

San Franciscans point with pride, however, to the rebuilding program that began almost immediately. Within two weeks, people were being hired to clear away rubble and begin the reconstruction process. Businesses and private individuals accepted the disaster as a challenge to build a new and better city. Amazingly, more than two-thirds of the buildings destroyed by the fire had been replaced in less than three years.

Further Reading: Arabian, Vatche. "The Great 1906 Earthquake and Fire," <http://www.webivore. com/tguide/quake/quake.html>; Bronson, William. *The Earth Shook, the Sky Burned: A Photographic Record of the 1906 San Francisco Earthquake and Fire*. San Francisco: Chronicle Books, 1986; Hansen, Gladys, Emmet Condon, and David Fowler Cameron, eds. *Detail of Disaster: The Untold Story and Photographs of the*

San Francisco Earthquake and Fire of 1906. San Francisco: Cameron & Company, 1989; London, Jack. "The Story of an Eyewitness." *Collier's*, 5 May 1906. Also at http://www.sfmuseum.org/hist5/jlondon.html; "The Scourging of San Francisco." In Bryce Walker and The Editors of Time-Life Books, *Planet Earth: Earthquake*. Alexandria, VA: Time-Life Books, 1982, pp. 62–73; Thomas, Gordon, and Max Morgan Witts. *The San Francisco Earthquake*. New York: Stein and Day, 1971.

GREEK FIRE

Greek fire is one of the earliest forms of incendiary weapons of major significance. An incendiary weapon is one that makes use of fire and flame to injure or kill enemies and to destroy their weapons of war. Greek fire material is said to have been invented by a Greek, Egyptian, or Syrian engineer named Callinicus. Virtually nothing is known about the life of Callinicus, including his place or year of birth or death. He is remembered today solely because of this one invention.

The Nature of Greek Fire

Modern scholars are not certain as to the composition of Greek fire. The recipe for making this weapon was carefully guarded by the Greeks for many centuries. It is thought to have contained some combination of quicklime (calcium oxide; CaO), saltpeter (potassium nitrate; KNO_3), and volatile petroleum oils, sulfur, resin, pitch, and/or other substances. The two critical components of this mixture are quicklime and a flammable material, such as sulfur or pitch. When quicklime comes into contact with water, it reacts with water to form calcium hydroxide, with the evolution of large amounts of heat.

$$CaO + H_2O = Ca(OH)_2 + \text{heat}$$

The heat produced is sufficient to ignite the sulfur, pitch, or other flammable material with which the quicklime is mixed. The presence of petroleum oils causes the mixture to float on water as it burns.

The Greeks also developed a delivery system for this invention that allowed them to deliver Greek fire to their enemy with relatively little danger to themselves. The invention consisted of a syphon through which the mixture of quicklime, sulfur, and other components was fired at an enemy ship. This system is sometimes said to be the earliest form of the modern flame thrower.

The first use of Greek fire occurred in 673 A.D. by Byzantine Greeks defending the city of Constantinople. At the time, the city was under attack by a Muslim fleet. In the harbor of Constantinople, the Greeks assembled a relatively small fleet of ships equipped with barrels of Greek fire. As Muslim ships approached the harbor, the flammable liquid was sprayed on them from hoses. As soon as the liquid hit water, it ignited and set fire to the enemy ships. There was no way for the attackers to extinguish the fires that were produced or to protect their ships from destruction.

Historians credit the use of Greek fire as a major deterrent to the spread of Muslim forces into Asia Minor and Europe. The new weapon had an impact on warfare at the time that is comparable to the explosion of the first nuclear weapons at the end of World War II. As with most weapons, Greek fire eventually became obsolete. The development of gunpowder and weapons able to hurl bombs across large distances finally made Greek fire an outmoded and ineffective battle weapon. *See also* INCENDIARY BOMBS.

Further Reading: Goudsblom, Johan. *Fire and Civilization*. London: Allen Lane, 1992, pp. 137–140; "Greek Fire," <http://www.greece.org/Romiosini/greek_fire.html>, 12/22/98; Partington, J.R. *A History of Greek Fire and Gunpowder*. Cambridge: Heffer, 1960.

GREENHOUSE EFFECT. *See* CLIMATE CHANGE

H

HALL OF FLAME

The Hall of Flame Museum of Firefighting is the world's largest museum devoted to the subject of firefighting. The museum was founded in 1961 by George F. Getz, Jr., in Lake Geneva, Wisconsin. The museum originated with the gift of a 1924 American LaFrance fire engine to Mr. Getz from his wife Olive Atwater Getz. Before long, the Getzs were collecting fire engines and firefighting apparatus from every part of the world.

As the Getz collection grew, the museum was moved from Lake Geneva to Kenosha, Wisconsin, and, in 1970, to Scottsdale, Arizona, its present location. The museum now occupies 47,500 square feet with more than 130 firefighting vehicles, thousands of smaller artifacts, and a library containing more than 6,000 holdings and over 50,000 graphics. Among the types of objects found in the museum are firefighting apparatus, uniforms, artwork, a model two-room safety house, and a variety of hands-on exhibits for children. In 1998, the museum added a new gallery, the National Firefighting Hall of Heroes, dedicated to firefighters who have died in the line of duty.

The Hall is supported by a membership of over 500 individuals, fire departments, and corporate bodies and is visited each year by more than 30,000 visitors. For further information, contact:

Hall of Flame
6101 East Van Buren Street
Phoenix, AZ 85008
(602) 275–3473
http://www.hallofflame.org/
e-mail: Webmaster@Hallofflame.org

The Hall of Flame is the largest of more than 300 fire museums in the United States and Canada. In addition, as of late 2000, there are 41 fire museums in Europe, 12 in Australia and New Zealand, one in Japan, and one in Ecuador. These museums range from the very modest—often no more than a room in a fire station—to larger buildings similar to the Hall of Flame.

Fire museums may have a number of functions. Some tell the story of an individual firehouse or a local or regional fire company. Others focus on the men and women who have served in a fire company in an individual community or a whole state. Others are devoted to fire equipment, particularly the kinds of fire engines used in a town, city or state. Finally, many museums combine these functions, often with the additional goal of teaching the general public, especially children, about the history and goals of firefighting.

Some examples of other fire museums in the United States include the following:

African-American Firefighter Museum
1401 South Central Avenue
Los Angeles, CA 90021
(213) 744–1730
http://www.lafd.org/museum2.html

Specializes in the history of the role played by African-Americans in the history of firefighting in the United States.

American LaFrance Memory Lane
11710 Statesville Blvd.
Cleveland, NC 27013
(704) 278–6200
http://www.freightliner.com/corp/lane/
Devoted to fire trucks built by American LaFrance.

American Museum of Firefighting
117 Harry Howard Avenue
Hudson, NY 12534
(518) 828–7695
http://www.regionnet.com/colberk/
amermusefire.html
A museum founded in 1925 to preserve fire engines and firefighting memorabilia, now covering 21,328 square feet of exhibit space.

Baltimore Equitable Society Fire Museum
21 North Eutaw Street
Baltimore, MD 21201-1794
(410) 727–1794
http://www.baltimoreequitableins.com/
history/museum.html
Tells the story of one of the first fire insurance companies in the United States and its role in firefighting in the Baltimore area.

Central Ohio Fire Museum & Learning Center
240 N. Fourth Street
Columbus, OH 43215
(419) 464–4099
http://fire.ci.columbus.oh.us/museum.htm
A facility designed to teach about the history of firefighting while offering hands-on learning experiences for children who want to learn more about the work of firefighters.

Oklahoma State Firefighters Museum
2716 NE 50th Street
Oklahoma City, OK 73111
(405) 424–1452
http://www.okstfirefighters.org/
about_us.htm
A museum that houses offices of the state firefighters association and includes a memorial to firefighters who have died in the line of duty.

Further Reading: "Fire Museum Network," <http://firemuseumnetwork.org/>

HALONS

Halons are chemical compounds that contain carbon, hydrogen, bromine, and chlorine and/or fluorine.[1] The most common use of halons is as fire extinguishers. Although many different halons are theoretically possible, only a relatively few have been produced commercially. Among the most common of these halons are halon-1211, halon-1301, and halon-2402.

Each digit in the name of a halon stands for the number of atoms of carbon, fluorine, chlorine, and bromine, respectively. For example, halon-1211 contains one carbon atom, one bromine atom, one chlorine atom, and two fluorine atoms. The chemical formula for this compound is $CBrClF_2$.

To a moderate extent, halons suppress fires by such traditional physical means, such as blanketing the fuel, thereby removing its access to oxygen, and reducing the temperature of the burning fuel. Their primary mode of function, however, is chemical. When most fuels burn, hydrogen in the fuel is oxidized:

$$4H \text{ (from fuel)} + O_2 = 2H_2O$$

Halon molecules interfere with that process. Bromine atoms in the halon molecule react with hydrogen atoms in the fuel, as indicated by the following chemical equation:

$$CBrClF_2 + H^+ \text{ (from fuel)} = CClF_2^+ + HBr$$

As hydrogen atoms are removed from the fuel, combustion is reduced, and the fire is extinguished.

During the 1960s and 1970s, halons became popular worldwide in both stationary and portable fire extinguishing systems. For example, most fire extinguishing systems in commercial aircraft once used halons as the active agent.

During the late 1970s, however, evidence began to accumulate that certain halogen-containing compounds, including the halons, were having detrimental effects on the ozone layer in the Earth's stratosphere. The mechanism by which this action occurs has been hypothesized as follows:

Chlorine or bromine atoms are released from chlorofluorocarbons (CFCs), halons, and other halogen-containing compounds when exposed to solar energy in the atmosphere. For example:

$$C_2Cl_2F_4 \text{ (CFC-114)—solar energy—}$$
$$= C_2ClF_4 + Cl^*$$

and

$$CBrClF_2 \text{ (halon-1211)—solar}$$
$$\text{energy—} = CClF_2 + Br^*$$

Free chlorine or bromine atoms then attack ozone molecules, converting them to diatomic ("normal") oxygen molecules. With bromine, for example, the reaction is:

$$Br^* + O_3 = BrO + O_2$$

Concern about this type of damage led to the Montreal Protocol on Substances That Deplete the Ozone Layer, signed in 1986. Under terms of that agreement, production of halons was banned as of 1 January 1994.

The problem for many industries has been that no totally effective replacement for halons has yet been discovered. Many partially acceptable alternatives are available, however. Among these alternatives are water, carbon dioxide, hydrochlorofluorocarbons (HCFCs), chlorofluorocarbons (CFCs), and a variety of dry chemicals. Under the Significant New Alternatives Program (SNAP) of the 1996 Clean Air Amendments Act, the U.S. Environmental Protection Agency has prepared lists of partially acceptable alternatives for each of the most popular halons. These lists are available on-line at http://www.epa.gov/ozone/title6/snap/snap.html. At the same time, chemical researchers are attempting to find new products that will be as effective as fire extinguishing agents as the halons, but that will not have the harmful environmental effects of those compounds.

Note

1. Halon (with a capital *H*) is also the tradename for a specific polymer of carbon and fluorine produced by the Allied-Signal Chemical Company. That compound is not discussed here.

Further Reading: *1998 Report of the Halons Technical Options Committee.* [Nairobi]: United Nations Environment Programme, 1999; "Questions and Answers on Halons and Their Substitutes," <http://www.epa.gov/ozone/title6/snap/hal.html>.

HEARTH FIRE

A hearth is the floor or the area in front of a fireplace, usually made of brick, stone, cement, or concrete. The term *hearth* comes from the Latin word for "focus," and is etymologically related to the word *heart*.

Hearth and Home

In the earliest stages of human history, fire was available only from natural sources, such as lightning or volcanic eruptions. Without the knowledge as to how to make fire, humans had to collect, transport, and protect fire obtained from such sources. As a consequence, fire was a resource that required constant attention.

The usefulness of fire also added to the necessity of protecting it. Humans used fire to heat their homes, cook their meals, provide light at night, and guard against wild animals. The loss of fire was not just an inconvenience; it was a potential disaster for early families. For these reasons, humans needed to find a way to protect and nourish the fires they obtained, at first, from natural sources and, later, learned how to build themselves. Fireplaces were probably one of the earliest devices invented for this purpose.

By necessity, the fireplace and its hearth became the center of family life. Family members gathered around the fire to stay warm and to cook their meals. They were literally and figuratively bound to each other by the fireplace, the hearth, and the fire it contained.

The role of the hearth fire was so critical in family life that many customs eventually became associated with its maintenance. For example, it was considered bad fortune in some cultures if the fire were allowed to go out. In addition, the beneficial qualities of the fire were often acknowledged by certain traditions and rituals, such as passing one's hands through the smoke of the fire to gain protection from illness and bad luck.

Given the importance of fire in the lives of early humans, it is not surprising that hearth fire itself sometimes took on mystical or religious meaning. Among the Ainus of Japan, for example, the hearth became regarded as a shrine that held a fire goddess. In many cultures, the distinction between hearth fire as a practical tool for heating and cooking became inextricably interwoven with the hearth fire as a sacred and venerated object.

Hearth and Community

Over time, the hearth fire grew to have a broader significance. It became the core not only of family life, but also of communal life as well. At many times and in many places, a central hearth fire was maintained by priests or other individuals whose sole responsibility may have been to ensure that the eternal flame for which they were responsible was never extinguished. Among the best known groups with this kind of responsibility were the Vestal Virgins of ancient Rome. It was their responsibility to protect the everlasting flame in the Temple of Vesta that was regarded as the heart and protector of the Roman nation.

The hearth fire thus assumed enormous mystical powers in many societies. As late as the nineteenth century, the American writer Nathaniel Hawthorne was propounding the patriotic appeal of "Pro aris et focus" (For altar and hearth). In his 1843 essay on "Fire Worship," Hawthorne said that this call was historically "the strongest appeal that could be made to patriotism." He went on to say that:

> The holy Hearth! if any earthly and material thing—or rather, a divine idea, embodied in brick and mortar—might be supposed to possess the permanence of moral truth, it was this. All revered it.

Hearth Fire Today

The practical significance of the hearth fire died out as alternatives to the fireplace for heating and cooking were developed. Homes that have fireplaces today do so not because they are such a good source of heat (they seldom are), but because they still provide a focus for family activities.

With the disappearance of the utilitarian fireplace has also come a diminution in many parts of the world of the hearth as a symbol of community and common beliefs. Although orators may still occasionally rally their listeners around calls for "home and hearth," the day when hearth fires could serve as a familial, communal, or religious bond are probably now long gone.

Further Reading: "Customs of the Hearth Fire," <http://www.dalriada.co.uk/Archives/hearth.htm>, 07/31/99; "Fire," <http://www.power.inms.nrc.ca/bamji/Fire.htm>, 09/30/99; Frazer, J.G. *Balder the Beautiful: The Fire-Festivals of Europe and the Doctrine of the External Soul*, 2 vols. London: Macmillan and Company, 1913, passim; Pyne, Stephen J. *Vestal Fire: An Environmental History, Told through Fire, of Europe and Europe's Encounter with the World*. Seattle: University of Washington Press, 1997, pp. 49–78; Seton, Nora Janssen. *The Kitchen Congregation: Gatherings at the Hearth*. New York: Picador, 2000; Victorin-Vangerud, Nancy M. *The Raging Hearth: Spirit in the Household of God*. St. Louis: Chalice Press, 2000.

HEAT

Heat is a form of energy that results from the motion of particles in a body. Heat is one of the three factors needed for combustion—the other two being a fuel and an oxidizing agent, such as oxygen. Combustion occurs when the fuel and oxidizing agent are raised to a temperature sufficient to cause ignition of the fuel.

Caloric Theory of Heat

Relatively little scientific attention was paid to the fundamental character of heat until the eighteenth century. At that point in time, the general understanding was that heat was a fluid that flows from a warmer body to a cooler one. The fluid was generally given the name of *caloric*. This characterization of heat was largely satisfactory at the time because it fit the obvious experimental data.

Flaws in the caloric theory of heat began to appear in the mid-1700s, however, especially with the work of the Scottish chemist Joseph Black (1728–1799). Black reasoned that, if the caloric theory were correct, he should be able to measure weight changes as heat flowed from one body to another. That is, the warmer body should lose weight as it loses caloric, and the cooler body should gain weight as it gains caloric. In a series of carefully conducted experiments, Black was unable to observe any weight changes in such instances.

The caloric theory was not destroyed by these experiments, however, because there was always the possibility that caloric was a weightless fluid. Scientists at the time (and even now) did not necessarily have problems imagining a fluid-like material that has no weight. As a result, the caloric theory survived Black's research, albeit in a somewhat modified form.

Such was not to be the case with another group of experiments conducted in the late 1790s by the American-British-French physicist Benjamin Thompson, Lord Rumford (1753–1814). Rumford's interest in the nature of heat was spurred by his observation of the process by which cannons are made. In this process, a boring tool is turned against a cylindrical block of metal until a hole is made in the metal. Rumford observed that a very large amount of heat was produced in this process, so much heat that the metal had constantly to be cooled with water.

The traditional explanation of this phenomenon was that caloric in the metal was released by the boring process, causing the heating of the metal. The problem that Rumford noted, however, was that there didn't seem to be any limit to the amount of caloric that could be released from the gun metal. He commented that so much caloric seemed to be produced by the boring process that, were it to be restored to the metal, the metal would melt. But, of course, the metal was solid to begin with. The only conclusion from this observation was that more caloric was being taken out of the metal than it contained to begin with!

To investigate this phenomenon further, Rumford switched from a typical pointed bore to a blunt boring device. When he used this device on the metal, no hole was made in the cylinder and no shavings were produced, of course, but the metal continued to grow warmer and warmer as long as the boring proceeded.

Rumford finally concluded from his research that the notion of caloric was untenable. He suggested, instead, that heat was the result of some form of motion. That is, he thought that the energy represented by the motion of the boring tool against the metal was converted into heat energy. He had, in a simple form, stated the modern definition of heat.

The Nature of Heat

Perhaps the most important single change in pre-Rumfordian and modern notions of heat is that we no longer think of heat as "something" that resides in a body. In fact, scientists talk about heat only in terms of the flow of energy from one body to another, the gain of heat by one body and the loss of heat by another body. The total heat energy in a body cannot be measured, although the flow of heat energy from one body to another can be.

The amount of heat gained or lost by a body depends on three factors: the mass of the body, its specific heat, and the temperature change it undergoes, or:

$$\Delta Q = mc\Delta T,$$

where ΔQ is the gain or loss of heat energy, m is the mass of the body, c is the specific heat capacity of the body, and ΔT is the temperature change.

The specific heat capacity of a body is simply a measure of how easily a body gains or loses heat. For example, a fair amount of heat must be added to water to raise its temperature by 1°C. By convention, the specific heat capacity of water is arbitrarily set at 1.000 calories per gram per degree Celsius. That means that it takes 1.000 calories of heat to raise the temperature of one gram of water by one degree Celsius. The *calorie* is a common unit of measure for heat changes. Most other substances have specific heat capacities less than water. The table below lists some typical values.

Specific Heats	
Substance	*Specific Heat Capacity (cal/g•°C)*
iron	0.11
copper	0.093
aluminum	0.22
lead	0.0306
wood	0.4
mercury	0.033
benzene	0.41
air	0.25
helium	1.24

Sources of Heat

Heat can be generated from other forms of energy in a number of ways. For example, mechanical energy can be converted to heat energy by friction. When two objects are rubbed against each other (as in the Rumford experiment), mechanical energy is changed into heat energy.

Fire is an example of the conversion of chemical energy into heat energy. The molecular bonds that make up the molecules of a fuel can be broken, releasing energy in the form of heat.

Electrical energy can also be converted to heat energy. When an electrical current passes through all but the very best of conductors, it encounters resistance and some electrical energy is converted to heat energy.

Transmission

Heat can be transmitted from one body to another body in three ways: by conduction, convection, and radiation. Conduction occurs when two bodies of different temperature are in direct contact with each other. The heat energy, often in the form of the movement of molecules of the warmer body, is transmitted directly to the molecules of the second body. Both bodies remain motionless, but energy is transferred from one to the other.

In convection, heat is transferred as matter itself flows from one region to another. For example, a flame applied to the bottom of a container of cold water provides energy to the water adjacent to the flame. As the water warms, it becomes less dense and is pushed upward by cooler, denser water around it. Eventually a current develops in which warm water rises to be replaced by cold water. Heat is transferred from one part of the container to another part by this process.

Radiation involves the transfer of heat through space, with or without the presence of matter. If you place your hand below a heat lamp, it will be warmed by heat transferred from the lamp by means of radiation.

Heat versus Temperature

Heat and temperature are easily confused. The concept of heat refers to the total amount of thermal (heat) energy contained within a body, while temperature is a measure of the average kinetic energy (molecular motion) of the body. It is possible for two containers of water, one holding 10 cubic centimeters and the other, 1,000 cubic centimeters, to be at the same temperature, say 20°C. Although both the molecules of water in both containers have the same average kinetic energy, the large container holds more water and, thus, has a larger total amount of heat energy.

Effects

Heat produces a number of effects on matter. Many of those effects are caused by the fact that thermal energy from a warm body causes the molecules of a cooler body to move more rapidly and, therefore, to move farther apart from each other. For example, most objects expand when they are heated. Expansion occurs because molecules of the heated object move faster and move farther apart. Thermal energy also causes objects to evaporate and to boil. The primary difference between a gas (or a vapor) and a liquid is that the molecules of the former are moving more rapidly and are farther apart.

Thermal energy can also cause chemical reactions, such as combustion, to occur more rapidly. A piece of coal stored at room temperature reacts very slowly with oxygen of the atmosphere. That reaction occurs so slowly that it would be impossible to observe any changes in the coal during a person's lifetime. But heating the coal causes the process of oxidation to take place more rapidly. As a general rule of thumb, raising the temperature by 10°C doubles the rate at which a chemical reaction occurs. If coal is heated from room temperature (20°C) to red heat (more than 500°C), the time required for the reaction between coal and oxygen is vastly reduced. It is under such circumstances that an object catches fire.

The temperature to which an object must be raised before it will catch fire is known as its *kindling point*. The kindling points of some common objects are shown below.

Substance	Kindling Point
carbon	400°C (750°F)
phosphorus	34°C (93°F)
sulfur	240°C (470°F)

A somewhat different measurement is used to determine the temperature at which gases and vapors catch fire. This measurement is known as the *ignition temperature* of the material. The ignition temperatures of some common gases and vapors are listed below.

Substance	Ignition Temperature
acetaldehyde	185°C (365 °F)
acetone	700°C (1,300 °F)
benzaldehyde	180°C (360 °F)
benzene	740°C (1,400 °F)
ethanol	558°C (1,040 °F)
hydrogen	580–590°C (1,080–1,090 °F)
isopropyl alcohol	590°C (1,090 °F)
kerosene	295°C (560 °F)
methane	650–750°C (1,200–1,400 °F)

Applications of Heat Theory in Fire Sciences

As should be obvious, the general theory of heat and heat transfer is of significant importance for those who deal with forest, urban, intermix, and other types of fires. For example, fire scientists have developed models of the way fires spread in a tall building, in a forest, and in other settings. In most cases, such fires are transferred by means of convection currents as very hot masses of air move from one location to another. Models of fire behavior must, therefore, take into account the physics of heat transfer by convection.

On the other hand, heating by conduction is of importance in understanding how fire moves through a particular material, whether it be the walls of a building or the bark of a tree. Again, the conductivity of various materials is quite different, so that the effects produced by fires in different settings will be very different.

Heating by radiation must also be taken into consideration in understanding the behavior of urban, forest, or other kinds of fires. During a forest fire, for example, fuels may gain heat to a significant extent by the radiation produced by other burning fuels near them. *See also* FLAMES; FURNACES; ROMAN BATHS.

Further Reading: Note: Most introductory high school and college physics texts provide a good introduction to the subject of heat. Fuchs, Hans U. *The Dynamics of Heat.* New York: Springer Verlag, 1996; Halpern, Alvin M. *Schaum's Outline of Preparatory Physics: Mechanics and Heat.* New York: McGraw-Hill, 1994, Volume 1; Von Baeyer, Hans C. *Warmth Disperses and Time Passes: A History of Heat.* New York: Random House, 1999.

HELL

In Christian theology, hell is generally thought to be a place where devils and evil spirits live and where the damned are sent after their death. The concept of hell varies widely among denominations. In some cases, hell is considered to be a real place, often located at the center of the Earth, where the damned are condemned to spend all eternity suffering from unspeakable torments, the worst of which is an unending fire. In other denominations, hell is thought to be an abstract condition in which a person's soul is forever separated from God.

The word *hell*, itself, is probably derived from Old Norse, where it was the name of the Queen of the Underworld, Hel. The word is used thirty-one times in the King James Version of the Old Testament of the Bible, and twenty-three times in the New Testament. Later translations of the Bible use the term less often, and in some of the newest versions, the word does not appear in either the Old or New Testament. This trend may represent some of the problems associated with translating both the word itself and the ideas that it represents.

Old Testament

In the Old Testament, the word *hell* is always a translation of the Hebrew word *se'ol*, or "sheol." The closest translation of that word is probably just "grave," a place to which all people—rich and poor, good and evil—go when they die. As the word is incorporated into the Christian Bible, however, it becomes transformed. Eventually it is taken to be a region that consists of two halves. One, *Abraham's bosom* was designed to receive the righteous who died before Christ appeared on Earth. When he was crucified and arose from the grave, he took to Heaven with him those who had originally been assigned to Abraham's bosom.

The other part of sheol, in Christian theology, is a place of torment for sinners. It includes the Lake of Fire (mentioned in the Book of Revelations 21:8), where evildoers are doomed to spend eternity burning in the flames of the Lake.

New Testament

The word *hell* in the New Testament is translated from three original terms. The first of these, hades, is used ten times and is the Greek equivalent of *sheol*. Thus, it is used in a fairly neutral sense, simply as a resting place for those who have died.

A second source of the word *hell* is the Greek word *gehenna* which, in turn, came from a similar Hebrew term. Gehenna is the root source for *hell* on twelve occasions in the New Testament. During Old Testament times, Gehenna was a region outside the walls of Jerusalem used as a city dump. Wastes in the dump burned continuously and not even the worst rains could put out the fires that burned there. One reason for the persistence of the fires was that brimstone (sulfur) was added to the dump to make sure that bodies of dead animals and criminals burned completely. It is probably this concept of hell that many Christians in the past and today have had foremost in their minds. The Bible warns in a number of places that those who are not saved will endure an eternity of burning in the flames of hell, which corresponds to the eternally burning fires of Gehenna.

The third source term for *hell* in the New Testament is the Greek word *tartaroo*, which is used only once and refers only to a place where fallen angels are sent. It makes no reference to the fate of humans after death.

The precise interpretations of Biblical references to hell (or, more precisely, to the source words for this term) have been the subject of extensive discussion and debate among scholars. There is considerable disagreement, for example, as to whether bodies burn forever in hell or whether bodies *and* souls burn forever. There are questions, also, as to whether evildoers who have died have already gone to hell or whether they reach that destination only after the Second Coming.

Some Christian denominations have taken a very literal view about the location and nature of hell. They cite Biblical passages to show that hell is located at the center of the Earth and that it is a place in which the damned are consumed by flames. One of the most frequently mentioned passages is the story told in which a rich man in hell asks for relief from the torments of hell. He looks upward from hell, sees Abraham and Lazarus and cries:

> Father Abraham, have mercy on me, and send Lazarus, that he may dip the tip of his finger in water, and cool my tongue, for I am tormented in this flame. (Luke 16)

Similar descriptions of hell can be found in the New Testament. In Matthew (13:42), for example, Jesus warns that sinners "shall [be] cast into a furnace of fire: there shall be wailing and gnashing of teeth." The Book of Revelations also gives vivid descriptions of the flames to be encountered in hell. In Chapter 20, verse 15, for example, the writer warns that, "Whosoever was not found written in the book of life was cast into the lake of fire."

Some theologians also provide support for the belief that hell can be found at the center of the Earth in the Bible. They cite Matthew (12:40), for example, which says that "For as Jonas was three days and three nights in the whale's belly, so shall the Son of man be three days and three nights in the heart of the Earth [in hell]." Jesus' fate is actually foretold in the book of Ephesians (4:9), which tells that he is now ascended, but "that he also descended first into the lower parts of the Earth." For many modern Christians, scientific reports that temperatures at the center of the Earth exceed 5,000°C and the eruption of hot lava from volcanoes confirm their view that a hot and fiery hell exists at the Earth's center. In Christian denominations where literal views of hell such as these are accepted, the possibility of being sent to hell after death can be a terrifying fate.

Other Christian denominations regard hell in more symbolic terms. They believe the term refers to a state in which the soul may remain for some period of time after death, during which it is purified by the love of God.

Further Reading: Camporesi, Piero. *The Fear of Hell: Images of Damnation and Salvation in Early Modern Europe.* Translated from Italian. Cambridge: Polity Press, 1990; Fudge, Edward William, and Robert A. Peterson. *Two Views of Hell: A Biblical and Theological Dialogue.* Downers Grove, IL: Intervarsity Press, 2000; McGee, Matthew, "Hell Part 2: The Differences between Hades and the Lake of Fire," <http://www.matthewmcgee.org/helwords.html>, 02/28/99; "In the Afterlife: Sheol, Heaven and Hell," <http://webstu.messiah.edu/~vb1151/home.htm>, 03/02/99; Pollock, Constance, and Dan Pollock, eds. *Visions of the Afterlife.* Waco, TX: Word Books, 1999; Rumford, Douglas J. *What about Heaven and Hell?* Wheaton, IL: Tyndale House Publishers, 2000; Seymour, Charles Steven. *A Theodicy of Hell.* Dordrecht, Netherlands: Kluwer Academic Publishers, 2000; "Two Common Views of 'Hell'," <http://members.aol.com/hunting444/hell.html>, 02/28/99; "What Is Hell?" <http://www.pernet.net/~cosmo1/hell.htm>, 03/02/99.

HEPHAESTUS

Hephaestus is the god of fire in Greek mythology. Because of his skill in working with fire, he also became the god of forging, smithing, and other fire-related occupations. In Roman mythology, a comparable god of fire was called Vulcan.

A variety of tales are told about Hephaestus' life. According to one version, he was born to a virgin Hera, wife of Zeus. This story corresponds in some respects to the story of Jesus' birth of a virgin Mary. Other storytellers claim, however, that Zeus was indeed the father of Hephaestus, and Hera's story was concocted in order to explain why the child was born before Zeus and Hera were married. That explanation has been known to be used in other circumstances also.

In any case, Hephaestus was unusual among the gods in that he was deformed. He was said to be "ill-made and lame in both legs" by some writers. In fact, he is often depicted as a muscular man leaning on a crutch. The cause of Hephaestus' deformity is subject to debate. According to some stories, he was born with his handicap. According to other stories, he was injured when his mother (or, in other versions, his father) became angry at the child and hurled him to the Earth. He landed on the island of Lemnos in some stories or in the sea, in others.

Upon reaching the Earth, Hephaestus was saved by two sea nymphs who raised him as their own child. Over time, Hephaestus developed remarkable skill in metalworking and produced the finest weapons, pieces of jewelry, ornamental objects, and pieces of furniture ever seen. When his mother learned of his skills, she demanded that he be returned to Olympus. Stories disagree as to how he was finally convinced to return to his home, but he eventually did so.

Hephaestus was also involved in the death of Prometheus, the figure from Greek mythology who brought fire to humans. When Zeus discovered that Prometheus had stolen fire from Olympus to give to humans, he ordered Hephaestus to nail Prometheus to Mount Caucasus as punishment for his crime. Prometheus was later released, by Hephaestus himself, according to some accounts, or by Heracles, according to others. *See also* GODS AND GODDESSES.

Further Reading: Andrews, Tamra. *Legends of the Earth, Sea, and Sky*. Santa Barbara, CA: ABC-CLIO, 1998, pp. 80–82, passim; "Encyclopedia Mythica," <http://www.pantheon.org/mythica/articles/>, 07/13/99; Goudsblom, Johan. *Fire and Civilization*. London: Allen Lane, 1992, pp. 97, 110, 120; "Greece: Myth and Logic." In Pierre Grimal, ed. *Larousse World Mythology*. New York: G. P. Putnam's Sons, 1965, pp. 96–175.

HUNTING WITH FIRE

Hunting with fire involves any activity in which the pursuit, capture, and/or killing of animals is accomplished using fire. Fire hunting was one of the earliest forms of hunting known to humans. In the earliest stages of human culture, people were faced with the problem of hunting down creatures much larger than themselves, such as mammoths, mastodons, and wild horses. The discovery of fire greatly increased the ability of humans to hunt such animals.

For example, hunters during the Early Paleolithic Age are known to have used torches and bonfires to corner animals before attacking them with rocks, spears, and other weapons. Another practice was to set fire to grass and shrubs in order to drive animals toward a precipice. When the frightened animals fell or jumped off the cliff, they were killed by hunting partners waiting at the bottom.

Some primitive cultures today still use fire hunting to pursue and kill their prey. Among Aborigine tribes in Australia, for example, brush fires may be set to drive kangaroos, wallabies, and other game in the direction of hunters who are hidden from view. Members of the Lango tribe of Uganda use a similar practice. They set fire to a section of grassland, driving their game out of hiding and into the hands of a circle of warriors surrounding the area.

Fire hunting was probably an important practice among early cultures because of the necessity for cooperation among hunters. Each person had his or her specific task, and everyone had to work together to successfully complete the hunt.

Further Reading: Editors of *Life*. *The Epic of Man*. New York: Time Incorporated, 1961, Chapter 1; Pyne, Stephen J. *Vestal Fire: An Environmental History, Told through Fire, of Europe and Europe's Encounter with the World*. Seattle: University of Washington Press, 1997, pp. 27–30, passim.

I

IGNIS FATUUS

Ignis fatuus is a meteorological phenomenon in which flickering lights of unknown origin appear over marshes, usually on windless nights shortly after sunset. The lights always appear to have a pale color, such as blue, pink, or green, and never are pure white. As an observer attempts to approach the lights, they seem to float away into the distance. In some cases, the lights seem to move quickly from one place to another, changing direction in an instant. This behavior has led some observers to believe that sightings of supposed unidentified flying objects (UFOs) may actually be cases of ignis fatuus.

The term *ignis fatuus* comes from a Latin term meaning "foolish fire." The phenomenon is also known more commonly as will-o'-the-wisp, jack-o'-lantern, fox fire, friar's lantern, and corpse candle. It is usually regarded as an ominous sign that has become a part of the folklore of many countries. According to one story, the lights are burning coals being carried about on the Earth by dead souls who have been rejected by Satan and banished from hell.

Many explanations have been given for the phenomenon, none of which has been widely accepted. Perhaps the most common explanation is that the lights come from burning methane gas ignited by spontaneous combustion when it comes into contact with hydrogen phosphide. Both gases are produced by the decomposition of dead organic matter. The problem is that similar phenomena are not observed in other situations where organic matter is decaying and would be expected to produce something like ignis fatuus.

Another possible explanation for ignis fatuus is that the light is produced by the phosphorescent radiation given off by decaying plant and animal material. At this point in time, no single explanation appears to be satisfactory for the formation of ignis fatuus.

IGNITION

Ignition is the act of starting a fire. At one time in the distant past, humans had no knowledge of methods by which fire can be produced. They had to rely on natural sources, such as lightning and volcanoes, to obtain the fire they needed and wanted. Archaeologists believe that humans first learned how to make fire sometime between 1.5 million and 500,000 years ago. The first objects clearly designed for fire making have been found in Neolithic sites dating to about 5,000–10,000 years ago.

Natural Sources

By far the most common natural source of fire in most parts of the world is lightning. Lightning is a flow of high-voltage electricity which, when it strikes a combustible object, can set it afire. Archaeologists hypothesize that early humans took advantage of lightning-caused fires to ignite their own camp-fires and torches. For thousands of years, the primary means of starting a fire was by using an existing fire in the form of a torch, a camp-fire, or some other ongoing flame.

Another natural source of fire is associated with volcanic eruptions. When hot lava is emitted from a volcano, it commonly sets fire to vegetation in its path.

A less common source of natural fire is rock slides. The friction produced when rocks and boulders rub against each other may produce enough heat to ignite moss, dry leaves and grass, or other flammable materials in the vicinity.

At one time, the friction caused by tree branches' rubbing against each other was also thought to be a cause of fire. But there is little or no evidence to support this idea.

Early Artificial Sources of Ignition

One of the earliest methods for igniting fire developed by humans was probably the simple action of rubbing two objects together, such as two sticks or a stone and a piece of flint (hard quartz). In the first case, friction may produce sufficient heat to ignite a combustible material, such as dry leaves or grass or punk (usually made from dried fungi). In the second case, friction produces a spark which may also be used to ignite the material. Once the combustible material has started to glow, fanning or blowing on it will produce a full-blown flame.

Over the centuries, humans invented a variety of tools to more efficiently produce the friction needed to ignite a fire. Many of these tools involved the friction between two pieces of wood, one a hardwood and one a softwood. Heat produced by this friction was used to ignite a combustible material. An example of such a tool is a *fire drill*, probably first used in prehistoric Africa. A fire drill consists of a drilling stick and a board that contains a depression. The stick is held vertically with one end in the depression and then rotated back and forth by a twisting motion of the hands. Friction between the drilling stick and the board produces sufficient heat

to ignite punk or some other combustible material placed in the depression of the board.

Many variations of the fire drill have been developed. For example, the ancient Egyptians developed a form of the fire drill known as a *bow drill* in which the string of a bow was wrapped around the vertical stick used to start the fire. Pulling on the bow with a back-and-forth motion caused the vertical stick to rotate as in the hand-operated fire drill, but the action was easier and produced a stronger force on the horizontal stick.

Australian aboriginals developed yet another form of friction device called a *fire saw*. The fire saw consisted of a flat board that could be rubbed back and forth across the rough surface of a second flat piece of wood, rock, or porcelain. Again, friction between the flat board and the second material produced enough heat to ignite a combustible material placed on the board.

Fire-making tools like these are now used primarily by hikers, campers, and others who choose not to carry or may not have available to them more modern fire-making devices, such as matches and lighters. Many websites can be found on the Internet offering books, pamphlets, classes, and other forms of instruction on the use of fire drills, bow drills, and other primitive forms of fire-making devices.

Until the seventeenth century, most fires were started from torches or flames that were kept burning or by one of the friction techniques described above. Matches are also a fire-making device that makes use of the heat generated by friction. A match is a strip of pasteboard or wood that contains a combustible material and an oxidizing agent at its tip. When scratched on a surface, a chemical reaction takes place between these materials and a flame is produced. Matches produce a flame much more easily than bow drills or similar devices, they are easy to carry around, and they are very inexpensive. The invention of the match was one of the great inventions in human history since it made it possible for anyone to ignite a fire with virtually no effort at all.

The earliest matches were often quite dangerous to use. In some cases, the reaction that occurred was more of the nature of a small explosion than a simple fire. Match making improved over time, however, and the ability of anyone to light a fire safely almost anywhere with a match is now simply taken for granted.

Modern Ignition Devices

Today, a range of ignition devices and systems are available for starting different kinds of fires. For example, a cigarette lighter is a friction device in which a piece of steel is made to rub against a special alloy known as *misch metal*. When friction occurs, the misch metal gives off a spark that ignites a lighter fuel, such as butane, absorbed in the lighter wick. In principle, the system is little different from the one used by prehistoric peoples when they rubbed a rock against a piece of flint in order to produce a spark.

Most people are familiar with the ignition system used to light the gas stove found in many homes. In older models of this system, a very weak flame called a *pilot light* is kept burning near each burner on the stove. When the stove is turned on, a flow of gas is released to one of the burners, which is ignited by the burning pilot light. An improvement on this design is to replace the pilot light by an electrical sparking device located next to each burner on the stove. When the burner is turned on, an electrical spark is produced at the same time that a flow of gas is released to the burner. The spark ignites the gas, lighting the burner.

A variety of methods are available for igniting fireworks. Two of the most common are called a *black match* and *touch paper*. A black match consists of a string that has been impregnated with a metallic powder. When the string is touched with a match, it begins to burn rapidly, igniting the fireworks to which it is attached. Touch paper is made simply by soaking some absorbent material, such as a paper towel, in a saturated solution of potassium nitrate and then allowing the material to dry. When the dry material is touched with a match, it catches fire and burns very quickly.

One of the most familiar ignition systems is the one used in internal combustion engines. Such engines operate when a spark plug ignites a mixture of gasoline and air inside a cylinder. The voltage required to produce such a spark is very large, about 120,000 volts. That voltage is produced by an ignition coil attached to the spark plug. The ignition coil acts as a step-up transformer, converting the voltage available from the vehicle itself (about twelve volts) to the very high voltage needed by the spark plug.

The ignition system used in diesel engines operates differently from that in a gasoline engine. In both engines, the fuel-air mixture is compressed before being ignited. In both cases, compression heats the mixture although, in a gasoline engine, not to a temperature high enough to cause combustion. By contrast, the compression of the fuel-air mixtures in a diesel engine generates enough heat to ignite the mixture without the need for some outside ignition system.

Research is constantly being carried out to find new and more efficient methods of ignition. One of the most promising goals of such research today is the user of laser beams for ignition. For example, scientists have now found that they are able to ignite a specific fuel, such as methane, by focusing a laser beam tuned to the frequency of the methane molecule. This technique is being studied as a less expensive, more efficient method of igniting fuel mixtures in power plants and other facilities in which gaseous fuels are burned.

Laser beam technology is also being studied as a method for igniting propellants used by military units. Traditionally, medium- and large-size guns are fired by first igniting a detonator or by setting off an electric spark. Research has shown, however, that the same results can be achieved by aiming a laser beam to ignite the charge, while at the same time providing greater dependability and safety to those working with the guns. *See also* ETERNAL FLAMES; FIREFIGHTING; FOREST

FIRE; INTERMIX FIRE; LIGHTNING; MATCHES; SMOKEY BEAR; SPONTANEOUS COMBUSTION.

Further Reading: "Combustion: Laser Ignition of Natural Gas Mixtures Accomplished," <http://www.gri.org/pub/abstracts/4912.html>; Cooke, Lawrence S., ed. *Lighting in America: From Colonial Rushlights to Victorian Chandeliers.* New York: Main Street/Universe Books, 1975, pp. 133–136; "Earth Connection," <http://www.earth-connection.com/adult.html>, 08/15/99; "Fire Building," <http://www.snipercountry.com/fire.html>; Frazer, J.G. *Balder the Beautiful: The Fire-Festivals of Europe and the Doctrine of the External Soul,* 2 vols. London: Macmillan and Company, 1913, Chapter 3; "The Ignition System," <wysiwyg://91/http://autorepair.about.com/>; Lyons, John W. *Fire.* New York: Scientific American Books, 1985, Chapter 7; Rossotti, Hazel. *Fire.* Oxford: Oxford University Press, 1993, Chapter 3.

INCENDIARY BOMBS

Incendiary bombs are weapons, usually dropped from aircraft, whose primary purpose it is to set fires on the land and buildings on which they fall. The genesis of the true modern incendiary bomb awaited the development of dependable carriers from which to drop them; that is, modern aircraft.

History

During World War I, the Germans used lighter-than-air ships (zeppelins) and airplanes to release incendiary bombs on their targets. The bombs were very primitive, consisting of a thermite core and cotton cloth dipped in tar and encased in a bucket-shaped metal container.

Thermite is a mixture of iron oxide (Fe_2O_3) and powdered aluminum. It has remained one of the most popular ingredients in the manufacture of an incendiary bomb. When the thermite is ignited, a chemical reaction occurs that does not require the presence of oxygen. In this reaction, temperatures as high as 2,200°C can be reached, a condition under which many metals, including steel, begin to soften and melt.

The original thermite-cotton-tar mixture proved to be unsatisfactory because it was difficult to deliver accurately and did not always ignite, and German scientists developed more efficient forms of the incendiary bomb. In one of these, gasoline and paraffin were mixed in a container and ignited to produce a hot, long-burning fire. The second development became known as the *electron bomb*. It consisted of a mixture of thermite and magnesium as the charge, encased in a magnesium container.

The appeal of this weapon is that magnesium burns with a very hot white flame. When the thermite in the bomb begins to burn, it sets fire to the magnesium with which it is mixed and then to the magnesium casing itself. The weapon is very efficient, and effective models weighing less than two pounds were developed by the Germans. They were not, however, used in World War I.

In the period between World War I and World War II, there was a dramatically different level of interest in incendiary bombs in the United States and Germany. American military strategists were more interested in working on other forms of weaponry, such as high explosive bombs, while the Germans continued to develop their incendiary arsenal, especially the electron bomb.

Shortly after the onset of World War II, the U.S. Army was to realize the error of its ways. The Germans bombarded London with both electron bombs and with much larger oil bombs weighing more than 100 kilograms (200 pounds). The electron bombs were highly effective since they tended to set fire to any wooden structure on which they landed. The oil bombs were also productive since the burning liquid they released spread outward across a wide radius, igniting fires in every direction. The property damage and loss of life caused by German incendiary bombs in London were enormous. Some of the most horrible stories of World War II involve the destruction produced by this horrible weapon.

By mid-1941, the U.S. Army had begun research on its own incendiary weapons. Those weapons were of three types: thermite, electron, and phosphorus bombs. Phosphorus bombs contained a few pounds of white phosphorus, a material that burns vigorously when exposed to air. When a phosphorus bomb explodes, it scatters tiny pieces of phosphorus over a large area. Phosphorus fires can be extinguished fairly easily with water, but as soon as a piece of phosphorus dries off, it begins to burn once more.

In 1942, a new type of incendiary bomb was developed that revolutionized incendiary warfare. That invention was napalm, a combination of aluminum metal and three or more fatty acids. Military strategists were excited about the possibility of using napalm in incendiary bombs because the material burns very slowly, but with a very hot flame. Before long, napalm bombs were being used widely in both the European and Pacific theaters of war.

Incendiary bombs were especially effective against Japan because so many buildings there were built of cardboard, wood, or other highly flammable materials. In one of the largest bombing raids of the war, a fleet of 334 B-29 bombers flying at an altitude of 5,000–9,000 feet dropped more than 1,500 tons of incendiary bombs on Tokyo on the night of 9–10 March, 1945. Officials estimated that more than fifteen square miles of land were devastated, more than 80,000 people were killed, and more than one million were left homeless.

Japanese Response

The Japanese air force was not able to mount an effective counterattack against the U.S. mainland in retaliation for the Tokyo firebombings. However, it had devised a form of incendiary attack of its own only a few months earlier. The first of more than 9,000 balloons had been launched from Japan on 3 November 1944, aimed at the U.S. mainland. Each balloon carried one antipersonnel bomb and two incendiary bombs. The Japanese counted on the prevailing westerly winds to blow these balloons across the Pacific Ocean and over the North American continent.

The Japanese used a variety of incendiary bombs, including both electron bombs and oil bombs like those launched by the Germans against Great Britain. The Japanese also built bombs consisting entirely of thermite as well as phosphorus bombs. White phosphorus is a highly combustible material that ignites spontaneously in air. A phosphorus fire tends to emit burning particles in every direction, igniting new fires wherever they land.

The Japanese balloon bombs were a nearly complete failure. No more than about a thousand of the bombs ever reached the North American continent. Those that did so did almost no damage. Six members of the Mitchell family were killed while removing an incendiary bomb from the woods near Lakeview, Oregon, on 5 May 1945, but they are the only known casualties to result from the Japanese attack.

Since World War II, incendiary weapons have been included in the military arsenal of most large nations. These weapons are relatively easy to build, easy to deliver, and highly destructive. They were used by the United States in the Korean Conflict (1950–1953) and the Vietnam War (1950–1975). In the latter conflict, napalm bombs were used extensively, not only on inhabited areas and against enemy troops, but also to burn away forested areas in an effort to deprive Viet Cong soldiers of cover from U.S. troops.

Further Reading: "Balloon Bomber," <http://www.af.mil/news/airman/0298/bombsb.htm>, 01/05/99; Fisher, George J.B. *Incendiary Warfare.* New York: McGraw-Hill, 1946; Kleber, Brooks E., and Dale Birdsell. *The Chemical Warfare Service: Chemicals in Combat.* Washington, DC: Office of the Chief of Military History, United States Army, 1966, Chapter XVII; Museum of the City of San Francisco, "Bombs: What to Do and When to Do It," <http://www.sfmuseum.org/war/bombs3.html>, 01/05/99; Partington, J.P. *History of Greek Fire and Gunpowder.* London: W. Heffer and Sons, 1961; Pyne, Stephen J. *Fire in America: A Cultural History of Wildland and Rural Fire.* Seattle: University of Washington Press, 1997, pp. 390–403; Stockholm International Peace Research Institute. *Incendiary Weapons.* Cambridge, MA: MIT Press, 1975.

INCENDIARY WEAPONS. *See*

Flame Thrower; Flares; Greek Fire; Incendiary Bombs; Napalm; Smoke Screen; Tracer Bullets

INFERNO

Inferno is one of three major sections of Dante Alighieri's classic poem, *La Divina Commedia* (*The Divine Comedy*). Dante is thought to have begun the poem in about 1307 and finished it in 1321, just before he died. The other two sections of the poem are *Purgatorio* and *Paradiso*. Each section consists of 100 cantos and includes a total of more than 14,000 lines.

The poem tells of Dante's journey through the various regions of hell, to the rim of purgatory, and then on to heaven, where he briefly has a glimpse of God Himself. The poet is accompanied on the first part of his journey by the Roman poet Virgil and on the last part, by his lady Beatrice.

In *Inferno*, Dante describes the nine regions of hell to which sinners can be sent, each more terrible and each buried deeper in the Earth. The upper regions are reserved for those whose sins are less terrible (lust and gluttony, for example, and children who were never baptized ("limbo"), while the lowest regions contain those guilty of sins such as treason, heresy, bestiality, and fraud.

From its title, one might expect the poem to be filled with visions of fire and flames, but the term *inferno* is used by Dante to refer to more general forms of punishment that are beyond human understanding. The lowest level of hell, in fact, is not covered by fire but by the coldest cold one can imagine. The cold represents the sinners' total absence of love and separation from God. It is here that Satan himself resides.

Fiery punishments are, however, an essential part of Dante's hell. They do not appear until Canto 10 and are reserved for those guilty of the worst sins. In Canto 14, for example, Dante tells of sinners who are being exposed to a constant rain for fire that will never be put out. Whatever meaning Dante may originally have had in mind for the word, the term *inferno* has now become a synonym for a raging fire. *See also* RELIGIOUS ALLUSIONS TO FIRE.

Further Reading: Alighieri, Dante. *The Divine Comedy*, translated by John Ciardi. New York: W. W. Norton, 1977; Alighieri, Dante. *The Divine Comedy*, translated by Robert Pinsky. New York: Farrar Straus & Giroux, 1995; "The Divine Comedy by Dante Alighieri," <http://www.divinecomedy.org/>, 08/31/99; Payton, R.J. *A Modern Reader's Guide to Dante's Inferno*. New York: Peter Lang Publishing, 1992.

INTERMIX FIRE

An intermix fire is a fire that occurs at the boundary, or *interface*, between a wilderness area and a built-up area containing human habitations. Over the past two decades, intermix fires have become arguably the most serious new fire management problem facing the United States and many other developed nations of the world.

Origins of the Problem

The fundamental reason that intermix fires have become an issue is the change in demographic patterns in many developed nations, such as the United States. For a period of about two centuries beginning in the eighteenth century, the general trend of population growth was in the direction from rural to urban. That is, fewer and fewer people in developed nations remained on farms and more and more moved to large urban areas. In the United States, as an example, urban areas occupied not quite 6 percent of all the land area of the United States in 1950. By 1990, the fraction had increased to more than 16 percent.

After 1950, however, a new demographic pattern began to emerge. People in developed nations began fleeing crowded inner cities and looking for new living space in a more pastoral setting, often on the edges of wilderness areas. Homes began to be built in areas that had their own fire history, one with which urban fire departments had little experience.

But just as people were migrating toward and into the wilderness, they were also incorporating the wilderness into their own homes and neighborhoods. To make their return to nature even more complete, they often used materials (fuels) from wilderness with which to build their homes and landscape their yards. They used the most flammable wood for shake roofs, and they planted trees and bushes especially adapted to wildfires to naturalize their yards.

At the same time, the very presence of humans in interface areas increased the likelihood of fires. Not only did the opportunities for arson increase, but also the simple day-to-day machines, power lines, railroads, automotive traffic, and recreational activities made fires both possible and inevitable. These patterns created opportunities for a new type of fire—the intermix fire—that had some of the worst characteristics of both wilderness and urban fires and that presented entirely new challenges for the firefighting establishment. For example, in the period between 1955 and 1979, 2,408 structures in California were destroyed by intermix fires. In the period from 1980 to 1993, a period only one-half as long, however, that number had more than tripled.

As the first intermix fires broke out, it became clear that neither the U.S. Forest Service, and its allies in the fight against wilderness fires, nor the city fire departments understood the new challenges they faced or the methods by which those challenges could be met. As historian of fire Stephen Pyne has written:

> Urban fire fighters save people; their cherished image is that of a lifesaver; their nightmare is a high-rise fire. Wildland fire fighters encounter forests, grasslands, and brush . . . they measure losses in acres, not lives or property. (Pyne 1997)

Intermix fires began to develop as a widespread national problem at the end of the 1980s. The most extensive of those fires swept through communities such as the Painted Cave area of Santa Barbara in 1990 and again in 1992; the Dude Creek area outside Payson, Arizona, in 1990; the Pattee Canyon region near Missoula, Montana in 1977; the Berkeley Hills area of Oakland, California in 1991; and the city of Malibu, California, in 1993. Many other parts of the United States and other developed countries were similarly affected. Intermix fires had begun to appear in such far-flung regions as the border between human communities and the bush in Australia, the boreal forests in Canada, and the Mediterranean biomes of southern Europe.

Fire Management in the Interface

The phenomenon of intermix fires has posed a whole new dilemma for the fire management profession in the United States and throughout the world. Wilderness and urban firefighting agencies are exploring new ways to work together to learn how to manage intermix fires more efficiently.

A key component of this new effort has been an approach known as *fire risk analysis*. The U.S. Forest Service has been using a fire behavior prediction system known as BEHAVE for three decades to predict fire risk. However, the system does not take into account the presence of lodging structures, such as private homes, nor does it distinguish between natural wilderness vegetation and plants used in gardens and landscaping.

As a consequence of the Berkeley Hills fires in Oakland in 1991, a consortium of federal, state, and private agencies in California developed a fire risk analysis program similar to BEHAVE for use with intermix fires. The program is based on a *fire-behavior triangle,* whose three vertices are topography, weather, and fuel.

Topography can dramatically affect the way a fire spreads through an area. Narrow canyons, for example, can act like chimneys, as they do in Santa Ana fires, and steep hillsides can increase the rate at which fuel preheats.

Weather also plays a role in determining fire hazards, with hot, dry, windy conditions greatly increasing the chance of a serious fire. Fires can create their own "weather" also as they increase both temperature and wind speeds that further intensify a blaze.

The nature and amount of fuel present in an area also determines the severity of an intermix fire. Studies have shown that the type of roofing used on a house accounts for as much as 75 percent of its risk of catching fire. Other structural factors include the type of siding, decking, and fencing used in the house. A study of the Painted Cave Fire near Santa Barbara, California, in 1990, found that nearly three-quarters of 701 buildings in the area with a nonflammable roof survived the fire, while only 19 percent with a flammable roof survived.

The two most important vegetation hazard factors around a home are the type and amount of surface fuel available and the amount of defensible space. Defensible space is defined is the space between a structure and a wilderness area in which vegetation has been cleared or modified to reduce possible fire risk. In the Painted Cave Fire, just over three-quarters of 542 buildings with a thirty-foot defensible space survived the fire, while only 38 percent without defensible space survived.

Fire officials understand that risk analysis systems will never prevent intermix fires from occurring, but such systems can decrease the chance that fires will begin in a given area and can increase the ability of firefighters to bring a blaze under control. Fire risk analysis systems are now being used in many parts of the country to promote certain types of legislation, to inform the methods by which urban and wilderness fire management agencies can work together, and to educate the general public about appropriate fire prevention techniques.

For example, the State of California now requires homes built in an interface region to have a defensible space of at least thirty feet in width. Homeowners are encouraged to clear even wider spaces, especially if their home is on a sloping lot. Educational programs also encourage homeowners to take defensive measures against intermix fires, such as keeping their roof and chimneys clean, installing spark arrestors (devices that prevent the escape of burning particles from a chimney), storing firewood away from the house, and controlling the vegetation around the house. *See also* BEHAVE; BERKELEY HILLS (CALIFORNIA) FIRE OF 1991; FOREST FIRES; SANTA ANA WINDS; WILDFIRE LANDSCAPING.

Further Reading: "Fire Danger in the East Bay Hills," <http://www-ehs.lbl.gov/hills/frd.htm>, 07/26/99; Pyne, Stephen J. *World Fire: The Culture of Life on Earth*. Seattle: University of Washington Press, 1997, pp. 269–295; Queen, Phillip L. *Fighting Fire in the Wildland/Urban Interface*. Bellflower, CA: Fire Publications, 1993; *Urban Wildland Interface Code*. Whittier, CA: International Fire Code Institute, 1997; "Wildfire Information," <http://www.prefire.ucfpl.ucop.edu/wildfire.htm>, 07/26/99.

INTERMOUNTAIN FIRE SCIENCES LABORATORY. *See* MISSOULA-INTERMOUNTAIN FIRE SCIENCES LABORATORY

INTERNAL COMBUSTION ENGINE

An internal combustion engine is a machine for converting the energy stored in chemical bonds of a fuel into mechanical energy. An internal combustion engine is similar in some fundamental ways to a steam engine. In both cases, the expansion and/or contraction of a gas or vapor causes a piston to move inside a cylinder. The up-and-down or back-and-forth motion of the piston may then be converted to a rotary motion by means of a crankshaft, which turns a wheel, a set of wheels, or some other mechanical devices.

The primary difference between an internal combustion engine and a steam engine is that the latter is an *external* combustion engine. That is, the gas or vapor needed to move the piston is generated outside the cylinder and then introduced into the cylinder, where it does work on the piston. In the internal combustion engine, the gas or vapor is generated inside the cylinder, above the piston head, after which it exerts a force on the piston.

History

Both the internal combustion engine and the steam engine have a common intellectual heritage in the research of the Dutch physicist Christiaan Huygens (1629–1695) and his assistant, Denis Papin (1647–1712). Huygens invented a device known as a *gunpowder engine* in which a small charge of explosives was ignited inside a cylinder. The force of the resulting explosion drove a piston through the cylinder. Huygens' device suggested a general method by which a piston-and-cylinder engine might operate, although the gunpowder engine was too dangerous and too inefficient to find any practical application.

Papin modified the design of Huygens' engine by using a small amount of water in the cylinder rather than gunpowder. When the water was heated to boiling, it changed to steam and pushed a piston upward in a cylinder. When the cylinder was cooled, the steam condensed, and atmospheric pressure pushed the piston back down again. Papin's engine was very slow and cumbersome, but it did work without posing the threat of injury to the operator.

It is somewhat remarkable that the experiments of Huygens and Papin led almost immediately to the development of external combustion (steam) engines, although internal combustion engines designed on the same principle did not appear for about 150 years. When that line of research did begin, inventors tried three different fuels for use in the internal combustion engine: gases, oils, and petroleum-based fuels (gasoline).

One of the first successful internal combustion engines was invented by the French engineer Étienne Lenoir (1822–1900) in about 1860. Lenoir's engine consisted of a single cylinder and piston in which illuminating gas was ignited by means of a spark produced by an induction coil. Lenoir's engine was, of course, very primitive, but it contained many of the elements of a modern internal combustion engine, and it was used successfully in a number of industrial operations.

Modern Internal Combustion Engines

A critical breakthrough in internal combustion engine technology occurred in 1876 with the invention of the four-stroke cycle by German engineer N.A. Otto (1832–1891). The so-called *Otto cycle* soon became the standard design for all high-power internal combustion engines.

The four stages of the Otto cycle include: (1) intake; (2) compression; (3) power; and (4) exhaust. During the initial stage, the intake valve in the cylinder opens, and a mixture of fuel and air is introduced into the cylinder from the carburetor. At this point, the piston is in its lowest position in the cylinder. In stage two of the cycle, both intake and exhaust valves close, and the piston moves upward in the cylinder to compress the fuel/air mixture. Stage three occurs when a spark from the spark plug atop the cylinder ignites the fuel/air mixture. The explosion that follows drives the piston downward in the cylinder again. In stage four of the cycle, the intake valve remains closed, the exhaust valve opens, and waste gases produced during ignition are ejected from the cylinder.

A number of variations on the standard four-stroke engine have been developed. One of the earliest of these modifications was the two-stroke engine, in which intake, combustion, and exhaust are all combined in a single up-and-down movement of the piston. Probably the main disadvantage of this type of engine is its reduced efficiency, compared to the four-stroke engine. Some unburned fuel remains in the cylinder after combustion in the two-stroke engine, reducing the efficiency of combustion in the next ignition cycle. Because of this disadvantage, two-stroke engines are used today only in specialized, usually small, appliances, such as lawn mowers and other garden tools. *See also* STEAM ENGINE.

Further Reading: Derry, T.K., and Trevor I. Williams. *A Short History of Technology from the Earliest Times to A.D. 1900*. New York: Dover Publications, 1993, Chapter 21; MacCoul, Neil, and Donald Anglin, "Internal Combustion Engine." In *McGraw-Hill Encyclopedia of Science & Technology*, 8th edition. New York: McGraw-Hill, 1997, Volume 9, pp. 342–349; "Internal Combustion Engine." In *The New Illustrated Science and Invention Encyclopedia*. Westport, CT: H.S. Suttman, 1989, pp. 1303–1309; Stone, Richard. *Introduction to Internal Combustion Engines*. Warrendale, PA: Society of Automotive Engineers, 1993; Taylor, Charles Fayette. *The Internal-Combustion Engine in Theory and Practice: Combustion, Fuels, Materials, Design*, revised edition. Cambridge, MA: MIT Press, 1985.

INTERNATIONAL ASSOCIATION OF FIRE CHIEFS

The International Association of Fire Chiefs (IAFC) was founded in 1873 as the National Association of Fire Engineers. The organization changed its name to the International Association of Fire Chiefs in 1926. It currently has a membership of about 12,000 fire chiefs and chief fire officers in forty countries around the world. Members are responsible for providing fire, emergency medical services (EMS), and other emergency services which, in the United States, cover about 70 percent of the population.

IAFC's mission statement states that the organization's goal is "to provide leadership to career and volunteer chiefs, chief fire officers, and managers of Emergency Services Organizations throughout the international community." The association attempts to achieve this goal by providing information, educational services, and other programs for the enhancement of the profession's members.

The work of IAFC is carried out largely through seven sections, eleven committees, and a task force on Chief Certification. Among the sections are those on Apparatus and Maintenance, EMS, Federal/Military, and Volunteer Chief Officers. Examples of IAFC committees include those on Fire Prevention, Hazardous Materials, Health and Safety, Risk Management, and Terrorism.

An important function of IAFC is its publications program, which produces books, manuals, and other materials on fire department issues. Examples of such products include the *Fire Service Joint Labor Management Candidate Physical Ability Test*, *A Comprehensive Analysis of the OSHA Respiratory Standard*, and *EMS Transport Systems*. The association also administers a number of scholarships for firefighters and their children, including the Arends Scholarship at Utah Valley State College, the Friendship Veterans Fire Engine Association Scholarship, the NFPA Fire Safety Educational Memorial Fund, and the Texas Commission on Fire Protection Scholarship.

For further information, contact:
International Association of Fire Chiefs
4025 Fair Ridge Drive
Fairfax, VA 22033
(703) 273–0911
http://www.iafc.org

INTERNATIONAL ASSOCIATION OF WILDLAND FIRE

The International Association of Wildland Fire (IAWF) is a nonprofit organization with over 2,100 members in eighty-one countries. Its primary mission is to promote the understanding of wildland fire and its management. The association serves as a clearinghouse for technical and professional information on wildland fires. In order to achieve this objective, it maintains an extensive library and software database that includes more than 60,000 magazine and journal articles and a list of over 100,000 researchers, managers, consultants, libraries, and agencies involved in forest fire management.

The association publishes *The International Journal of Wildland Fire*, a quarterly, peer-reviewed journal, and *Wildfire Magazine*, a monthly magazine concerned with equipment, training, safety, intermix fires, and human components of fire management. The association sponsors two conferences annually, one dealing primarily with research and fire ecology, and the other with fire management and safety. It also maintains a website with extensive information for both professionals and interested amateurs. A feature of special interest on the website is a Youth Menu with information, lessons, and activities on wildland fire for boys and girls.

For further information contact:
International Association of Wildland Fire
East 8109 Bratt Road
Fairfield, WA 99012–0328
(509) 523–4003
http://www.wildfiremagazine.com

THE INTERNATIONAL JOURNAL OF WILDLAND FIRE. *See*

INTERNATIONAL ASSOCIATION OF WILDLAND FIRE

INTIMIDATION FIRES

Throughout history, fire has been used as a way of threatening, punishing, persecuting, torturing, and executing individuals. In some cases, punishment by fire has been exacted by governmental agencies against their enemies. In other instances, it has been religious groups acting against the rebellious, disbeliever, or sinner. In most such cases, the threat of being tortured or killed by fire has been used to remove opposition or obtain agreement with a political philosophy or religious dogma. Fire has also been used by individuals or gangs to express their disapproval of some ethnic, religious, sexual, racial, or other group. Some examples of the use of fire for the purpose of intimidation are discussed below.

Firebombs

Firebombs can be made from devices as simple as a *Molotov cocktail*, a bottle filled with gasoline or some other flammable liquid into which a wick has been fitted. The wick is ignited just before the bottle is thrown. When the bottle lands, the flammable liquid ignites, causing a conflagration in any structure on which it lands.

More complex firebombs can also be made with everyday materials by following instructions in booklets or websites readily available to any person. Such bombs can be made more destructive in a variety of ways, such as by adding small nails to the bottle. In such "nail bottles," the nails are released when the bomb explodes, traveling outward in every direction like tiny projectiles.

Firebombs have been used widely over the past half century by individuals and groups wishing to express their disapproval of other people with whose beliefs or lifestyles they object. Common targets are gay men and women, Jews, and people of other religious beliefs. For example, a firebomb containing nails was thrown into a gay pub in the Soho district of London on 30 April 1999, causing extensive damage to the building, killing three people and injuring eighty more. The bombing was the third in two weeks carried out by a far-right group that called itself the

White Wolves. Earlier the group had released firebombs in two other areas of London with large East Asian populations.

Firebombs have been especially popular weapons among Palestinians against Israeli citizens and soldiers. During October 1998, for example, firebombs were used in attacks against Israelis on the Mount of Olives (5 October), in Eastern Jerusalem (7 October), and in the area of el-Aroub (31 October).

Some other examples of firebombing incidents over the past decade include the following:

- On 10 October 1990, an elderly couple in Phoenix, Arizona, was nearly killed when a firebomb intended for a teenager living near them hit their window instead.
- On 1 June 1997, angry students in South Korea used firebombs against riot police during a political demonstration against President Kim Young-sam.
- During the first week of December 1997, a rash of violent attacks in eastern France resulted in numerous fires and injuries and the death of one eighteen-year-old boy.
- In late June 1999, three synagogues in Sacramento, California, were firebombed by individuals who left behind anti-Semitic pamphlets and literature.
- On 23 September 1999, a teenager in San Jose, California, was charged with throwing a firebomb at the home of a Roman Catholic judge.
- In April and May of 2000, a series of firebombing incidents in Vientiane, the capital of Laos, was blamed on members of the Hmong ethnic minority who live in the northern part of the country.

Cross Burnings

Shortly after the Civil War ended, supporters of slavery adopted a new technique for frightening newly freed slaves and their supporters: cross burning. The usual practice was for groups of White citizens to construct a large wooden cross, wrap it in cloth, soak it with gasoline, and then erect the cross on property owned by a Black person or family. The cross was then set on fire. The group was most commonly associated with cross burning has long been known as the Ku Klux Klan.

Although no longer as widespread as it was in the late nineteenth and early twentieth centuries, cross burning has not disappeared from American society. During the late 1990s, the U.S. Justice Department brought charges based on cross-burning incidents in an average of two dozen cases per year. In recent years, cross burners have typically chosen Black families who have moved into previously all-White areas or biracial couples as the targets of their protests.

The U.S. Supreme Court has been called upon to rule on the legality of cross burning in only one case, *R.A.V. v. St. Paul* (505 U.S. 377). This case was based on the arrest of an individual, referred to in the case only as "R.A.V.," who had been arrested in 1990 for burning a cross on the lawn of a Black family in St. Paul, Minnesota. The arrest was based on St. Paul's so-called "Bias-Motivated Crime Ordinance." That ordinance prohibited the display of any symbol that could be expected to "arouse anger, alarm, or resentment in others on the basis of race, color, creed, religion, or gender." The U.S. Supreme Court overruled the Minnesota Supreme Court and upheld the original trial court in this case, agreeing with the latter that the St. Paul ordinance was overly broad and, therefore, an abridgment of the Free Speech amendment of the U.S. Constitution.

The Court's decision has not made it impossible, however, to arrest and convict individuals for cross burning. In most cases, a person who carries out such acts is charged with violation of a person's civil rights, using threat against the person on whose property the cross is burned, or violating a local or state hate crime ordinance.

Church Burning

As used here, the term *church burning* refers to arsonous attacks on churches whose members are primarily Black, as an attempt to intimidate or terrorize Black communities. During the civil rights movement in the United States in the 1960s, Black churches in the South were often targeted by White citizens who sought to frighten those working for the civil rights of African Americans. Perhaps the most notorious of these attacks was the bombing of the all-Black Sixteenth Street Baptist Church in Birmingham, Alabama, on 15 September 1963. A bomb was thrown into the basement of the church during morning services, resulting in the death of four young girls.

Beginning on 1 January 1995, a similar outbreak of arson attacks on Black churches appeared to have begun. In the next eighteen months, nearly forty fires of "suspicious origin" destroyed Black churches in North Carolina, South Carolina, Alabama, Georgia, Mississippi, and other southern states. Black and White citizens and political leaders alike

Ruins of a Charlotte, North Carolina church after a 1996 arson fire. © *Bob Padgett/ Archive Photos.*

began to fear that a new epidemic of church burning was underway. President Bill Clinton ordered the formation of a special task force including representatives of the U.S. Bureau of Alcohol, Tobacco and Firearms, the Federal Bureau of Investigation, and the U.S. Marshals Service to investigate these attacks.

The task force's investigation was inconclusive. It found that, overall, the number of church fires in the South for 1995 and 1996 did not significantly exceed the annual average. It did discover a handful of cases in which racial hatred was a motive for the church burnings. But it also found that many of the suspected arson cases were actually accidental fires having nothing to do with political or social issues. It also discovered that the rate of fires among Black churches and the rate of solved cases for such fires was not significantly different than it was for White churches.

Further Reading: Chalmers, David Mark. *Hooded Americanism: The History of the Ku Klux Klan*. Durham, NC: Duke University Press, 1987; *Church Burnings*. Hearings before the Committee on the Judiciary, U.S. Senate, 104th Congress, second session, 27 June 1996. Washington, DC: Government Printing Office, 1997; Hjelm, Norman A., ed. *Out of the Ashes: Burned Churches and the Community of Faith*. Nashville, TN: Thomas Nelson, 1998; "Rebuilding Burned Churches," <http://gbgm-umc.org/advance/Church-Burnings/index.html>, 10/06/99; Wade, Wyn Craig. *The Fiery Cross: The Ku Klux Klan in America*. New York: Oxford University Press, 1998.

IROQUOIS THEATER FIRE OF 1903

The most deadly single-structure fire in U.S. history occurred on 30 December 1903 in Chicago's Iroquois Theater. The fire broke out at about 3:15 P.M. during a matinee performance of the popular musical show, Mr. Blue Beard, Jr. The performance was sold out, with about 1,900 people in attendance. Although no exact count was possible, authorities set the number of deaths at just over 600. The majority of those killed were women and children. By contrast, only one out of more than 500 performers and stage hands lost his life.

In some ways, the Iroquois conflagration was totally unexpected. The theater had opened less than a month earlier. It had been built at a time when architects and contractors were well aware of the fire danger posed by theatrical buildings. A number of systems had been developed for reducing the risk of fire and of possible dangers posed by fires. For reasons that will probably never be known, the Iroquois lacked many of the systems that had become standard in most large performance halls and theaters.

Most new theaters of the time were equipped with a heavy asbestos or iron curtain that could be dropped from the proscenium arch in case of fire. The purpose of the curtain was to prevent smoke and fire from reaching the audience. At the Iroquois, however, the fire curtain failed completely. Inexperienced stage hands were unable to pull the curtain completely down, and the fire spread rapidly from backstage into the audience. By some accounts, the curtain was not even made of fireproof material.

The fire apparently broke out when a hot light backstage set fire to a velvet curtain. The use of fireproof materials had become standard in theaters by this time, but the Iroquois lacked such simple precautions. In addition, it had become standard practice to station firefighters backstage, equipped with fire extinguishers, hoses, and other firefighting equipment. At the Iroquois, firefighters were equipped with only two containers of a firefighting powder called Kilfyres. It was totally ineffective in slowing down the fire.

Theater personnel were also untrained in fire exit procedures. As soon as the fire broke out, patrons rushed to the small number of inadequate exits, which soon became jammed with dead bodies. A few people who were able to escape jumped to their death from upper windows or from fire escapes. Some of the fire doors had been built to open inwardly only, immediately trapping the first customers who tried to escape.

The fire was put out with remarkable speed by the Chicago Fire Department. It was completely under control in less than half an hour. Nearly all of the casualties had occurred within the first fifteen minutes of the fire. The theater building itself was largely undamaged by the fire. It reopened for business less than a year later and continued to function as a theater until 1925.

Further Reading: "CPL Chicago: 1903, December 30: Iroquois Theater Fire," <http://www.chipublib.org/004chicago/disasters/iroquois_fire.html>; Gunzel, Louis. *Retrospects: The Iroquois Theater Fire, La Salle Hotel Fire in Chicago*. Chicago: Theater Historical Society, 1993; *Lest We Forget: Chicago's Awful Theater Horror*. Chicago: Memorial Publishing, 1904; Randall, Frank A. *History of the Development of Building Construction in Chicago*. Urbana: University of Illinois Press, 1949.

J

JET ENGINES

A jet engine is a type of heat engine that produces a forward thrust by expelling hot gases or fluids from the rear of the engine. The principle on which jet engines work was first stated by Sir Isaac Newton (1642–1727) in his Third Law in about 1687. The Third Law states that for every action, there is an equal and opposite reaction. Perhaps the most familiar example of this law can be seen when air escapes from a toy balloon. As air leaves the balloon in one direction (the "action"), the balloon flies off in the opposite direction (the "reaction").

In the case of a jet engine, the "action" is the expulsion of hot gases produced by the combustion of a fuel (or, less commonly, by some other means) from the rear of the engine. The "reaction" is the forward movement of the engine, carrying along with it any object to which it may be attached.

Two general kinds of jet engines exist: the air-breathing jet engine and the rocket. In an air-breathing jet engine, some fuel (such as kerosene) is burned, using oxygen from the air which enters the front of the engine. A rocket operates without the need for oxygen from the atmosphere. Instead, it combines two chemicals—one, the fuel, and the other, the oxidizer—to achieve combustion. For example, a common pair of chemicals used to propel rockets consists of liquid hydrogen (the fuel) and liquid oxygen (the oxidizer).

Air-breathing Jet Engines

The simplest of all jet engines is the ramjet. In a ramjet, air enters the front of the engine and is used to burn a fuel in the combustion chamber in the center of the engine. Hot gases escape through the rear of the jet, providing a forward push for the engine. The engine casing is designed in such a way as to increase the pressure (and, therefore, the amount) of air within the combustion chamber, thereby increasing the efficiency of fuel combustion.

The operation of a ramjet depends entirely on the amount of air entering the front of the engine. At low speeds, therefore, it is a relatively inefficient engine. At higher speeds, however, air enters the engine much more rapidly, the rate of fuel combustion increases, and the engine's efficiency increases proportionately.

A more efficient form of the jet engine is called the turbojet. The basic mode of operation in a turbojet is the same as in a ramjet. The major difference is that the turbojet contains a compressor attached to a turbine in the front part of the engine. The compressor contains a series of metal blades that rotate around a central shaft. As air enters the front of the engine, it is compressed by the rotating fan blades and then sent on to the combustion chamber.

The compressor can be operated independently of the jet engine itself. That is, a turbojet aircraft can turn on its compressor before leaving the ground, immediately increasing the rate of fuel combustion within the engine. The engine's efficiency still depends on air speed, but to a much smaller degree than with a ramjet. The exhaust gases in a turbojet have, of course, the primary function of providing thrust for the engine. However, a portion of these gases are also diverted back into the engine, where they are used to operate the compressor.

All air-breathing jet engines are very inefficient. For example, only about a quarter of the oxygen that enters the front of such engines is actually used to burn fuel. The rest simply escapes out of the back of the engine with exhaust gases. One way of dealing with this problem is by adding an *afterburner* to the engine. An afterburner is simply an extension of the engine itself in which a second "dose" of fuel is sprayed into the escaping gases. This fuel reacts with some of the hot oxygen present in the exhaust gases, increasing the thrust and efficiency of the engine. Afterburners may increase by nearly 50 percent the thrust produced by a jet engine.

Rockets

The vast majority of rockets that have been developed and used are chemical rockets. They use combinations of fuel and oxidizer, such as liquid oxygen. Generally speaking, the fuel/oxidizer combination can be solid or liquid. An example of the former is a rocket propelled by a combination of powdered aluminum metal (the fuel) and ammonium perchlorate (the oxidizer), held together in a plastic resin known as the *binder*. An example of a liquid fuel is one consisting of liquid hydrogen and liquid oxygen.

The hot gases needed to propel a rocket can be produced in nonchemical ways also. In a nuclear rocket, for example, a fission reactor is used to heat a fluid, which is then expelled from the rear of the rocket. Some rockets operated by expelling hot charged particles—electrons or ions—rather than hot gases. Nuclear and ion rockets are relatively uncommon and tend to be used in specialized applications.

Further Reading: Boyne, Walter, Terry Gwynn-Jones, and Valerie Moolman. *How Things Work: Flight*. Alexandria, VA: Time-Life Books, 1990; Hill, Philip Graham, and Carl R. Peterson. *Mechanics and Thermodynamics of Propulsion*. Reading, MA: Addison-Wesley Publishing, 1992; Huenecke, Klaus. *Jet Engines: Fundamentals of Theory, Design, and Operation*. Stillwater, MN: Motorbooks International, 1998; "Jet Engine." In *The Rand McNally Encyclopedia of Transportation*. Chicago: Rand McNally, 1976; "Jet Engines." In *How It Works*. New York: Simon and Schuster, 1971; Karagozian, A. R. "Jet Propulsion." In Robert A. Meyers, ed. *Encyclopedia of Physical Science and Technology*. Orlando, FL: Academic Press, 1987; Rossotti, Hazel. *Fire*. Oxford: Oxford University Press, 1993, Chapter 11; Shaw, John M. "Jet Engines." In Frank N. McGill, ed. *Magill's Survey of Science: Applied Science Series*. Pasadena, CA: Salem Press, 1993; Sutton, George P. *Rocket Propulsion Elements: An Introduction to Engineering of Rockets*. New York: John Wiley & Sons, 1992.

JOURNAL OF WILDLAND FIRE, INTERNATIONAL. *See* INTERNATIONAL ASSOCIATION OF WILDLAND FIRE

K

KILN. *See* FIRING OF POTTERY

L

LAMPS AND LIGHTING

A lamp is a device for providing light. Until the invention of the incandescent electric light bulb in the 1800s, nearly every type of lamp used by humans was operated by means of a flame. Today, gas and oil lamps, candles, and other types of flames are generally used only for decorative or emergency lighting. Most nonelectric forms of lighting include torches, candles, or lamps that use some liquid or gaseous fuel, such as kerosene or natural gas (methane).

Early Lamps

Scientists can only speculate about the earliest forms of lighting used by primitive humans. Simple camp fires may well have been the first source of light used to brighten a cave or a dark evening.

It seems likely that the first devices specifically used for lighting were simple torches discovered when humans observed the way a burning branch gives off light. Even today, some primitive tribes use torches made of reeds, branches, or other plant materials, tied or twisted together, and soaked in a combustible liquid, such as oil, pitch, or resin. Young children in modern societies sometimes soak cattail heads in kerosene and light them to use as a primitive form of torch.

Torches of these designs are poor sources of light and give off a lot of smoke. The first improvement on torches appeared in about 18,000 B.C. among cave dwellers in Western Europe. These improvements consisted of hollowed-out stone cups or clay-lined wicker baskets into which oil or tallow was poured. Moss or some other kind of spongy plant was used as a wick in these primitive lamps.

Over time, humans found ways to make lamps a more efficient source of light. For example, lamps were often encased in containers made of glass or oiled paper to prevent breezes from blowing out the flame. The first use of lamps in Greece can be traced to about the seventh century B.C. In fact, the word *lamp* itself comes from the Greek word *lampas*, for "torch." By about 500 B.C. both Greeks and Romans were using lamps made of a covered container holding oil with an opening for holding a wick.

One of the most ingenious of the early lamp designs was one invented by Leonardo da Vinci (1452–1519). The flame in da Vinci's lamp was enclosed in a glass cylinder which was surrounded, in turn, by a globe made of glass and filled with water. This design not only protected the flame from being blown out, but also enhanced the brightness of the light produced.

Regardless of the materials and shape of a lamp, the principle by which they supplied light was always the same. The combustible liquid was soaked up by the wick. When the liquid reached the top of the wick, it was ignited by means of a fire. Heat from the fire caused the liquid to evaporate and to begin to burn. The wick itself usually charred, but it was not supposed to actually catch fire.

Candles

Historians believe that candles were invented by the Egyptians in about the third millennium B.C. They consisted of a piece of tallow or beeswax with a wick running through the core. In appearance, they may not have been very different from the candles we use today. They also operated on the same principle in which heat from the burning wick melts wax, causes it to evaporate, and then ignites it at its top.

There has been remarkably little change in candle technology over the past five millennia. One important improvement was made in the seventeenth century with the discovery that spermaceti wax, obtained from whales, was much superior to tallow or beeswax for the manufacture of candles. Spermaceti wax is harder than tallow or beeswax, so it does not soften in warm temperatures. It also burns with less smoke than tallow and is less expensive than beeswax. Two centuries later, an even better material, paraffin, was discovered. Paraffin is one of the materials obtained from the fractional distillation of crude oil.

Some changes were made in the way wicks were made also. At one time, strings, threads, or fibers were simply twisted around each other. Later, it was found that braiding fibers made for a more permanent and efficient wick.

Today, the vast majority of candles are used for decorative or celebratory purposes. They are valued not so much for the light they produce as their artistic design, colors, aromas, and other properties.

Later Developments in Lamp Technology

Many of the lamps used throughout Europe and the early United States up to the eighteenth century were simple lamps consisting of little more than a long wick. A popular form of this lamp was the rushlight. A rushlight was made by dipping small rushes in grease or kitchen fat. After the rushes had dried, they could be lit to produce a relatively dim lamp that was used primarily to supplement the light of a fireplace.

The most popular forms of lamps in early America were oil lamps copied after those used in Europe at the time. They often consisted of nothing other than a bowl to hold oil and a wick. Among the best known of these lamps was the "Betty" or "Betsy" lamp. The origin of the name of these lamps is not known, but is thought to have been derived from the word *better* because they were "better" than lamps previously available.

A Betsy lamp consisted of three parts: a bowl to hold oil, a spout in which the wick could be inserted, and a long handle that permitted the lamp to be carried about or hung from a peg on the wall.

One of the first substantial changes in lamp design came in 1784 when a Swiss-English chemist by the name of Aimé Argand invented an oil lamp with a round wick, set above a circular container. With this design, air was able to flow over the wick from both outside and inside, increasing the efficiency with which combustion occurred. The addition of a glass chimney and metal reflecting plates around the lamp also increased the amount of light given off by the lamp.

Inventors continued to make modest improvements in the design of oil lamps through the second half of the eighteenth and the early nineteenth century. The fundamental problem with such lamps, however, was the lack of a really satisfactory fuel. At first, the fuel used in lamps was primarily the particular kind of oils available in the area, such as olive oil or sesame oil. Later, other kinds of liquid fuels, such as spermaceti oil, paraffin oil, or kerosene, became more popular.

Introduction of Gas Lamps

The crucial breakthrough in lamp technology, however, came with the conversion from liquid to gaseous fuels. The earliest suggestions for using flammable gases as fuels in lighting devices can be traced to the late 1660s. However, practical devices that burned gas rather than liquid did not actually appear until more than a hundred years later.

One of the first of these devices was invented by the French engineer Philippe Lebon (1767–1804) in about 1791. Lebon found that he could produce a flammable gas by the destructive distillation of wood. He called it *wood gas*. Destructive distillation is a process by which a material is heated in the absence of air. Rather than catching fire (since no oxygen is present), the material breaks down into simpler components, some of which may be flammable. The destructive distillation of wood had been a familiar process for centuries since it was the mechanism by which charcoal was made.

Lebon received a patent for a lamp that operated using wood gas in 1799, by which time he had used the invention to light at least one home in Paris and its gardens. He was prevented from further developing his invention, however, when he was killed during a robbery attempt in 1804.

The Scottish mechanic William Murdock (1754–1839) pursued a somewhat different approach to the manufacture of gas lamps and had greater success than did Lebon. Murdock heated soft coal, rather than wood, in a container from which air was excluded. The destructive distillation of soft coal results in the formation of coke, a thick liquid called *coal tar*, and a flammable gas known as *coal gas*. Coal gas is primarily methane and other hydrocarbons, and is a relatively good fuel. It was eventually used for a variety of heating, lighting, and other purposes.

Murdock developed a system for producing coal gas and then piping it to homes, factories, and other structures where it could be used for lighting. His first customer was a cotton mill in Salford, which installed 900 gas lamps to illuminate its factory, a private road, and a private residence owned by the company. Murdock adapted Argand's oil lamp design for use with his gas lamps.

Later Developments in Lamp Technology

Gas quickly became widely popular for a variety of lighting systems, such as homes, offices, factories and street light systems in a community. Gas lamps were popular for only a relatively short period of time, however. Once Thomas Edison had invented the electric light bulb in 1879, most lighting systems switched over from gas to electricity. The primary exception to this was applications in which portable lighting was needed as, for example, in the mining of coal and other minerals. The need for such lamps was essentially unaffected by the development of incandescent lights, and development continued in the design of gas and liquid lamps.

One of the most important inventions made was that of a pressure lamp. In a pressure lamp, fuel is forced out of a reservoir under pressure into the wick. One of the earliest pressure lamps was called the "Moderator Lamp." It was invented by a Frenchman known today only by the name of Franchot. Pressure in the lamp was produced by a spring that drove a piston through the fuel container, forcing the fuel upward into the wick.

One other invention that changed the nature of the pressure lamp was that of the mantle. A mantle is a thin gauze-like material made of some substance that glows brightly when heated. The possibility of using a mantle in a lamp was first conceived by the Austrian chemist Karl Auer, Baron von Welsbach (1858–1929). Welsbach hypothesized that lamps could be made to glow more brightly if they included some material that, when heated, gave off more light than the flame of the lamp itself. He tried a number of substances in his effort to find such a material.

Welsbach finally discovered a nearly ideal material to use for such a purpose. He impregnated a very thin, cylindrical, cotton fabric with a solution of thorium nitrate containing a small amount of cerium nitrate. When the fabric was ignited, the cotton itself burned away, leaving behind a very thin ceramic material consisting of thorium and cerium compounds. When this material was placed around the flame of a pressure oil lamp, it glowed with a brilliant white light.

The concept developed by Welsbach is essentially the same one used in modern pressure lamps. When a person buys a mantle for such a lamp, the first thing that must be done is to ignite the mantle. When the mantle has completely burned up, a thin, crusty-like material is left behind. It is this material that is then suspended within the lantern and that gives off light when the lamp is operated.

Many people today are familiar with pressure lamps since they are widely used for camping and other outdoor activities. They are sold under trade names such as Coleman, Tilley, Petromax, Primus, Evening Star, and AGM.

Further Reading: *A Brief History of the Origin and Use of Coleman Lamps and Lanterns*. Wichita, KS: The Coleman Company, 1980; Cooke, Lawrence S., ed. *Lighting in America: From Colonial Rushlights to Victorian Chandeliers*. New York: Main Street/Universe Books, 1975; Derry, T.K., and Trevor I. Williams. *A Short History of Technology from the Earliest Times to A.D. 1900*. New York: Dover Publications, 1993, Chapter 17, passim; Hobson, A. *Lanterns that Lit Our World: Book 2*. New York: Golden Hill Press, 1997; "How a Gas Lantern Works," <http://www.howstuffworks.com/gas-lantern.htm>, 09/08/99; Meadows, C.A. *Discovering Oil Lamps*. Princes Risborough: Shire Publications, 1994; "Pressure Lamps Unlimited," <http://ourworld.compuserve.com/homepages/awm/history.htm>, 09/07/99; Travers, Bridget, ed. *World of Invention*. Detroit: Gale Research, 1994, pp. 361–362.

LANTERNS. *See* LAMPS AND LIGHTING

LAVOISIER, ANTOINE LAURENT

Antoine Laurent Lavoisier is generally regarded as the "Father of Modern Chemistry." One reason that he is accorded this honor is that he was the first person to enunciate the modern theory of combustion.

Biographical Background

Lavoisier was born in Paris on 26 August 1743. He was born into a well-to-do family and received all the benefits of a good upbringing, financial security, and an excellent education. He studied law and was granted a license to practice as an attorney in 1764.

He was also very interested in scientific subjects, however, and did some of his first original experiments in the same year he was accredited in the law.

Not surprisingly, Lavoisier eventually became involved in governmental politics. He was a member of many boards and commissions and was active in the development of a new lighting system installed in many cities and towns. In 1786, Lavoisier invested money in a financial firm known as the Ferme Général, a private company whose job it was to collect taxes for the French government. Although he was never involved in the actual collection of taxes, Lavoisier's affiliation with the firm was eventually to prove disastrous to his future.

Lavoisier's life would appear to have been headed for a bright and successful future were it not for political turmoil stirring in the nation at the time. When the French Revolution broke out in the 1790s, Lavoisier had already made enough enemies to seal his doom. The most important of those enemies was Jean-Paul Marat, whom Lavoisier had blackballed for membership in the French Academy of Sciences in 1780. In the early days of the revolution, Marat accused Lavoisier of plotting against the public welfare and demanded his execution. Marat was actually assassinated before any action was taken on his complaints, but his accusations had achieved their goal. Lavoisier was tried as an enemy of the people, found guilty, and guillotined on 8 May 1794. He was buried in an unmarked grave in Paris, almost certainly the most tragic death of the revolution.

Scientific Achievements

Although Lavoisier was interested in experimental research, he was far more important as a theorist. He lived during a period when the science of chemistry was just beginning to develop and badly needed someone who could provide a sense of organization and cohesiveness for the field. Lavoisier was that man. For example, he published a book in 1787 in collaboration with Claude Louis Berthollet (1748–1822) and Antoine François Fourcroy (1755–1809) entitled,

Methods of Chemical Nomenclature. The book cut through the confusion of previous naming systems and outlined a clear and functional system of naming chemical elements and compounds.

In some ways, Lavoisier's most important contribution was in his explanation of the process of burning. Historians of science have pointed out that each field of science appears to have some crucial problem that must be resolved if its early development is to go forward. In the case of physics, it was the question of motion; in the case of astronomy, it was a matter of heliocentrism; and in the case of biology, it was the circulation of the blood. In the case of chemistry, it was combustion.

As an example, the Dutch physician Hermann Boerhaave (1668–1738) once wrote that:

> If you make a mistake in your exposition of the Nature of Fire, your error will spread to all the branches of physics, and this is because, in all natural production, Fire . . . is always the chief agent. (Hermann Boerhaave. *Eléments de Chimie.* Paris: Chardon, fils, 1752, vol. I, 144.)

Since the first dawnings of chemical research at the beginning of the seventeenth century, burning had been explained by the phlogiston theory, a theory that supposed that burning involved the loss of a substance (phlogiston) during the process of combustion. The theory explained many phenomena associated with combustion, but created some new problems of its own.

In 1778, Lavoisier had devised a new theory of combustion. He based this theory on a number of experiments he claimed to have conducted himself, although he certainly relied also on the discoveries of other pioneer chemists of the time. For example, Lavoisier reported on a number of experiments in which he heated objects in air, weighing them both before and after combustion. His attention to the quantitative aspects of the experiments was both unusual for the time and crucial for the success of his work. He showed that in every case, an ob-

ject gained weight after it had been heated. He explained this fact by assuming that the object being heated combined with oxygen of the air to form a "calx," a compound that we would today call an oxide.

Lavoisier's explanation appears to us today, of course, to be simple and obvious. But it was not so received by many of his contemporaries. Most of those scientists had grown up believing in the phlogiston theory, and it was not easy for them to reject everything they had believed before. In fact, it was a number of years before the oxidation theory of combustion became generally accepted in the chemical community and the phlogiston theory was finally laid to rest.

Further Reading: Asimov, Isaac. *Asimov's Biographical Encyclopedia of Science & Technology.* 2nd revised edition. Garden City, NY: Doubleday & Company, 1982, pp. 222–226; Donovan, Arthur. *Antoine Lavoisier: Science, Administration, and Revolution.* Cambridge: Cambridge University Press, 1996; Holmes, Frederic L. *Antoine Lavoisier: The Next Crucial Year.* Princeton, NJ: Princeton University Press, 1998; Holmes, Frederic L. *Lavoisier and the Chemistry of Life: An Exploration of Scientific Creativity.* Madison, WI: University of Wisconsin Press, 1989; Holton, Gerald, and Duane H.D. Roller. *Foundations of Modern Physical Science.* Reading, MA: Addison-Wesley Publishing Company, 1958, Chapter 16; Ihde, Aaron J. *The Development of Modern Chemistry.* New York: Dover Publications, 1984, Chapter 3; Partington, J.R. *A Short History of Chemistry.* London: Macmillan and Company, 1937, Chapter 7; Wightman, William P.D. *The Growth of Scientific Ideas.* New Haven, CT: Yale University Press, 1951, Chapter 16.

LEARN NOT TO BURN®

Learn Not to Burn® (LNTB) is a fire safety education program developed by the National Fire Protection Association in 1979. It is now taught in more than 50,000 classrooms in the United States. The program teaches twenty-two key fire safety behaviors at four different learning levels: preschool, K–2, 3–5, and 6–8. It is designed to be taught as part of regular classroom work at all of these grade levels, with lessons designed for use in language arts, math, history, art, and science.

The LNTB preschool program is available in English, French, and Spanish and is designed to teach eight basic fire and burn prevention behaviors. Short and simple lessons are designed to involve children in activities related to fire safety. Each lesson is accompanied by a fire safety song provided on a cassette tape.

The K–8 sequence of lessons introduces an additional fourteen fire safety behaviors at increasing levels of complexity. Materials for these lessons include reproducible activity sheets, discussion topics, and evaluation forms. Lessons for the curriculum are provided in a set of three resource books for grades K–2, 3–5, and 6–8. These resource books are available in English, Spanish, and French.

Additional information about the curriculum can be obtained from:

National Fire Protection Association
1 Batterymarch Park
P.O. Box 9101
Quincy, MA 02269–9101
(800) 344–3555

LEOPOLD REPORT

The Leopold Report was a report made by a committee of naturalists, wildlife biologists, wildlife managers, and experts in other fields of wildlife management to the Secretary of the Interior in 1963. The committee consisted of A. Starker Leopold, S.A. Cain, C.H. Cottam, Ira N. Gabrielson, and T.L. Kimball. The committee was formed in 1962 in an effort to answer the question: "How shall excess wildlife populations be controlled in our national parks?"

Philosophy of the Report

The report was of major significance in the history of National Park Service policies because it clearly defined for the first time the importance of viewing the plants, animals, and abiotic components of a wilderness area as a whole and complete ecosystem. Every feature of such an area was inextricably bound up with every other feature, and all elements were interconnected with each other. This view was a dramatic divergence from traditional policies that dealt with a particular spe-

cies or group of species of plants or animals one at a time.

The committee noted that relatively few national parks or national wilderness areas were large enough to maintain self-regulating ecological systems. In those regions, however, the committee emphasized the importance of allowing natural systems to come into equilibrium with each other. Human interventions should occur rarely and only for very special reasons.

In smaller areas, the committee said, some form of management was necessary in order to restore ecological systems to their natural and normal condition. The report pointed out that:

> Most biotic communities are in a constant state of change due to natural or man-caused processes of ecological succession. In these "successional" communities, it is necessary to manage the habitat to achieve or stabilize it at a desired stage.

The report proposed an apparently simple goal for the Park Service. A park, it said, should be maintained "as nearly as possible in the condition that prevailed when the area was first visited by the white man. A national park should represent a vignette of primitive America." The authors of the report recognized the magnitude of this recommendation. They called the implications of the above statement "stupendous."

The Report's View on Fire

The Leopold Report was of fundamental importance to National Park Service fire policies and practices. Prior to its release, the Service had been following a policy of fire prevention and suppression for decades. Professional land managers and the general public alike had come to believe that fire in a national park or other wilderness area was "bad" and to be prevented or suppressed. The report emphasized that fire is a normal and natural part of every ecosystem, and it can not be excluded from a system without bringing about dramatic disruptions. It continued the line of reasoning above to point out that "fire is an essential management tool to maintain East Africa open savanna or American prairie."

The Leopold Report committee spoke out strongly about the role of fire, both natural and human-made, in achieving the goal of returning wilderness areas to their "natural" state. It made the point that, "Of the various methods of manipulating vegetation, the controlled use of fire is the most 'natural' and much the cheapest and easiest to apply." This recommendation was followed by more specific suggestions as to the kinds of fires that should be used in different wilderness habitats under the control of the National Park Service.

Further Reading: "The Leopold Committee Report: Wildlife Management in the National Parks." *American Forests*, April 1963, pp. 32–35.

LIGHTHOUSES

A lighthouse is a tall building containing a bright light that provides a warning to mariners. Lighthouses are usually built at the entrance to a harbor, along a rough coast, or at some other point on a shoreline where hazards to navigation may exist.

History

The earliest method for warning mariners probably consisted of nothing more than an open fire built on top of a bluff or high hill.

As maritime traffic became more important, however, the need for more sophisticated warning towers became obvious. There is some evidence that lighthouses were being built in North Africa as early as 1000 B.C. The purpose of these structures was to guide travelers on religious journeys. The first true lighthouse of which we have any information is one built in about 660 B.C. to guide mariners through the Greek Isles.

The oldest lighthouse of which we have definite knowledge was the Pharos of Alexandria, built in about 290 B.C. The Pharos was built by order of the Egyptian pharaoh Ptolemy Soter to protect ships entering the harbor of Alexandria. It was thought to be at least 85 meters (280 feet) tall and was regarded as one of the Seven Wonders of the Ancient World. Light was produced by wood fires and then reflected off polished metal sheets. The Pharos light is thought to have been visible for a distance of about 50 kilometers (30 miles).

Lighthouse building reached a high stage of development during the Roman Empire. Dozens of structures were erected on the Atlantic and Mediterranean coasts, from Great Britain to Palestine. In fact, the oldest existing lighthouse in the United Kingdom was built by order of the Emperor Caligula at Dover Castle in 90 A.D. During the Middle Ages, however, lighthouse construction decreased significantly. In fact, many existing lighthouses were torn down during the period.

With significant increases in maritime traffic during the fifteenth century, however, the rate of lighthouse building began to increase once again. Every nation in Western Europe who was at all dependent on ship-

Engraving of the Pharos of Alexandria, the oldest known lighthouse, which was one of the seven wonders of the ancient world. © *Historical Pictures Archives/CORBIS.*

ping—including Great Britain, France, Germany, Spain, and Sweden—constructed warning towers along their coasts and at the entrance to their harbors and ports. An important breakthrough in construction methods occurred in 1756 when English engineer John Smeaton (1724–1792) designed the first stone lighthouse to be built on Eddystone Rock near Plymouth. Stone lighthouses were not only sturdier, but also less likely to catch fire than their wooden predecessors.

The first lighthouse built in the United States was the Boston Light, erected on Little Brewster Island at the entrance to Boston Harbor in 1716. Today, there are about 10,000 lighthouses in the United States.

Lights

To provide an effective warning system, a lighthouse requires two major components: a source of light and a means for magnifying (or intensifying) that light. Prior to the invention of the incandescent lamp in the 1860s, light was provided by the burning of wood, coal, fuel oil, or candles. In many cases, the systems used were quite simple. Many lighthouses had nothing more complicated than a large metal basin or lamp in which one of these fuels could be burned.

Wood, coal, and oil had the advantage of providing brighter lights than did candles. But the flame they produced was less steady and inclined to be more smokey than that of a candle. Many efficient lighthouses, then, obtained their light from burning candles. Smeaton's Eddystone Rock lighthouse, for example, derived its light from twenty-four candles, which was said to be visible by telescope for distances of up to twenty kilometers.

One of the most popular systems for producing light was the burning of preheated oils. The oils were first warmed on a stove and then carried to a lamp where they were set afire. This system was both clumsy and dangerous, but it provided one of the most efficient forms of lighting during the late eighteenth and early nineteenth century.

The increasing availability of petroleum products in the early 1800s provided a better fuel for lighting: kerosene. By the 1870s, kerosene had become the fuel of choice at most lighthouses around the world. At first, the kerosene was preheated and converted into a vapor, which was then burned to produce light. Later, pressure burners were invented in which kerosene vapor was forced through a nozzle and then ignited.

After 1900, yet another fuel became popular: acetylene. Acetylene gas burns with an intense white light and became the fuel of choice in many lighthouses until well into the twentieth century. As late as 1980, for example, about eighty automatic lighthouses along Australian coasts were still being powered by acetylene lamps.

By the end of the nineteenth century, electrical power was becoming available in most parts of the developed world. Most new lighthouses use some form of electricity to generate light. At first, carbon arc lamps were used, although Thomas Edison's tungsten filament incandescent light bulb soon became the source of choice in the vast majority of lighthouses.

Magnification

Until the mid-eighteenth century, only the most primitive forms of light magnification were available. A highly polished sheet of metal, for example, could sometimes be used to increase the intensity of a light, but the improvement was usually minimal.

The first major improvement in methods for the magnification of light came in 1752 when an English scientist, William Hutchinson, invented a parabolic reflector. A parabolic reflector is a surface from which light rays are reflected in such a way that they all leave the surface in a straight line. Hutchinson's invention solved the perennial problem with all earlier lights that a large fraction of the light produced by fire radiated away in directions that were not visible to mariners and, therefore, of no value in the warning system.

The greatest breakthrough in lighthouse magnification systems occurred in the early 1820s with the invention of a new type of lens by French physicist Augustin Jean Fresnel (1788–1827). The so-called *Fresnel lens* or

Fresnel prism consists of a complex system of glass prisms, arranged so as to focus light by both reflection and refraction. The Fresnel lens resembles a beehive in appearance and is placed around a fire light source. It was capable of producing the most intense beam ever available from a lighthouse. By the mid-nineteenth century, Fresnel lenses had been installed in many lighthouses throughout the world. They remained the magnification system of choice everywhere until the 1920s, when new forms of incandescent lighting became available.

By the close of the twentieth century, the role of lighthouses in providing navigational warnings had changed significantly. In the first place, most lighthouses had become fully automated. Lighthouse keepers, who had once played crucial roles in operating and maintaining lights, had become virtually obsolete. In the second place, most warning systems no longer depend primarily on beacons of light for their guidance. Instead, mariners can use satellite and global positioning systems to determine their position with a high degree of accuracy.

Further Reading: Guiberson, Brenda Z. *Lighthouses: Watchers at Sea*. New York: Henry Holt, 1995; International Association of Marine Aids to Navigation and Lighthouse Authorities. *Lighthouses of the World*. Springfield, TN: Globe Pequot Press, 1998; Jones, Ray. *American Lighthouses*. Springfield, TN: Globe Pequot Press, 1998; "The Lighthouse Society of Great Britain," <http://www.btinternet.com/~k.trethewey/>.

LIGHTNING

Lightning is a powerful discharge of electrical charge in the atmosphere. Lightning is the single most common cause for wildfires throughout the world. By some estimates, up to 44,000 thunderstorms occur daily throughout the world with more than 100,000 lightning bolts striking the Earth on an average day.

Whether or not a lightning bolt starts a fire depends on a number of factors, including the energy of the lightning strike, the time it takes for the lightning bolt to complete its journey, the kind of fuel available in an area, the current weather, and the topography of the region. In general, the fires started by lightning tend to be small and to die out fairly quickly. Some important exceptions occur in those areas in which dry fuel has been allowed to accumulate over long periods of time, as when vigorous programs of fire suppression have been in force.

Lightning in Mythology

Allusions to lightning in mythology are less common than allusions to fire. Perhaps the best known of these allusions is associated with Zeus, the chief god in Greek mythology. In leading the battle of the Olympians against the Titans, Zeus is said to have advanced on the enemy, hurling flashes of lightning that set fire to the surface of the Earth.

In many cases, ancient myths associate lightning with thunder and rain and appoint a single god to rule over all of these weather phenomena. The Zapotec people of ancient Mexico, for example, worshipped Cocijo, the god of lightning and rain. The god's name was apparently transformed among later Mayan stories to Chak, also called the god of lightning and rain.

Scientific Explanation

The process by which lightning forms is not completely understood. The fundamental requirement for lightning to occur is that a charge difference must exist between different parts of a cloud. That is, one part of a cloud must develop an electrical charge that is negative compared to that of a second part of the cloud.

One theory as to how a charge separation develops is that water droplets become ionized as they freeze to form ice crystals. Ionization is the process by which atoms or molecules break apart into two charged particles, one positive and one negative. Experiments have shown that negatively charged ions are formed and migrate to the warmer portions of a droplet, while positively charged ions migrate to the cooler part of the droplet. As rapidly moving updrafts of air move through a cloud, they may sweep the positively charged ions upward, leaving the negatively charged ions in lower parts of the cloud.

However, when charge separations develop in a cloud, they represent an unstable state. The normal tendency in such a circumstance is for negatively and positively charged ions to rejoin to form a neutral particle. The problem is that air is a poor conductor of electricity, and the potential difference between positive and negative ions must be quite large before they are able to overcome the electrical resistance of air. In general, the potential difference must exceed about 3,000 volts per meter for a current to flow from one part of a cloud to a second part. When the potential difference has reached this value, the current flow occurs very rapidly with an enormous release of energy. This release of energy heats air and water molecules along its path, accounting for the flash of lightning, and causes air to expand rapidly, accounting for the sound of the thunder.

The charge differential within a cloud can also be neutralized by the movement of negative ions (electrons) from the cloud to the Earth's surface. The potential difference required for this to occur can be of the order of a few millions, or even hundreds of millions, of volts. When that limit is reached, a lightning stroke may travel from the base of the cloud to the ground.

The phenomenon of lightning striking the Earth is quite complex and has been studied in great detail by means of motion pictures. Each single flash of lightning actually consists of three to four separate strokes, separated from each other by about fifty milliseconds of time. Soon after the lightning flash has struck the Earth, a discharge occurs in the reverse direction, from Earth to the cloud. This reverse movement of electrons is called a *return stroke*. All other factors being equal, the longer the return stroke in a lightning burst, the greater the chance that fire will break out.

Lightning and Forest Fires

Lightning is an important source of ignition of forest fires historically because, for many centuries, it was the only way in which humans were able to obtain fire. Before they learned how to make fire on their own, hu-

mans had to wait for a lightning strike (or some other natural form of ignition) to get the fire they needed and wanted. Throughout history, lightning has continued to be an unrelenting and dependable cause of natural wilderness fires. This fairly obvious fact has at times been ignored by humans responsible for fire management whose goal it has been to prevent or suppress wildfires entirely.

In some areas, where forests, brush, grasslands, and other sites for wildfires have been totally eliminated (as in some parts of Europe), lightning can have only limited success in igniting new fires. But in an area where vegetation still grows, lightning fires are inevitable.

For example, during the long period in American history when prevention and suppression were the guiding principles of forest management, large amounts of fuel (dead and dying trees and brush, for example) accumulated. When a lightning storm finally struck this fuel, huge fires broke out. Such was the case in the great Yellowstone Park fires of 1988. Fire managers now realize that, since lightning fires are inevitable, some form of prescribed burning is essential at least in some areas under some circumstances. It is only in this way that lightning can be deprived of the fuel it needs to generate really massive conflagrations. *See also* FOREST FIRES; IGNITION.

Further Reading: Agee, James K. *Fire Ecology of Pacific Northwest Forests.* Washington, DC: Island Press, 1993, Chapter 2, pp. 207–209; Burby, Liza N. *Electrical Storms.* New York: Rosen Publishing Group, 1999; Lutgens, Frederick K., and Edward J. Tarbuck. *The Atmosphere: An Introduction to Meteorology,* 6th edition. Upper Saddle River, NJ: Prentice Hall, 1995, Chapter 10; Pyne, Stephen J. *Vestal Fire: An Environmental History, Told through Fire, of Europe and Europe's Encounter with the World.* Seattle: University of Washington Press, 1997, pp. 19–24, passim; Uman, Martin A. *All About Lightning.* New York: Dover Publications, 1986; Viemeister, Peter E. *The Lightning Book.* Cambridge, MA: MIT Press, 1972; Whipple, A.B.C. *Planet Earth: Storm.* Alexandria, VA: Time-Life Books, 1982, Chapter 5; Wright, Henry A., and Arthur W. Bailey. *Fire Ecology: United States and Southern Canada.* New York: Wiley-Interscience, 1982, passim.

LITERATURE

Fire is the subject of a countless number of books. In many instances, the primary focus of these books is fire itself. A review of the books listed in the "Further Reading" sections at the end of many entries in this book provides a good introduction to such books. The books by John W. Lyons (*Fire*, 1985), Hazel Rossotti (*Fire*, 1993), Johan Goudsblom (*Fire and Civilization*, 1992) and the various books on fire by Stephen Pyne are examples of works that attempt to deal with many aspects of fire.

In other cases, an author focuses on some specific aspect of fire. Some examples include James K. Agee's *Fire Ecology of Pacific Northwest Forests* (1993), John L. Bryan's *Fire Suppression and Detection Systems* (1982), J.G. Frazer's *Balder the Beautiful: The Fire-Festivals of Europe and the Doctrine of the External Soul* (1913), and Margaret Sullivan's *Firestorm! The Story of the 1991 East Bay Fire In Berkeley* (1993).

The titles listed as references in this encyclopedia do not begin to encompass the works in which fire is a primary theme. Other examples of books that focus on fire itself include the following:

- Christopher Duffy's *Fire & Stone: The Science of Fortress Warfare, 1660–1860* (Greenhill Press, 1996), describes some of the ways in which fire was used as a military weapon over a two-century period.
- Norman MacLean's *Young Men & Fire* (University of Chicago Press, 1993), tells the story of the Mann Gulch fire in Montana that claimed the lives of thirteen young smoke jumpers on 5 August 1949.
- Kenneth Werrell's *Blankets of Fire: U.S. Bombers over Japan during World War II* (Smithsonian Institution Press, 1998), provides an introduction to the use of incendiary weapons during the war against Japan.

References to fire also appear in the titles of many books as a symbol of passion, commitment, desire, obsession, or some other powerful emotion. John Van Houten Dippel's, *Bound upon a Wheel of Fire: Why So Many German Jews Made the Tragic Decision to Remain in Nazi Germany* (Basic Books, 1996), tells of the lives of six successful German Jews who stayed in their native country even when it became clear that Hitler's anti-Semitic policies threatened their lives. Carson Van Lindt's, *Fire & Spirit: The Story of the 1950 Phillies* (Marabou Publishing, 1998) narrates the story of Philadelphia's National League baseball team's apparently hopeless struggle to win the 1950 league championship.

The imagery of fire as a symbol of passion and commitment appears frequently in works that tell of the struggle of women, Blacks, and other minorities to find success in American society. Among the most famous of these books is James A. Baldwin's, *The Fire Next Time* (Dial Press, 1963). Baldwin uses the image of fire in a variety of ways in the book. He warns that the racial injustices existing in the United States in the 1960s are so great that they will ignite a fire in the hearts of Blacks that may eventually lead to riots in which real fires will be prominent. Other books that reflect Baldwin's theme include Henry Mayer's, *All on Fire: William Lloyd Garrison and the Abolition of Slavery* (St. Martin's Press, 1998) and Nelson Peery's *Black Fire: The Making of an American Revolutionary* (New Press, 1995).

A number of books on religious topics also call on fire as a symbol of passion and commitment. Heather Choate Davis selects a familiar religious symbol for the title of her book, *Baptism by Fire: The True Story of a Mother Who Finds Faith during Her Daughter's Darkest Hour* (Bantam Doubleday Dell Publishers, 1999). The book explains how her daughter's near-death seizure and subsequent brain surgery brought her a new faith in God. *The Bliss of Inner Fire: Heart Practice of the Six Yogas of Naropa* (by Lama Thubten Yeshe, Wisdom Publications, 1998) also calls on the imagery of fire in the heart to describe fundamental rituals in Tibetan yoga.

Annelise Orleck's *Common Sense & a Little Fire: Women and Working-Class Politics in the United States, 1900–1965* (University of North Carolina Press, 1995) follows the lives of four immigrant women activists who started in garment sweatshops to rise to influence positions in the American labor movement, largely because of the "fire in their heart" for better lives for their fellow women.

Fire has also long been a ubiquitous theme in world fiction. One of the most famous works in English literature in which fire has a prominent role is Rudyard Kipling's *The Jungle Book* (1893). In the story "Mowgli's Brothers," for example, the boy-wolf Mowgli overcomes the wolf pack's efforts to kill him by overturning the firepot and threatening his "brothers" with a firebrand:

> Mowgli thrust his dead branch into the fire till the twigs lit and cracked, and whirled it above his head among the cowering wolves. . . . He strode forward to where Shere Khan sat blinking stupidly at the flames, and caught him by the tuft of his chin. . . . "Up, dog!" Mowgli cried. "Up, when a man speaks, or I will set that coat ablaze!" (University of Virginia Library, Electronic Text Center, p. 41)

Fire is a major theme in another animal-related book, *Bambi*, written by Felix Salten in 1926 and made into a Walt Disney movie in 1942. Probably the most dramatic scene in that animated classic was a forest fire that killed Bambi's mother.

Fire continues to be a popular theme in contemporary fiction. For example, author Don Winslow was an arson investigator for fifteen years before becoming an author. Many of his works feature his fictional counterpart, Jack Wade, also an arson investigator called in to solve difficult problems involving suspicious fires. The book review periodical, *Booklist* calls one Winslow work, *California Fire and Life* (1999) "tough as nails and entertaining as hell."

One of the great popular literary and film successes of the early 1990s was Tom Wolfe's *Bonfire of the Vanities*. Wolfe calls on the famous story of the Dominican monk Girolamo Savonarola's rise to power in Florence in 1497 to highlight the rise of materialism in modern America. The novel and film tell the story of a Wall Street financier, Sherman McCoy, who, with his mistress, makes a wrong turn and ends up in the South Bronx late one evening. After running down a young Black man who they believe is attacking them, their lives begin to unravel.

LONDON FIRE OF 1666. *See* THE GREAT LONDON FIRE OF 1666

LOOKOUT TOWERS

A lookout tower is an elevated structure from which observations can be made. In these lookouts, an observer is watching for fires so that firefighters can be dispatched to the affected region.

The use of fire lookout towers goes back at least as far as the Roman Empire. The primitive firefighting brigades established by Caesar Augustus in 32 B.C., for example, relied on reports from watchmen stationed at high points in the city to know when fires developed. The use of such watchmen continued for most of the next 1,800 years.

Fire lookout towers became especially useful in another setting in the late nineteenth century: the national and state forests. Rising concerns about the conservation and preservation of the nation's forest resources led to the creation in 1907 of the U.S. Forest Service (USFS). One of the duties assumed by the USFS was firefighting in the 132 million acres of forests under its care. The task of watching over this much land was, of course, overwhelming. The only reasonable approach developed for monitoring the forest lands was the development of lookout towers at strategic locations in the forest. The first lookout towers were built in the 1910s and were little more than simple cabins built at high points in a forested region. More sophisticated structures were soon being built, however, and at an accelerating rate.

Construction of a lookout tower was not an easy task. Building materials and commu-

Fire lookout station. *Courtesy of Bureau of Land Management at the National Interagency Fire Center.*

World War II again interrupted the development of fire lookout towers and introduced an even more important change. Techno-logical developments resulting from the military effort made lookout towers much less crucial to the protection of forest reserves. Fire-fighters began to rely on observations made by aircraft and satellites. They were also able to replace human observers with sophisticated fire detection equipment, such as infrared (heat) detectors that could locate fires that were not visible to the human eye.

nication equipment had to be hauled many miles through forested areas and up the side of mountains. Still, after a hiatus caused by World War I, lookout towers were being built at the rate of two to four per year in each ranger district in the U.S. West by 1930. The towers were built in such a way that some overlap was achieved, with no area of forest land being left without observation.

A variety of architectural styles were used for lookout towers. These structures had to include living space for tower occupants (the "fire finder"), storage space, and a room from which observations could be made. One of the most popular structures was called the D-5 (named after a California fire district in which they were first built) that consisted of a cupola with windows on all four sides built on top of a cabin that also contained four windows in all directions.

The period of maximum lookout construction occurred during the days of President Franklin Roosevelt's Civilian Conservation Corps. As one of the Corps' major responsibilities, more than 60,000 miles of trails and 600 lookout towers were constructed.

Other factors also led to a reduction in lookout tower construction and maintenance. The USFS found it more difficult to find men and women who were willing to give up a "normal" life to live twenty-four hours a day, week by week, in a lookout tower. Also, the threat of lawsuits by individuals who fell from or were injured in lookout towers raised questions as to the economic feasibility of these structures.

A report issued by the USFS in 1974 noted that the highpoint of the lookout development in the United States occurred in 1953 when 5,060 permanent structures existed. After that time, older, less efficient, and more dangerous structures were being taken down to be replaced by more modern technologies. *See also* NATIONAL FIRE PROTECTION ASSOCIATION.

Further Reading: Kresek, Ray. *Lookouts of the Northwest.* Fairfield, WA: Ye Galleon Press, 1984; Spring, Ira, and Byron Fish. *Lookouts: Firewatchers of the Cascades and Olympics.* Seattle: The Mountaineers, 1981.

M

MALTESE CROSS

The Maltese Cross is a form of the cross also known as a cross formée, or Cross Pattee-Nowy. It has long been taken as a symbol of the firefighting profession, representing its commitment to the prevention and extinguishing of fires and the saving of lives.

The emblem is thought to have been adopted from a similar symbol used by the Knights of St. John of Jerusalem, also known as the Knights Hospitallers. The Knights were originally a charitable organization formed in the eleventh century to help the sick and poor. They established a number of hospitals and hospices to serve these individuals. Late in their existence, the Knights also took on military responsibilities, providing assistance to Crusaders fighting for the conquest of Jerusalem. At one point, they also took part in battles on the Island of Malta, after which the cross is named.

The association of Maltese Cross and firefighting arose because the Knights Hospitallers were often involved in the fighting of fires. Attacking armies at the time often used a variety of incendiary materials, such as burning sulfur, oil, and petroleum products. Members of the Knights were called upon to save lives and properties by doing battle with fires set by these incendiary weapons.

MATCHES

A match is a short, slender piece of wood or pasteboard tipped with a mixture that catches fire when rubbed on a specially prepared surface.

History

Humans have long wanted to have ways of setting fire to materials, whether they be camp fires, forest fires, or cigarettes. The earliest device for doing so was probably just a simple torch. The objects that we now know as matches were probably not invented until after 1000 A.D. Marco Polo reported seeing the use of matches during his visit to China in the late 1200s. But the first clear evidence we have for the invention of such a device can be traced to an invention by the Irish-English physicist and chemist Robert Boyle (1627–1691) in 1681. Boyle's device consisted of two parts. One was a piece of paper coated with phosphorus, and the other was a wooden splint with a sulfur tip. When the wooden splint was drawn through a crease in the paper, it caught fire.

The invention of the first modern match is usually credited to the British pharmacist John Walker (1781–1859) in 1827. Walker coated the tip of a wooden splint with a mixture of potassium chlorate and antimony sulfide. When the match was struck on a surface, a chemical reaction between the two compounds produced a spark and a flame. Unfortunately, the spark was sometimes quite pronounced, and the match holder ran the risk of being showered with sparks from the match.

Walker called his matches *Congreves* in honor of Sir William Congreve, inventor of a military rocket. A competitor of Walker's soon produced his own version of the Congreve and named his matches *Lucifer* because of the foul smell they produced when ignited.

Most early matches used some form of phosphorus on their striking tip. Yellow phosphorus has a very low kindling temperature and is easily ignited simply by rubbing it against a rough surface. Unfortunately, yellow phosphorus is highly toxic, and workers who produced matches coated with phosphorus often developed a terrible disease known as *phosphorus necrosis*, or "phossy jaw." It was not until the 1840s that inventors found ways to use red phosphorus in matches, an allotrope of the element that does not ignite as easily, but is much safer to work with and use.

Modern Matches

The first modern matches were produced in the 1840s by European chemists who found that a compound of phosphorus, phosphorus sesquisulfide, was ideal for use on matches. It has a low kindling temperature, but is not nearly as toxic as yellow phosphorus.

During the early part of this century, match making in most of the world was dominated by a single Swedish company, owned by Ivar Kreuger (1880–1932), known as the "Match King." Kreuger developed an extensive system of match making that started with the growing of trees from which to make the wooden splints to the actual sale of the matches themselves.

Until the mid-nineteenth century, most matches were "strike-anywhere" matches. The tips of these matches were covered with a material that would ignite when rubbed against almost any rough material: wood, concrete, and even rough cloth. Then, in 1844, the first *safety match* was invented by the Swedish chemist Gustave E. Pasch.

Pasch's safety matches would ignite only when they were rubbed against a specific kind of surface. Since this invention, the vast majority of matches in use have been safety matches. They possess the important property that they are highly unlikely to ignite accidentally, as can a "strike-anywhere" match.

The modern "strike-anywhere" match consists of three parts: the wooden splint, a head made of antimony sulfide and potassium chlorate, and a tip made of phosphorus sesquisulfide. When the match is rubbed on a rough surface, the tip ignites and sets fire to the head, which then causes the wood splint to burn. The wooden splint is usually impregnated with ammonium phosphate, which reduces the rate at which it burns.

The safety match, by contrast, consists of only two parts: the wooden splint and the head. The head usually consists of a mixture of antimony sulfide and potassium dichromate. The striking surface, which is kept separate from the match itself, is usually coated with red phosphorus. When the wooden splint is rubbed on the striking surface, the red phosphorus briefly catches fire from the heat generated by friction. The burning phosphorus then ignites the materials on the head of the match, which burst into flame.

The modern matchbook was invented in 1892 by an attorney from Philadelphia, Joshua Pusey. In Pusey's original model, the matches and the striking surface were folded together inside a matchbook. With this design, it was relatively easy for the matches and striking surface to rub against each other and to ignite spontaneously. Today, the striking surface is placed on the back of the matchbook to prevent this kind of accident. *See also* IGNITION.

Further Reading: Beaver, Patrick. *The Match Makers*. London: Henry Mellard, 1985; Cross, M.F., "A History of the Match Industry," *Journal of Chemical Education* 18 (1941): 116–120, 277–282, 316–319, 380–384, 421–431; Goudsblom, Johan. *Fire and Civilization*. London: Allen Lane, 1992, pp. 170–172; "Match," in Bridget Travers, ed. *World of Invention*. Detroit: Gale Research, 1994, pp. 406–407; Derry, T.K., and Trevor I. Williams. *A Short History of Technology from the Earliest Times to A.D. 1900*. New York: Dover Publications, 1993, pp. 551–553.

METALLURGY

Metallurgy is the science and art of working with metals. The field of metallurgy is usually divided into three major parts: extractive metallurgy, physical metallurgy, and process (or manufacturing) metallurgy. Extractive metallurgy includes all those processes by which metals are separated from their ores and then refined. Physical metallurgy involves the analysis of metals and the study of their physical and chemical properties. Process metallurgy is the set of processes by which a raw metal is converted into some useful objects, such as a propeller or pan.

History of Metallurgy

Metallurgy is one of the oldest arts known to humans. Early humans almost certainly came across pieces of gold, copper, iron, and other metals found naturally in the Earth's surface or in the remains of meteors. At first, these metals were probably used in their native form or they were simply hammered into simple objects. Among the earliest metal objects found by archaeologists are pins, hoops, beads, and other decorative and useful objects dating to the fifth millennium in Egypt. Copper pipes were also used to deliver river water to the homes of wealthy Egyptians.

Probably the first true metallurgical process to be developed was annealing. Annealing involves the heating of a metal so that it becomes soft enough to work more easily. Annealing is suitable for use with only a small number of metals, such as copper and gold, which have low melting points. Its use was limited also because early artisans had no techniques for producing very hot flames that would be needed to soften most metals, such as iron.

One of the first chemical processes discovered was smelting, the process of melting an ore in order to separate the metal it contains from impurities with which it is mixed. In some cases, melting alone is sufficient to achieve this objective. Ores of gold and copper, for example, often contain the pure metal mixed with sand and other earthy materials. When the mixture is heated, the metal melts and separates from impurities.

In other instances, smelting may be associated with a chemical change, such as oxidation or reduction. Primitive metallurgists probably first observed the reduction of metallic ores accidentally. When an ore such as iron oxide is heated over a charcoal fire, a chemical change occurs in which the oxide reacts with the charcoal to form pure metallic iron:

$$2Fe_2O_3 + 3C \text{ (charcoal)} = 4Fe + 3CO_2$$

It probably required little imagination for these artisans to adapt this natural chemical change for their own use in extracting metals from their ores.

Oxidation also occurs naturally when ores are heated. The most common reaction is one in which a sulfide ore is converted into the comparable oxide. In the case of lead sulfide, for example:

$$2PbS + 3O_2 = 2PbO + 2SO_2$$

This change, known as *roasting*, is a necessary first step when a metal occurs as a sulfide ore. After roasting, the oxide formed can then be converted to the pure metal by the reduction process described above.

It appears that the simple chemical changes involved in smelting were widely used, even if they were not understood from a scientific standpoint, as early as the fourth millennium B.C. In fact, the production of bronze by the use of such techniques was already in wide use during the period between 4000 and 3000 B.C., an era that is often called the *Bronze Age* because of the ubiquity of bronze objects.

Further developments in metallurgy depended on the invention of tools and techniques for producing hotter fires. Probably the most important of these inventions was the bellows. The bellows was used to blow air across a fire at a high rate of speed. The increased availability of oxygen allowed the fire to burn at a higher temperature, making possible metallurgical processes that do not occur at lower temperatures.

The first furnaces using bellows began to appear as early as 1800 B.C. One of the first products of such furnaces was wrought iron, a form of iron that contains about 3–5 percent earthy impurities known as *slag*. Wrought iron forms naturally and easily when iron ores are heated to the melting point of iron (about 1,500°C). As metallic iron is produced, it mixes with sand and other impurities present in the ore. Wrought iron can be worked quite easily, although it is too soft to be used for some purposes. Nonetheless, the discovery of techniques for making iron were so successful that the metal soon replaced bronze as the most widely used metal. That step marked the beginning of another age in human history now known as the *Iron Age*.

The Iron Age saw the development of other metallurgical techniques designed to improve the quality of iron products. One of these techniques is quenching, the process of plunging a hot piece of iron into cold water. Quenching makes significant changes in the physical properties of a metal. In the case of iron, it increases the strength and hardness of the product.

Cementation methods were also developed during the Iron Age. Cementation is a process by which a piece of one metal (iron, at this point in history) is coated with a second metal. The most common cementation method involves heating a piece of iron to a high temperature and then placing it into a powdered metal of some other kind. The powdered metal melts and adheres to the outer surface of the iron. Cementation also adds certain desirable properties to a metal, such as corrosion resistance to iron.

Metallurgical techniques developed during the Iron Age remained largely unchanged for about 2,000 years. One important step forward was the development of more efficient furnaces during the fourteenth century. For example, the continuous-shaft furnace, the precursor of the modern blast furnace, first appeared in Germany in 1323. The continuous-shaft furnace made possible the production of a new form of iron, *cast iron,* that is harder and stronger than wrought iron. With cast iron, a whole new class of weapons and implements could be produced.

Metallurgy experienced a spurt of growth after 1700 as the Industrial Revolution began. The demand for iron and steel products increased the need for new and better forms of these materials. At the same time, as machinery and techniques improved, better methods for making metals became available. One of the most significant advances early in this period was the use of coke, rather than charcoal, for firing blast furnaces. An English inventor by the name of Abraham Darby is often credited for inventing the first successful coke-fired blast furnace at Shropshire in 1709. However, such furnaces did not become widely used until after the middle of the century. By 1760, there were only seventeen such furnaces in all of Great Britain. By 1775, that number had grown to thirty-one and by 1790, to eighty-one. At that time, coke-fired ovens were by far the most common type of iron- and steelmaking furnaces in the country.

Techniques of Metallurgy

Today, extractive, physical, and process metallurgists have a host of techniques and procedures available to them. A complete discussion of those techniques is beyond the scope of this article. The following examples are only illustrative of the variety of methods now in use.

Casting. Casting is the process by which a molten metal is poured into some type of mold, giving it a desired shape. Casting is a very old process and was employed as soon as artisans found ways to heat a metal to a temperature high enough to cause it to melt. Many types of molds can be used in the casting process. One of the most common types is a sand mold. A sand mold consists of sand shaped into the design needed for the final metal product. Liquid metal is poured into the mold and allowed to solidify. The sand mold is then destroyed and the metal product retrieved.

A variation of sand mold casting is called shell mold casting. In shell mold casting, the mold consists of a thin layer of sand bonded with a thermosetting resin. After the mold has been filled with liquid metal and the metal cooled, the mold is broken.

In some cases, metal molds are made for casting. The advantage of a metal mold is that it can be reused many times. The disadvantage is that the metal mold must be made of some material that will not react or bond with the liquid metal poured into it.

Forging. Forging is a process by which a hot metal is hammered, rolled, extruded, or otherwise deformed by pressure. Forging is one of the oldest metallurgical techniques known, although a variety of modifications have been developed for use in the manufacture of products of specific shapes.

Powder metallurgy. Powder metallurgy is a process by which tiny metallic particles are forced together to form a particular product. The particles are commonly formed by melting a piece of metal and then spraying a jet of inert gas through the melt. The spray breaks down the metal into very small particles and cools them to produce a fine powder. The powder can then be sprayed, pressed, or otherwise worked into some particular shape. Powder metallurgy is useful in the manufacture of products with delicate, irregular, or very small parts.

Some professional organizations of interest include the following:

Society for Mining, Metallurgy and Exploration
8307 Shaffer Parkway
Littleton, CO 80127
Tel: (303) 973–9550; (800) 763–3132
Fax: (303) 973–3845
http://www.smenet.org
e-mail: sme@smenet.org

American Institute of Mining, Metallurgical, and Petroleum Engineers
Three Park Avenue
New York, NY 10016–5998
Tel: (212) 419–7676
Fax: (212) 419–7671
http://www.aimeny.org/
e-mail: aimeny@aimeny.org

Ancient Metallurgy Research Group
Professor A. M. Pollard

Department of Archaeological Sciences
University of Bradford
Bradford, England BD7 1DP
Tel: +44 (0) 1274–383531
Fax: +44 (0) 1274–385190
http://www.brad.ac.uk/acad/archsci/depart/resgrp/amrg/amrginfo.html.
e-mail: p.budd@bradford.ac.uk

Canadian Institute of Mining, Metallurgy, and Petroleum
3400 de Maisonneuve Blvd., W, Suite 1210
Montreal, Quebec, Canada H3Z 3B8
Tel: (514) 939–2710
Fax: (514) 939–2714
http://www.cim.org/
e-mail: cim@cim.org

Further Reading: Brandt, Daniel A. *Metallurgy Fundamentals*. Tinley Park, IL: Goodheart-Wilcox Company, 2000; Chandler, Harry. *Metallurgy for the Non-Metallurgist*. Materials Park, OH: ASM, International, 1998; Smallman, R.E., and R.J. Bishop. *Modern Physical Metallurgy and Materials Engineering: Science, Process, Applications*. Woburn, MA: Butterworth-Heinemann, 2000; Walker, John R. *Modern Metalworking*. Tinley Park, IL: Goodheart-Wilcox Company, 2000.

METAPHORS

The words *fire, flame, smoke,* and other combustion-related terms are often used in contexts that have nothing to do with physical fire itself. The following are examples of the many ways in which fire-related words are used as metaphors:

Fire ants are members of the genus *Solenopsis,* specifically *S. richteri* (the black fire ant) and *S. invicta* (the red fire ant). Fire ants get their name from the painful burning sensation produced by their bite.

Fire blight is a plant disease caused by the bacterium *Erwinia amylovora*. It gets its name because the bacterium causes a plant's leaves and fruit to take on a brownish or blackish appearance, as if they have been burned.

Firebrand was originally used to describe any burning piece of wood, such as one that might be used to start other fires. Today, the term is also applied to individuals who have strong feelings and opinions on some topic and set out to stir up or "set fire" to others.

Firebrats are insects that belong to the order *Thysanura*. They get their name from the fact that they are able to withstand very high temperatures.

Firefly is an insect (especially members of the family *Lampyridae*) that give off a glow in the dark. The glow resembles a small flame, although it is produced by a chemical mechanism very different from oxidation.

Fire opal is a form of the gem opal that gets its name from the brilliant yellows, oranges, and reds it displays. The colors are so bright that they remind one of a flame and give the impression that the gem might shine in the dark, which it doesn't.

Firestone is an older name for a type of pyrite (iron sulfide) once used widely for starting fires by striking against another material.

Firethorn is any plant belonging to the genus *Pyracantha*, given its name because of the brilliant red and orange berries it produces after flowering.

Fireweed is a plant that gets its name from the fact that it moves quickly into an area that has been covered by fire and begins to germinate and grow quickly. The taxonomic name for fireweed is *Epilobium angustifolium*.

Fireworms are certain types of worms and larvae that get their name because they are luminescent. One example of fireworm is *Odontosyllus*, a member of the *Syllidae* family, found in coral reefs of tropical waters.

Ring of Fire is a narrow belt of active volcanoes that roughly describes a circle around the perimeter of the Pacific Ocean. About three-quarters of the known active volcanoes on Earth occur within the Ring of Fire, also known as the *Circle of Fire*.

MINE FIRES AND EXPLOSIONS

Mining has always been one of the most hazardous of all occupations. Although risks exist in surface mines, underground mines are far more dangerous places to work. Throughout the centuries, untold numbers of men, women, and children have been killed in explosions, fires, subsidence events, and other mine accidents.

History

At one time, mine fires and explosions were responsible for hundreds and even thousands of deaths each year in the United States, with comparable numbers being reported in other countries. In the first decade of the twentieth century, for example, the number of coal mine fatalities in the United States exceeded 2,000 per year, with the great majority of that number resulting from fire and explosion. The largest single disaster during that period was an explosion at the Number 6 and Number 8 mines at Monongah, West Virginia, in which 361 men and boys were killed.

The worst recorded fire-related coal mine disaster occurred in the Courrières coal mine in France in 1906, in which 1,100 workers were killed. Explosions and fires taking more than 100 lives were not uncommon throughout Western Europe, Canada, and Australia prior to the mid-twentieth century. Other terrible disasters occurred at the Oaks Colliery (coal mine) in Great Britain on 12 December 1866 (361 killed), the Senghenydd Universal Mine in Wales on 14 October 1913 (439 killed), and the Radbold mine in Germany's Ruhr region in 1908 (360 killed).

Sources and Causes

The most common source of mine explosions and fires is methane gas, also known as *fire damp*. Methane is produced underground by the decay of organic materials and can be found in any underground mine, metallic or nonmetallic. It is far more common in coal mines, however, where it almost always occurs along with coal. Methane normally ignites and explodes at a concentration of only 5.5 percent. Under special conditions, however, it may become explosive at concentrations of about 0.5 percent.

Coal dust is a second source of explosions and fires. Coal dust is produced inevitably by the normal processes of coal mining, such as cutting and moving blocks of coal. It poses a high risk because the very large surface area of coal dust particles dramatically increases the rate of combustion. Combinations of methane and coal dust can pose even more serious threats than either source by itself.

Fires and explosions can be set off by any number of different materials or simple acts involved in the mining process, including faulty electrical equipment; sparks produced from machinery; smoking by workers; the use of explosives to open up coal seams; friction from wire ropes; and accidental fires from flammable liquids, such as gasoline or kerosene.

Progress of a Mine Fire

Coal mine fires are particularly difficult to extinguish because of the abundance of fuel available. Once a fire begins, it has all the methane and coal it needs to continue burning for very long periods of time. When the fire breaks out in a surface mine, it can usually be attacked by traditional firefighting methods, such as the use of water or other fire-suppressing materials. But fires in underground mines present special problems. In most cases, there is no realistic way to get firefighting materials into the mine itself. Even flooding a mine is not effective, since the heat produced by the fire is sufficient to vaporize water pumped into the mine.

One famous example of the way mine fires may develop is the fire that broke out near Centralia, Pennsylvania, in 1962. The fire began when city officials approved the burning of trash in a landfill adjacent to an abandoned coal mine. Before anyone realized what had happened, the fire had worked its way downward into the abandoned coal seams. One of the characteristics of underground fires is that they can travel in any direction, following the fuel on which it feeds. Despite all efforts by local, state and federal authorities, the fire continued to rage for more than thirty-five years. The federal government and the Commonwealth of Pennsylvania together spent nearly $100 million fighting the fire and relocating residents of Centralia. Officials finally decided to allow the fire to burn itself out.

Fires such as the Centralia fire pose two major threats to people living in the area. First, carbon monoxide is always produced during any fire. In an underground fire, the quantities of oxygen available for combustion may be greatly reduced, and the amount of carbon monoxide produced, greatly increased. The gas can quickly accumulate and reach toxic levels in homes and buildings above and around the burning coal seam.

A second risk associated with mine fires is subsidence. Subsidence occurs when underground fires burn away huge areas of coal. The cavern thus produced is often not able to support the weight above it, and its ceiling collapses. Earth, houses, buildings, trees, roads, cars, and people may fall into the newly opened hole.

For many years, many residents of Centralia attempted to stay in their homes, dealing with these risks on a day-to-day basis. Every family purchased a carbon monoxide monitor to warn them of high concentrations of the deadly gas. Eventually the risks became too great, however, and in 1996 the Commonwealth of Pennsylvania forced the thirty-two families still living in Centralia to sell their houses and move to new locations.

Large-scale mining disasters continued to occur in the United States until well into the 1960s. In 1968, for example, an explosion and fire at the Consol Number 9 mine at Farmington, West Virginia, killed 78 men. That disaster was a major factor leading to the Federal Coal Mine and Health Safety Act of 1969 (FCMHSA), which provided for regular inspections of coal mines, with serious penalties for any mine lacking adequate safety systems. Since the adoption of the FCMHSA, the number of deaths and injuries from mine fires and explosions has dropped significantly. The total number of coal mine fatalities from *all* causes between 1993 and 1997, for example, was 208.

Further Reading: *Coal Mine Fire and Explosion Prevention*. Washington, DC: United States Department of the Interior, Bureau of Mines, Information Circular 8768, 1978; DeKok, David. *Unseen Danger*. Philadelphia: University of Pennsylvania Press, 1986; "Mining Disasters—An Exhibition," <http://www.msha.gov/DISASTER/DISASTER.HTM>; Mitchell, Donald W. *Mine Fires: Prevention, Detection, Fighting*, 3rd edition. Chicago: Intertec Publishing, 1996; Pomroy, William H., and Annie M. Carigiet. *Analysis of Underground Coal Mine Fire Incidents in the*

United States from 1978 through 1992. Washington, DC: U.S. Department of the Interior, Bureau of Mines, 1995.

MINER'S LAMP. *See* DAVY SAFETY LAMP

MISSOULA-INTERMOUNTAIN FIRE SCIENCES LABORATORY

The Missoula-Intermountain Fire Sciences Laboratory is a division of the Rocky Mountain Research Station of the U.S. Forest Service. The laboratory was established as the Northern Forest Fire Laboratory in 1960 largely as a consequence of a wildfire in Helena National Forest in Montana in which fifteen smoke jumpers were killed.

The purpose of the laboratory is to conduct basic research on the control and use of fire. Researchers working at the facility conduct experiments in the field as well as in the laboratory itself. Although their primary focus is on fire issues in the thirteen states of the Intermountain West and Alaska, they are also concerned with fire problems on a nationwide basis.

Work of the laboratory is currently divided among three units: fire behavior, fire effects, and fire chemistry. The fire behavior work unit conducts basic and applied research on the behavior of wildland fires that can be applied by land managers in prescribed burns, fire suppression, and prefire planning. The fire effects work unit deals with the effects of fire on forests, rangelands, wetlands, and wilderness ecosystems. One of its goals is the development of guidelines for the management of ecosystems following wildfire. The fire chemistry work unit focuses on the chemistry of combustion processes and the emissions produced by wildfires and controlled burns.

For further information, contact:
Missoula-Intermountain Fire Sciences Laboratory
5775 W. Highway 10
Missoula, MT 59802
(406) 329–4886

http://www.xmission.com/~rmrs/staffs/labs/missifsl.html.

MOTION PICTURES AND VIDEOS

Given its history, role, and symbolism in human civilization, it would be surprising if fire were not a common theme in motion pictures and television films. And such has been the case.

Historic Fires

Great fires have themselves provided the basis for some of the best known films. The 1936 MGM historical spectacular, *San Francisco,* told of the catastrophe that followed the earthquake that struck that city on 18 April 1906. An all-star cast that included Clark Gable, Spencer Tracy, and Jeanette MacDonald made *San Francisco* one of the classic fire films of all times.

Two years after the release of *San Francisco,* 20th Century Fox issued another fire disaster film, *In Old Chicago.* This picture told the story of the great Chicago fire of 1871, with Alice Brady in the role of Mrs. O'Leary. Her two sons were played by two matinee idols of the day, Tyrone Power and Don Ameche. Both sons had fallen in love with a character played by Alice Faye, providing a romantic subplot to the fire story. The climax of the film, featuring the fire itself, was regarded as a masterpiece of special effects for the time.

Even the story of the "Triangle Factory Fire of 1911" found its way to the screen, albeit the television screen. In 1979, Don Kirshner produced a made-for-television movie telling of the holocaust that swept the Asch Building and killed 145 workers in less than half an hour.

Fire Films

Conflagrations in tall buildings have provided a recurrent theme for motion pictures and television dramas. Perhaps the best known and the most critically praised was the 1974 film *The Towering Inferno.* Featuring such stars as Steve McQueen, Paul Newman, William Holden, Faye Dunaway, Fred Astaire,

A scene from the movie *San Francisco* showing the city after the earthquake and fire of 1906. © *Archive Photos.*

Richard Chamberlain, and Dabney Coleman, *Inferno* became one of the box office hits of 1974. The film tells the story of a fire that breaks out in a 138-story skyscraper in San Francisco. The drama is based on a fundamental reality faced by modern firefighters, namely the problems encountered in dealing with fires in the upper floors of high buildings. In *Inferno*, the city fire department's equipment is unable to reach the upper floors of the new building. The challenge of rescuing people in these floors provides much of the drama for which the film was famous.

Two made-for-television films provide a reprise of the *Inferno* story. One was, *Fire! Trapped on the 37th Floor,* aired in 1991. The film was of particular interest because it was based on an actual disaster that occurred in a Los Angeles skyscraper in 1988. The theme of the movie is not primarily the fire itself, but the efforts of firefighters to save trapped

office workers on floors far beyond the reach of its mechanical equipment. The film starred Lisa Hartman, Peter Scolari, and Lee Majors as the Los Angeles Fire Department supervisor in charge of fighting the fire.

Terror on the 40th Floor was a second made-for-television film released in the same year as *The Towering Inferno*. The focus of the film is not as much on the fire itself as on the musings of seven men and women trapped by the blaze and convinced that they are about to die.

The drama of oil well fires has provided the inspiration for a number of films. One of the earliest of these was *Hellfighters*, released by Universal Studios in 1968. *Hellfighters* tells the story of a team of troubleshooters who travel around the world trying to control oil fires. It highlights the personal and romantic problems facing members of the team who are called to leave home at a moment's notice to pursue one of the most dangerous occupations in the world. The film featured John Wayne, Vera Miles, Jim Hutton, and Katharine Ross. The 1986 film, *Oceans of Fire,* told a complex story of five ex-convicts working on an oil rig off the coast of South America. The rig explodes and catches fire, and the five men are faced with fighting the fire by themselves.

Two foreign films also focused on oil fires in 1978. They were, *Oil,* an Italian film, and, *City on Fire,* produced in Canada. *Oil* told of the heroic efforts of firefighters to subdue a huge oil fire in the Sahara Desert, while *City* relates the tale of a medical team to assist victims of an out-of-control oil fire, ignoring the risk to their own lives. Shelley Winters and Barry Newman played a nurse and doctor, respectively, in the film.

Some of the earliest films dealing with fire combined two or more tried and true elements of a successful story. The 1937 Warner Brothers picture *She Loved a Fireman,* is an example. The story relates only tangentially to fire in that the male lead is a firefighter. He falls in love with the sister of his brigade captain, much to the captain's distress. The story reaches a happy end when the fireman saves the captain in a fire and wins the love of the sister.

Forest fires have also been the subject of motion pictures. An early (1930) 20th Century Fox production, *Under Suspicion,* told of a woman unjustly accused of a crime and an officer in the Royal Canadian Mounted Police trapped in a raging forest fire in Jasper National Park. The story ends happily when both escape and catch the real criminal in the bargain.

Further Reading: Bleiler, David, ed. *TLA Film and Video Guide 2000–2001: The Discerning Film Lover's Guide.* New York: Griffin Trade Paperback, 1999; Craddock, Jim, ed. *Videohound's Golden Movie Retriever 2001.* Detroit: Visible Ink Press, 2000; Matlin, Leonard. *Leonard Matlin's Movie and Video Guide 2001.* New York: Plume, 2000.

MUSICAL ALLUSIONS TO FIRE

Fire and flames are mentioned in the titles of many musical selections, ranging from clas-

sical to popular music. In some instances, the musical selection is associated with some fire-related event, as is the case with George Frideric Handel's, *Music for the Royal Fireworks.* Handel (1685–1759) composed this selection to accompany a fireworks display conducted in London's Green Park on 27 April 1749 to celebrate the signing of the Treaty of Aix-la-Chapelle.

In some operas and ballets, fire is an integral part of the story being presented. In the finale of Richard Wagner's *Götterdammerung,* for example, Brünnhilde throws herself onto the funeral pyre containing the body of Siegried, her lover. Wagner (1813–1883) wrote *Farewell and Magic Fire Music* to accompany this dramatic scene.

A 1901 opera by Richard Strauss (1864–1949) is based on an intriguing Flemish legend, *The Extinguished Fires of Audenarde.* In this story, a hermit with magical powers tries to kiss a young girl. When she rejects him, he creates a "fire famine," that results in the extinguishing of all fires in the village. The hermit restores fire to the village only when the girl repents and gives him a kiss.

In many cases, musical selections have been written about actual fires or fire ceremonies. *The Beltane Fire,* by Peter Maxwell Davies (1934–), for example, was inspired by a very old Celtic tradition in which fires were built on or about the first of May each year. Maxwell completed this piece in 1995 and referred to it as a "choreographic poem for orchestra." It consists of orchestral selections, along with dances such as *First General Dance, The Greulik's Dance*, and *Sowing Dance.* The final scene of the piece is itself called *The Beltane Fire.*

In the majority of cases, the connection between a particular piece of music and fire is not known, is not entirely clear, or has only been surmised by critics. Symphony #59, by Franz Joseph Haydn (1732–1809), for example, has long been known as *The Fire Symphony,* although no one knows how it got this name or who chose it. Some critics point out that the symphony was written in 1769, shortly after the completion of his opera, *Die Feuerbrunst (The Conflagration* or *The*

Burned House). Critics think that both compositions may have been inspired by a terrible fire that swept through the town of Eisenstadt, where he was employed. The fire destroyed many of Haydn's original manuscripts.

For some composers, fire has had a symbolic meaning that inspired their work. One of the most famous works of Russian composer Alexander Scriabin (1872–1915), for example, is his symphonic poem, *Poem of Fire* (*Prometheus*). Throughout his life, Scriabin was, according to biographer Jason Martineau, consumed by "the cosmic, symbolic, and transcendent." *Poem of Fire* was only one of many works in which he tried to express his search in musical terms. Martineau points out that Scriabin actually saw himself as a Prometheus character because of the revolutionary approaches he used in his own composing.

Another composer who has incorporated philosophical interpretations of fire into his music is the Belgian composer Nicholas Lens (1959–). A press release from Sony Classics has described Lens' *Flamma Flamma: The Fire Requiem* as a "wildly original work, boldly synthesizing the Western spiritual concept of a requiem mass for the dead with death rituals and ceremonies from non-Western cultures." Lens uses fire as the unifying theme in this conception, fire as a tool of life, a metaphor for passion, and an agent for transformation in nature.

Another adaptation of ceremonial fire to music is "The Ritual Fire Dance," by Manuel de Falla (1876–1946). This selection is part of a one-act ballet, *El Amor Brujo* (*Love, the Sorcerer*), written in 1915. The piece is taken from an actual dance done at midnight by Andalusian gypsies to drive off evil spirits.

For some musical compositions, the connection with fire intended by the composer is not clear. Russian composer Igor Stravinsky (1882–1971), for example, gave no indication as to why he named his 1908 composition *Feu d'artifice* (*Fireworks*). Richard Kassel (*Baker's Dictionary of Music*, 1997) surmises that the title reflects the na-

ture of the music itself, which he describes as "sparks of rhythmic fire."

Stravinsky's *Feu d'artifice* has had its influence on modern composers. Scottish composer Oliver Knussen (1952–) has said that the work inspired his own *Flourish with Fireworks*, written for Michael Tilson Thomas and the London Symphony Orchestra in 1988.

One composer who has adapted the ages-old theme of earth, air, water, and fire is Danish composer Willy Stolarczyk (1945–). On 2 May 1996, his monumental *Symphony for 96 Pianos: Earth, Air, Fire, and Water* was performed at Koldinghus castle in Denmark. The piece describes the rise of a storm (thunder in section one; lightning in section two) that results in a "dance of fire" involving dozens of pianos, four out-of-tune pianos, a player piano, a toy trumpet, and two humans dressed in full suits of armor trying to dance the tango.

Another Danish composer who has taken fire themes as the basis of his work is Niels Rosing-Schow (1954–). Rosing-Schow's first major success was the 1989 opera, *Brand* (*Fire*), which he later adapted as *Echoes of Fire*, a concert piece for seven instruments.

On occasion, the connection between "fire" or "flame" in a title and the composition itself is not clear even to critics. *Death and Fire: A Dialogue with Paul Klee*, by Chinese composer Tan Dun (1957–) is an example. Dun wrote the selection after viewing a collection of Klee's abstract paintings at the Museum of Modern Art in New York. The composer has pointed out that his piece should not be viewed as programmatic music; that is, music specifically connected with the real world, intended to sound like something or designed to make the listener think of some specific event or object. Instead, Dun uses his music to try to understand the thought processes that inspired Klee to produce the paintings.

Fire has been a common theme in popular music also. In the vast majority of cases, song writers have used the idea of fire and flames to describe feelings of love and passion. An early example of this device was a song made popular by the Horace Heidt Orchestra, *I Don't Want to Set the World on Fire* ("I Just Want to Start a Flame in Your Heart"), written by Eddie Seiler, Sol Marcus, Bennie Benjamin, and Ed Durham in 1940.

A decade later, singer Georgia Gibbs made popular another fire-related song by A. G. Vilodo (translated by Lester Allen and Robert Hill), *Kiss of Fire* (1952). In this song, the singer tells her lover that "I touch your lips and all at once the sparks go flying. . . . I know I must surrender to your kiss of fire."

Fire and ardor remained a theme of popular songs through the rest of the century. In 1967, one of the most popular hits by Jim Morrison and The Doors was *Light My Fire*, in which the main theme was "Come on baby, light my fire . . . try to set the night on fire." In 1977, *You Set My Heart on Fire*, written by Jack Robinson and James Bolden, included the lines, "Oh I need somebody to burn me in the flame of love. . . . Oh pretty baby, I'm burning, burning at your flame of love."

In many instances, the fire and passion theme is not very highly developed in popular songs. Occasionally, however, a composer will explore in somewhat more detail the analogy between the characteristics of a fire and human love. In 1982, for example, the group REO Speedwagon performed *Keep the Fire Burnin'*, in which they ask the object of the song text to "Keep the fire burnin', Let it keep us warm. . . . Let us not stop learnin', We can help one another be strong."

In 1977, Bob Seger also explored a somewhat more coarse aspect of the passion and fire theme in his song *The Fire Down Below*. He talks about streetwalkers "old Rosie" and "hot Nancy" and their male clients who include both "the rich man in his big long limousine," "the poor man," "the banker," "the lawyer," and "the cop," all watching for the women and all with one thing in common: "They got the fire down below."

In only rare cases do popular song writers use fire imagery to portray some emotion or idea other than love and passion. In one of the Marshall Tucker Band's greatest hits, for example, composer George McCorkle talks about the fire on the mountain that fascinates and calls immigrants to the American West. He tells that "there's fire on the mountain, lightning in the air, gold in them hills, and it's waiting for me there." The song is fundamentally one of sadness and disappointment, however, as the composer concludes the song with a confession of failure in that, "Now my widow, she weeps by my grave . . . shot down in cold blood by a gun that carried fame, all for a useless and no good worthless claim."

A few modern songwriters have used realistic images of fires in their works. The Bloodhound Gang's *Fire Water Burn* (1996) is a diatribe against society that focuses on a burning building. "The roof . . . is on fire," the song begins. But, the group concludes, "We don't need no water. Let the m _ _ _ _ r burn. Burn, m _ _ _ _ r, burn." *See also* "THE FIREBIRD."

Further Reading: Tyrrell, John, ed. *The New Grove Dictionary of Music and Musicians*, 2nd edition. New York: Grove's Dictionaries, 2000.

MYTHOLOGICAL ALLUSIONS TO FIRE. *See* BONFIRE; GODS AND GODDESSES OF FIRE; HEPHAESTUS; LIGHTNING; ORIGINS OF FIRE; PHOENIX; PROMETHEUS; RELIGIOUS ALLUSIONS TO FIRE; VESTA

N

NAPALM

Napalm is a substance used in chemical weapons developed early in World War II as a substitute for an early type of incendiary weapon made of gasoline and natural rubber. A search for the new material became necessary when early Japanese conquests during the war eliminated nearly all sources of natural rubber for the Allied nations.

Invention

Napalm was invented for use in flame throwers, which had originally used ordinary gasoline as a fuel. But pure gasoline cannot be controlled very well and is dangerous to those who use flame throwers. During World War II, members of the U.S. Army Chemical Corps developed a new flame-thrower fuel that consisted of gasoline mixed with latex (a form of natural rubber). This fuel was an improvement on traditional gasoline fuels because it was easier to control and adhered to materials on which it was sprayed. When rubber plantations in East Asia from which latex was obtained were lost in the early days of World War II, the U.S. military assigned the task of finding a substitute for latex to a group of scientists working at Harvard University, the Arthur D. Little Company, and the Nuodex Products Company.

The product developed by this team consisted of the product of aluminum metal combined with three naturally occurring acids: oleic, palmitic, and naphthenic acids. Oleic acid is found in almost all natural fats and palmitic acid is obtained from coconut oil. Naphthenic acid is related to naphthalene, a product of the purification of petroleum. The mixture's name came from a combination of two of these ingredients: naphthenic acid and palmitic acid.

Military authorities were excited about the invention of napalm. The new material burns much hotter and more slowly than does gasoline. It also has a tendency to stick to objects on which it is sprayed. As a result, it is a far more destructive material than is gasoline and many other incendiary products. The material could also be made quite inexpensively and stored safely when proper precautions were observed. It rapidly became the fuel of choice for U.S. flame throwers.

Other Applications

Military strategists soon developed other applications for this "wonder weapon." In early 1945, they developed a bomb, called the M-69, with the capacity to carry large amounts of napalm. The bomb was used extensively in the late weeks of World War II in bombing raids against Japan. It was particularly effective because so many Japanese homes and buildings were made of paper, wood, or other combustible materials. Fires started by napalm bombs frequently spread quickly across large sections of a city, destroying structures, killing and injuring people, and devastating urban infrastructures.

Napalm also saw extensive use during the Vietnam War. By this time, researchers had developed a new formulation for the material. Napalm-B, as the new product was called, is a mixture of 21 percent benzene, 33 percent gasoline, and 46 percent polystyrene. It was used primarily as a defoliant, a substance that removes the leaves from trees and other plants. The U.S. military dropped millions of gallons of the incendiary material somewhat indiscriminately on the Vietnamese countryside, destroying plant life, homes, and human life.

Napalm is now a widely available weapon employed in a variety of local conflicts. For example, both participants in the Arab-Israeli wars following World War II have used the material. It is still used as a fuel in flame throwers and tank cannons, as well as in aerial bombs, missile warheads, and hand grenades.

Demolition

At the conclusion of the Vietnam War, the United States military was confronted with the problem of destroying vast quantities of napalm for which there was no immediate use. Many localities refused to allow napalm to be stored or incinerated in their area, and many refused napalm-carrying trains to pass through their perimeters. Many people living in these localities recalled photographs or stories of the widespread destruction caused by napalm-initiated fires during the Vietnam War, and they were concerned that accidents during storage or transporting of the chemical might result in similar disasters for their own communities.

As late as 1998, the controversy over what to do with left-over napalm was still going on. In April of that year, for example, the Southwest Division of the Naval Facilities Engineering Command announced that a train carrying 12,000 gallons of napalm was being shipped to the Pollution Control Industries (PCI) of East Chicago, Indiana. The plan had been for PCI to recycle the left-over napalm.

However, residents in communities through which the train was to travel raised strong objections to having the napalm pass through their area. PCI's neighbors also exerted pressure on the company not to handle the material. Eventually, PCI decided not to honor the conditions of its contract with the military, and the train was returned to California. Its destination there was the China Lake Naval Weapons Center 120 miles northwest of Los Angeles. The napalm has been stored for more than two decades at the Center, and it appears that it will have to remain there somewhat longer until alternative, less controversial, plans for its destruction can be developed.

Further Reading: "Napalm." In David Wallechinsky and Irving Wallace. *The People's Almanac, #2.* New York: Bantam Books, 1978, p. 548; "Napalm." In Bridget Travers, ed. *World of Invention.* Detroit: Gale Research, 1994, p. 437; "Napalm Back in California without Protest," <http://www.cnn.com/US/9804/20/napalm.train/>, 12/22/98; "Napalm Removal & Disposal Team," <http://www. efdswest. navfac.navy.mil/dep/env/Pages/napinfo.htm>, 12/22/98; United Nations. Group of Consultant Experts on Napalm and Other Incendiary Weapons. *Napalm and Other Incendiary Weapons and All Aspects of Their Possible Use.* New York: United Nations, 1973.

Discarded Napalm containers. © *Ross Pictures/CORBIS.*

NATIONAL FIRE ACADEMY

The National Fire Academy was established in 1975 as a division of the U.S. Fire Administration. The purpose of the Academy is to provide specialized training and advanced management programs designed to prevent the loss of life and property from fire through cooperative programs with local communities. Courses and other types of training sessions are held at the Academy's Emmitsburg, Maryland campus and at local sites throughout the United States. The Academy estimates that more than 1.3 million individuals have received training in one or more of its programs over the past quarter century.

Fire education classes at the National Fire Academy. *Courtesy of U.S. Fire Administration.*

Training is offered to any person involved with fire prevention and control, emergency medical services, or fire-related management activities. There is no cost for any of the training sessions offered by the Academy, and many students also qualify for stipends that cover the cost of travel, housing, and meals. A schedule of Academy classes is available on its website.

For further information, contact:

National Fire Academy
16825 South Seton Avenue
Emmitsburg, MD 21727–8998
(800) 238–3358, x1035
(301) 447–1035
http://www.usfa.fema.gov/nfa/

NATIONAL FIRE PREVENTION WEEK

National Fire Prevention Week is a week-long program of education designed to inform people about the dangers of fire and the steps that can be taken to prevent fires. The program is currently sponsored by the National Fire Protection Association (NFPA).

National Fire Prevention Week began in an abbreviated form as National Fire Prevention Day in 1911. The Fire Marshalls Association of North America (FMANA) decided to commemorate the Great Chicago Fire of 1871 with an educational program about fire safety. Prior to that time, the Chicago fire had been memorialized with festivities in Chicago, celebrating the city's survival and rebirth. The first National Fire Prevention Day was held on 9 October 1911.

A decade later, in 1920, President Woodrow Wilson issued the first National Fire Prevention Day proclamation. Five years later, his successor, Calvin Coolidge, extended that recognition to the first National Fire Prevention Week to run from October 4 to 10. Since that time, the event has covered a week-long span from the Saturday to Sunday in October that includes October 9.

Each year, NFPA selects a theme for National Fire Prevention Week. They have been as follows:

1957–1958	Don't Give Fire a Place to Start
1959	Fire Prevention is Your Job . . . Too
1960–1961	Don't Give Fire a Place to Start
1962	Fire Prevention is Your Job . . . Too
1963	Don't Give Fire a Place to Start
1964	Fire Prevention is Your Job . . . Too
1965	Don't Give Fire a Place to Start
1966–1972	Fire Hurts
1973	Help Stop Fire
1974	Things That Burn
1975–1976	Learn Not to Burn
1977	Where There's Smoke, There Should Be a Smoke Alarm
1978	You Are Not Alone!
1979–1980	Partners in Fire Prevention

1981 EDITH (Exit Drills in the Home)

1982 Learn Not to Burn—Wherever You Are

1984 Join the Fire Prevention Team

1985 Fire Drills Save Lives at Home at School at Work

1986 Learn Not to Burn: It Really Works!

1986 Play It Safe . . . Plan Your Escape

1988 A Sound You Can Live With: Test Your Smoke Detector

1989 Big Fires Start Small: Keep Matches and Lighters in the Right Hands

1990 Keep your Place Firesafe: Hunt for Home Hazards

1991 Fire Won't Wait . . . Plan Your Escape!

1992 Test Your Detector-It's Sound Advice!

1993 Get Out, Stay Out: Your Fire Safe Response

1994 Test Your Detector for Life

1995 Watch What You Heat: Prevent Home Fires!

1996 Let's Hear It for Fire Safety! Test Your Detectors

1997 Know When to Go: React Fast to Fire!

1998–2000 Fire Drills: The Great Escape!

NFPA suggests a number of activities and provides a number of resources for carrying out the objectives of National Fire Prevention Week. Local fire departments and individual communities also develop activities to be carried out during the week. Some materials available from NFPA include a fire safety video; a fire safety brochure (in English and Spanish); a babysitters' fire safety tips brochure; an activity book entitled *Count Riskalot*; a *Fight or Flight* video for the hearing impaired; campaign stickers, T-shirts, and campaign brochures on the Fire Prevention Week topic of the year; and video, posters, and brochures on specific topics, such as fire drills and fire extinguishers.

Each community designs Fire Prevention Week activities appropriate to its own setting. The Chesterfield (Virginia) County Fire Department, for example, offers a 911 simulator to teach children the use of this emergency number, information on planning a home escape, puppet shows, safety demonstrations, a fashion show, prizes, and free video record-

ings of children. Fire trucks are also parked near local businesses to provide information on safety plans and Fire Prevention Week. Families who develop and practice a home escape plan become eligible to win a Great Escape weekend sponsored by a local establishment. *See also* NATIONAL FIRE PROTECTION ASSOCIATION.

NATIONAL FIRE PROTECTION ASSOCIATION

The National Fire Protection Association (NFPA) is an international, nonprofit organization founded in 1896 to protect people, property, and the environment from destructive fire. The organization was founded by a group of five representatives of fire insurance and fire sprinkler companies in Boston. The primary purpose of that meeting was to establish a single set of formal standards for the installation of sprinkler systems. Over the past century, the organization has grown in size and outlook. Today, NFPA consists of 68,000 individual members and more than 100 organizations in the United States, Canada, and seventy other countries. Its national headquarters are located in Quincy, Massachusetts, with regional offices in Florida, Delaware, Kentucky, California, and Washington, DC.

Since 1935, the primary publication of the organization has been the *Fire Protection Handbook*, the "bible" of fire protection information and standards for the industry. NFPA also publishes a regular journal, *NFPA Journal*, formerly *Fire Journal*. The organization also publishes a more technical publication, *Fire Technology*.

One of the association's major contributions is a set of codes dealing with almost every aspect of fire safety, published as *National Fire Codes®*. These codes are developed and reviewed by 205 Technical Committees consisting of more than 5,300 volunteer members. They serve as the basis for fire protection practices in most cities, towns, businesses, and other entities. *See also* FIRE CODE; *FIRE PROTECTION HANDBOOK*; NATIONAL FIRE PREVENTION WEEK.

For further information, contact:
National Fire Protection Association
1 Batterymarch Park
P.O. Box 9101
Quincy, MA 02269–9101
(800) 344–3555
http://www.nfpa.org

Further Reading: "The National Fire Protection Association," <http://www.nfpa.org/home.html>, 07/11/99.

NATIONAL INTERAGENCY FIRE CENTER

The National Interagency Fire Center (NIFC) is a cooperative endeavor involving seven governmental agencies concerned with the fighting of wildfires: the U.S. Fish and Wildlife Service (USFWS); the National Park Service (NPS); the U.S. Forest Service (USFS); the Bureau of Land Management (BLM); the Bureau of Indian Affairs (BIA); the Office of Aircraft Services (OAS); and the National Weather Service (NWS).

History

NIFC was established during the early 1960s at a critical moment in U.S. fire history when long-held doctrines about fire management were being reassessed. Some fire authorities had begun to accept the fact that prescribed burning might, in fact, be an essential and unavoidable tool in the management of fire. As this revolution in thinking was taking place, a number of agencies outside the USFS began to rethink their roles in fire management and fire control and began to look for ways in which they could participate in decision making on fire management policy.

The proximate factor leading to the establishment of NIFC was a suggestion by the U.S. Bureau of the Budget that BLM organize a fire center at Boise, Idaho for the supervision of fire issues on lands for which it was responsible in the West. At about the same time, the USFS was looking for a site at which to locate a new air center for forest-fire suppression. Both agencies recognized the desirability of sharing physical facilities, and the joint center was opened for business in 1968. By that time, the U.S. Weather Bureau had also joined the consortium.

Integrating the efforts of government agencies each with its own fiefdom to protect could never have been expected to be easy. And such controversies did arise. For example, the story is told that the warehouse in which fire equipment was stored at the

The National Interagency Fire Center, Boise, Idaho. *Courtesy of Bureau of Land Management at the National Interagency Fire Center.*

center was divided down the middle by a yellow line. On one side of the line was BLM's equipment; on the other side, the USFS's equipment.

The drive for cooperation, however, eventually proved stronger than the demands of individual agencies. The BIA joined the consortium in 1967, and the USFWS completed the team three years later. In 1973, the Department of the Interior established the Office of Aircraft Services, which was located at the center, and a year later, the NPS became a partner.

Today, NIFC appears to be a highly successful organization in which all partners participate in a sharing of their firefighting resources, resulting in a more economical and more efficient firefighting operation.

Functions

NIFC has a number of important firefighting functions. Its National Interagency Coordination Center was established in 1975 to provide logistical support and intelligence for dealing with wilderness fires anywhere in the nation. Over time, their activities have expanded to include responses to other emergencies, such as floods, hurricanes, and earthquakes.

NIFC is also responsible for the development of new firefighting equipment. It prepares specifications for new fire engines and "slip-on units," requisitions the necessary materials, field-tests and approves new units, and trains operators for the equipment. The center is currently responsible for about 400 fire engines used in firefighting in the West.

The center also operates the Remote Fire Weather Support Group, a system that provides up-to-date weather information that covers all of the United States, but that focuses primarily on the Western states. Data from 1,150 Remote Automated Weather Stations is sent via satellite to Boise, where it is analyzed by computer. Periodic fire weather reports and maps are prepared and published in hard copy and on the Internet for fire management agencies everywhere in the country.

NIFC also has developed an important fire education and prevention component. It pro-vides a number of resources in hard copy and on the Internet. Examples include posters dealing with "Fire's Role in Nature" for various parts of the nation, a regular publication, *Wildfire News and Notes*, "NWCG [Pacific Northwest Wildfire Coordinating Group] Wildfire Prevention Guides," and an Internet site, "FireWise," with information designed to help homeowners who live in fire-prone areas.

For further information, contact:
National Interagency Fire Center
3833 S. Development Avenue
Boise, ID 83705
(208) 387–5457
http://www.nifc.gov/

NATIONAL PARK SERVICE

The National Park Service (NPS) is the federal agency responsible for administering 376 national parks, national historic parks, national monuments, national preserves, national military parks, national battlefields, national historic sites, and other locations set aside for conservation and preservation. The first national park to be created was Yellowstone National Park, established in 1872. Two decades later, Yosemite and Sequoia National Parks were added to the system (1890). Over time, the system was expanded by the U.S. Congress to include sites of every description in which the historic and natural assets of the nation were to be protected, preserved, and made available for public use and enjoyment. Today, NPS locations are found in every state except Delaware, as well as the District of Columbia, American Samoa, Guam, Puerto Rico, and the Virgin Islands. These sites cover a total of more than 83 million acres.

As with its sister agency, the U.S. Forest Service, NPS has long been faced with issues of fire administration on the lands for which it is responsible. Like other federal and state agencies, it passed through a sequence in its history that included obliviousness to fires on its lands to active suppression policies and practices about fire to a philosophy of pre-scribed burns and controlled natural fires.

An important turning point in modern NPS fire policy was the Leopold Report of 1963. In that report, A. Starker Leopold and his colleagues stressed the importance of viewing nature as a unified system in which each individual part was dependent on all other parts of the system. This view emphasized the role of fire in an ecosystem and strongly criticized previous NPS efforts to exclude fire from national parks and other natural areas under its control. The service created a Branch of Fire Management to reformulate its policies about fire management and to create a working document on which to base its fire management practices.

The story of the NPS's fire management practices over the past three decades is a complex and interesting one. In many instances, the service's policies and practices were not entirely clear, varied from park to park, and may or may not have been followed at all its locations. Overall, there was a general trend away from fire suppression, its policy for many decades, to a laissez-faire approach to fire, allowing natural fires to run their course in many cases.

This policy was finally tested in its most extreme form in the 1988 fires that swept through Yellowstone Park in June to September of that year. The combination of contrary policies of first fire suppression, then disregard for fires, combined with unusually hot and dry weather produced a series of conflagrations that caused the service to rethink once more its policies about the role of fire on lands for which it was responsible.

Today, fire management in the national park system is regulated by Chapter 4:14 of the National Park Service Management Policies. That document classifies fires as prescribed burns or natural fires. It also outlines the circumstances under the former that is to be used and the latter that is to be allowed to burn on its own or be brought under control with fire suppression techniques. *See also* FIRE MANAGEMENT; U.S. FOREST SERVICE; YELLOWSTONE FIRES OF 1988.

Further Reading: Carsten, Lein. *Olympic Battleground: The Power Politics of Timber Preservation*. Seattle: Mountaineers Books, 2000; "Homepage for the National Park Service,"

<http://www.nps.gov/>; "History of the NPS," <http://www.nps.gov/bibe/npshist.htm>; and "Fire Management Policies," <http://www.nps.gov/seki/fire/nps_mgt.htm>; Rettie, Dwight F., and Stewart L. Udall. *Our National Park System: Caring for America's Greatest Natural and Historic Treasures*. Champagne-Urbana: University of Illinois Press, 1995; Sellers, Richard West. *Preserving Nature in the National Parks: A History*. New Haven, CT: Yale University Press, 1997.

NAZI DEATH CAMPS

[This entry contains some graphic material that may not be suitable for all readers.] After achieving power in Germany during the 1930s, the National Socialist Party ("Nazis"), under Adolf Hitler, established a formal policy calling for the extermination of certain groups of "undesirables," a term that included Jews, gypsies, the physically and mentally handicapped, and homosexuals. In this campaign, that extended over more than a decade, the Nazis used fire as a powerful weapon in achieving their "final solution."

As an example, one of the early episodes in the Nazi campaign against Jews occurred on 9 November 1938. According to German officials, the incident was a spontaneous uprising by members of the Hitler Youth and members of the elite SS guard in reaction to the murder of a German Embassy official in Paris by a displaced Polish Jew. The mob violence that broke out throughout Germany and Austria was aimed primarily at synagogues, but Jewish homes and businesses were not spared. In many cities, Jewish neighborhoods were set afire and burned to the ground. So many windows were broken in the riots that the episode has come to be called *Kristallnacht*, or "Night of the Broken Glass."

Extermination Camps

By far the most memorable fire-related events associated with the Holocaust, however, were those that took place at the concentration and extermination camps created to eradicate Jews and other "undesirables." These camps were established as part of the "Reinhard project" beginning in about 1941. They were built in various locations throughout occupied Poland, including the towns of Auschwitz-Birkenau, Belzec, Chelmno, Majdanek, Sobibor, and Treblinka.

These camps were designed as holding and extermination sites. Men, women, and children were brought to the camps and either set to work at forced labor or killed. One of the practical problems with which camp administrators had to deal was the large number of dead bodies that accumulated each day. At first, the camps simply buried dead prisoners in large mass graves. Before long, however, this method of disposal became totally inadequate for the thousands of corpses produced each day.

At that point, disposal by cremation became a common practice at all camps. The usual procedure was to build large funeral pyres on which bodies were stacked. Cremation was not only a more efficient method of disposal, but it was also more effective in hiding evidence of the acts taking place at the camps.

Crematoria at Auschwitz

The use of cremation as a means of exterminating Jews and others reached its highest level of development at the Auschwitz camp. In 1941, Heinrich Himmler selected the German firm of Topf and Sohne of Erfurt to draw plans and oversee construction of a large crematorium at Auschwitz. The crematorium was eventually built at the nearby town of Birkenau, rather than at Auschwitz itself.

The project consisted of five separate buildings, each one a separate crematorium (or *kremum*) in and of itself. The design of each building was somewhat different, although the general operation of the five units was quite similar. Prisoners were first led into a large changing room capable of holding up to 1,000 people. The victims were told that they were about to take a shower and to note carefully the location of their belongings. When they were led into the "showers," they were instead exposed to a deadly gas, usually hydrogen cyanide. After a batch of prisoners had been executed, their bodies were transferred to the crematorium, where they were burned.

According to original plans for the crematoria, Building I was designed to cremate 340 bodies per day; Buildings II and III, 1,440 bodies per day; and Buildings IV and V, 768 bodies per day. Overall, the five buildings were expected to incinerate a total of 4,765 bodies each day.

These numbers varied considerably, however, as operators of the plant constantly looked for new and more efficient ways to carry out their work. The cremation ovens were originally designed to operate on coke gas, but operators soon found that the burning of bodies produced significant amounts of heat itself. Systems were developed by which heat produced during cremation was recycled, reducing the amount and cost of coke gas needed to keep the operation going. For example, victims might be fed into the oven in pairs, one of whom was relatively well-fed combined with one who was nearly dead from starvation.

Each crematorium was about 100 meters long by 50 meters wide. The five buildings contained a total of forty-six separate retorts (furnaces) that held three to five persons each. The cremation process took about half an hour, after which a retort would be shut down, cooled, cleaned, and then reheated for the next groups of bodies.

The crematoria at Auschwitz-Birkenau and other camps did not meet the expectations of their designers. They received such intense use that they began to experience damage soon after they were put into operation. Kurt Prufer, senior engineer at Topf and Sohne, testified in 1946 that the brick lining of the ovens deteriorated rapidly and had to be replaced over and over again. The problem was, he said, that the "the strain on the furnaces was colossal."

Late in the war, the crematoria were no longer able to keep up with the demand placed on them by camp supervisors. Most camps switched to the burning of corpses in pits. Dozens of prisoners were killed and then dumped into mass graves. They were then sprayed with alcohol, gasoline, liquid human fat, and other combustible materials and set afire. After cremation was complete, the pits and their victims were covered with dirt.

By late 1944, the German government recognized the possibility of defeat in the war. It began an aggressive campaign of destroying crematoria. In December, for example, women prisoners at Auschwitz-Birkenau were assigned to the destruction of Crematorium III. They were also employed to cover over cremation pits, plant new grass and trees, and restore the area to as natural a condition as possible. In spite of these events a very large amount of information about the ovens and their operation was eventually gleaned from aerial photographs, archaeological expeditions, and personal testimony of people involved with the operation of the crematoria.

Further Reading: Arad, Yitzhak. *Belzec, Sobibor, Treblinka: The Operation Reinhard Death Camps.* Bloomington, IN: Indiana University Press, 1999; "Auschwitz: Crematoria," <wysiwyg://31/http://www.nizkor.org/faqs/auschwitz/auschwitz-faq-07.html>; Czech, Danuta, and Jadwiga Bezwinska, eds. *Amidst a Nightmare of Crime: Manuscripts of Prisoners in Crematorium Squads Found at Auschwitz.* New York: Howard Fertig, 1992; Czech, Danuta, and Walter Laqueur. *Auschwitz Chronicle: 1939–1945.* New York: Henry Holt, 1997; "The Extermination Factory - Auschwitz," <http://www.virtual.co.il/education/education/holocaust/camps/ausch/ausch1.htm>; Gutman, Israel, and Michael Berenbaum, eds. *Anatomy of the Auschwitz Death Camp.* Bloomington, IN: Indiana University Press, 1998; Swiebocka, Teresa, Jonathan Webber, and Connie Wilsack, eds. *Auschwitz: A History in Photographs.* Bloom—ington, IN: Indiana University Press, 1993.

NEED-FIRE

Need-fire is a form of fire celebration once commonly used by people to combat disease, plague, and other disasters. The term was often used to distinguish such fires from other celebratory fires that were set on regular schedules, such as new-fires and those that occurred during annual holidays. In some places, need-fires were called by other names, such as *living fires* (among Slavonic tribes) or *wildfires* (among Germanic tribes). Need-fires are still practiced today in some parts of the world, especially Scotland, Ireland, and Wales, although the distinction between need-fires and new-fires has largely been lost.

The origin of need-fires is not known, although authorities believe the practice extends back at least two or three thousand years. It appears to have originated in Europe and been associated primarily, if not exclusively, with that part of the world. The clearest historical records of need-fires appear to be those associated with the Celts and early Christian leaders of Ireland. There is a legend, for example, that St. Patrick set a need-fire on top of Slane Hill near Dublin on 1 May 433. In so doing, he usurped a privilege that had long belonged to the pagan kings of the Celts and, thereby, replaced those kings and their religion with that of Christianity.

Perhaps the most common occasion for a need-fire was the outbreak of disease among either humans or cattle. In such cases, a need-fire was built, and the affected people or animals were forced to walk through the smoke. The ceremony was based on the belief that the smoke from the need-fire would cure people or animals of the disease that afflicted them.

Precise rules were followed in building a need-fire. In most instances, the fire had to be started by rubbing two pieces of wood together. Stone and flint were generally not allowed as a means of igniting the fire. Complicated mechanisms were often constructed with which to ignite the need-fire. In the Slavic countries, for example, the mechanism often consisted of two poles about eighteen inches high with a roller attached between them. When the poles were rotated, the roller rubbed against a piece of softwood coated with tar. Friction between the woods set the tar on fire.

In such cases, special rules had to be followed concerning who would be allowed or required to start the fire. When more than one person was involved in starting a fire, for example, both individuals might have to have the same Christian name. In the western islands of Scotland, eighty-one married men were involved in igniting the need-fire by rubbing two heavy planks against each other. In Serbia, the need-fire was sometimes ignited by two children, a boy and a girl, between eleven and fourteen years of age working completely naked in a dark room.

Need-fires have survived until the present day in some parts of Scotland and Ireland. In most cases, they have changed their character somewhat and become part of an annual event known as Beltain, which takes place on the first day of May. The tradition of need-fires was recently remembered in a musical produced by David and Ed Mirvish entitled *Needfire: Passion of the Heart*. The musical was first presented in Toronto in early 2000. It consisted of songs and dance celebrating Canada's Celtic heritage. *See also* GODS AND GODDESSES; RELIGIOUS ALLUSIONS TO FIRE.

Further Reading: "Beltane Customs," <http://www.tartans.com/beltaine.html>, 08/05/99; Frazer, J.G. *Balder the Beautiful: The Fire-Festivals of Europe and the Doctrine of the External Soul*, 2 vols. London: Macmillan and Company, 1913, Volume 1, pp. 269–300; Frazer, Sir James George. *The Golden Bough: A Study in Magic and Religion*. New York: The Macmillan Company, 1958, Chapter 62; Pyne, Stephen J. *Vestal Fire: An Environmental History, Told through Fire, of Europe and Europe's Encounter with the World*. Seattle: University of Washington Press, 1997, pp. 68–78.

NERO. *See* BURNING OF ROME

NEW-FIRE

New-fire is a ritual practiced in many cultures in which all flames from the preceding year are extinguished and a new-fire is lit. The history of using new-fires goes very far back into human history, and the reasons for conducting this practice are not always clear. It appears that new-fires are sometimes designed to encourage the Earth to begin renewing itself. These fires are often lit at the winter solstice, when the Earth receives the least amount of sunlight, or at the vernal equinox, the beginning of spring.

New-fire also appears to be an opportunity for humans to renew and refresh themselves. They are designed to encourage people to leave the previous year behind them, to ask forgiveness for sins they have committed or offenses made to gods or spirits, and to promise to be better individuals in the year to come.

Anthropologists have found examples of the new-fire tradition in cultures throughout the world. Among the Incas, for example, the new-fire was lit at the summer solstice by using the sun's rays focused by means of a polished, concave mirror reflected on sheep's wool. The Zuni Indians of New Mexico kindled a new-fire twice a year, once at the winter solstice and once at the summer solstice. The fire was kindled by priests who rubbed two sacred sticks together.

The kindling of new-fire was an important ceremony that was carried out only by certain individuals following very specific instructions. For example, Zuni priests are not allowed to blow on smoldering embers produced by their fire drill since their breath would profane the gods at work in producing the fire.

New-fire was a custom among the Eskimo also. Once each year, on New Year's Day, a pair of Eskimo men passed through the village extinguishing all flames. A new-fire is ignited using rubbing sticks, and residents of the village relight their own lamps from the common ritual fire.

New-fire was celebrated in parts of China at the beginning of April when officials passed through the countryside carrying wooden clappers. They commanded that all fires be extinguished, announcing the beginning of a period known as *Han shih tsieh*, or "eating cold food." At the end of that period, a new-fire was ignited by focusing sunlight on dry moss.

New-fires were and are an important part of Christian worship in many regions. The most common time for these festivals has been at Easter time. All the members of a congregation are instructed to put out their own fire, and a new-fire is lit on the Paschal, or Easter, candle. Some authorities believe that the custom of having candles on our Christmas trees and lighting a yule log are modern-day remnants of this very old tradition. *See also* RELIGIOUS ALLUSIONS TO FIRE.

Further Reading: Campbell, Joseph. *The Masks of God: Primitive Mythology*. New York: The Viking Press, 1959, passim; Frazer, J.G. *Balder the Beautiful: The Fire-Festivals of Europe and the Doctrine of the External Soul*, 2 vols. London:

Macmillan and Company, 1913, Volume 1, pp. 120–146; Tylor, Sir Edward Burnett. *Religion in Primitive Culture*. New York: Harper Torchbooks, 1958, passim.

NUCLEAR WINTER

Nuclear winter is a term invented in 1982 by Paul Crutzen and John Birks to describe certain environmental effects that might occur as the result of a military action in which nuclear weapons are used on a large scale. The potential for such effects was explored on a theoretical basis by a group of researchers during the 1980s.

Mechanism

This analysis was based on some well-known environmental effects of nuclear weapons explosions of given magnitude. In general, such explosions result in the very rapid incineration of all combustible materials within a given range. That range depends on the size of the weapon. For example, a one megaton fusion bomb will totally incinerate all combustible materials within a circle of a fifty kilometer radius around the impact site.

One of the products formed during this type of explosion is soot. Soot consists of very fine particles of unburned carbon less than one micron (one micrometer) in size. Clouds that contain large quantities of soot have properties somewhat different from those of rain and ice clouds. In particular, they tend to reflect a larger fraction of sunlight back into space than do water clouds. At the same time, they tend to be transparent to thermal radiation given off by the Earth's surface.

The combination of these two factors produces an effect just the opposite of the natural "greenhouse effect" that occurs in the Earth's atmosphere. The greenhouse effect is a phenomenon in which solar energy passes through water clouds to reach the Earth's surface. Heat reflected off the Earth's surface is then captured by water clouds and prevented from escaping from the Earth. Overall, these two factors work together to warm the Earth's atmosphere.

Clouds containing large amounts of soot, however, would have just the opposite effect.

Heat would escape from the atmosphere more rapidly than it could be trapped. The average annual temperature of the Earth's atmosphere would begin to decrease, accounting for the term used to describe this phenomenon, *nuclear winters*.

Estimates of Cooling Effects

Researchers who have studied this problem estimate that a large-scale nuclear conflict could release as much as 10 million to 100 million metric tons of soot worldwide. Clouds holding this much soot might block out the sun for weeks or months, causing a drop in the temperature of the atmosphere by as much as 25°C (45°F). The scientists suggest that the cooling effect could actually last much longer, perhaps a period of years.

With the loss of sunlight and heat, vegetation on the Earth might begin to die off. If that were to happen, animal life also would be threatened. Mass starvation might spread throughout the planet. One researcher set the number of human casualties of hunger and disease in the first year of a nuclear winter at 3 billion. He then warned that "the future of civilization beyond that point would seem grim, with little infrastructure remaining to support a long-term recovery" (Turco, p. 461).

Further Reading: Crutzen, P.G., and J.G. Goldammer, eds. *Fire in the Environment: Its Ecological, Climatic, and Atmospheric Chemical Importance*. New York: John Wiley and Sons, 1993; Crutzen, P.J., and J. Birks, "Twilight at Noon: The Atmosphere after a Nuclear War." *Ambio* 11 (1982): 114; Harwell, M., and T. Hutchinson. *Environmental Consequences of Nuclear War, Volume II. Ecological and Agricultural Effects*. New York: John Wiley and Sons, 1985; Pittock, A. et al. *Environmental Consequences of Nuclear War, Volume I. Physical and Atmospheric Effects*. New York: Wiley, 1985; Sagan, Carl, and Richard P. Turco. *A Path Where No Man Thought: Nuclear Winter and the End of the Arms Race*. London: Century Press, 1990; Turco, Richard O. *Earth Under Siege: From Air Pollution to Global Change*. Oxford: Oxford University Press, 1997; Turco, R.O. et al. "The Climatic Effects of Nuclear War." *Scientific American*, 251 (1984): 33.

O

OAKLAND (CALIFORNIA) FIRE OF 1991. See BERKELEY HILLS (CALIFORNIA) FIRE OF 1991

OLD BELIEVERS

The Old Believers are a religious sect that broke off from the Russian Orthodox Church in the mid-seventeenth century. The reason for this decision, sometimes referred to as the Great Schism in Russian religious history, was the objection to a number of reforms instigated by the Orthodox metropolitan of Moscow, Nikon, in the early 1600s. Nikon's reforms were continued after his deposition, and the church continued to move away from traditional beliefs and organizational structure and to take on a more modern and more secular format.

One of the symbolic acts that Nikon abolished was the custom of plunging a candle into water to symbolize an important Biblical passage in Luke (3:16). In that passage, St. John preaches to his listeners that he will "baptize you with water." However, the Jesus who is to come "will baptize you with the Holy Spirit and with fire."

To Old Believers, the symbolism of baptism by fire was very significant. One dissident wrote a pamphlet in which he warned that Nikon's reforms were attempting to deny the faithful "the tongue of fire that had descended upon the apostles." A leader of the Old Believers, the Archpriest Avvakum, told his followers that fire was an essential part of their religion because of its ability to purge the body of sin and evil. He encouraged the faithful not to fear fire, but to "run and jump into the flames" happily since they would then be released from Earthly evils and transported to the bosom of God.

As it happened, both Avvakum and many of his followers were given the opportunity to take part in this experience. He was burned at the stake for heresy in 1682, and many of his followers voluntarily followed his example in following years. According to one authority, as many as 20,000 members of the sect committed suicide by burning themselves to death in the decade following Avvakum's death.

Pockets of Old Believers congregations survive in various parts of the world today. They tend to be found primarily in the region of ancient Rus (present-day Russia, Ukraine, and Byelorussia), in Latvia, and in Western Canada. Members of the sect believe that they are the last true Christians left on Earth. An interesting film on the Canadian Old Believers, *The Old Believers*, was produced in 1988 by Joe MacDonald, John Paskievich, and Ches Yetman, and directed by John Paskievich.

The Old Believers also play an important role in Modest Mussorgsky's opera *Khovanshchina*. Set in Russia in 1682, the opera tells the story of a plot by the Old Believers to overthrow Czar Peter and the government. The plot fails and the opera ends with a group of Old Believers locking themselves inside a building and setting fire to it. They sing of the joy and glory of meeting their death in the purifying flames of the fire.

Further Reading: Crumney, Robert O. *The Old Believers and the World of Antichrist: The Vyg Community and the Russian State, 1694–1855.* Madison, WI: University of Wisconsin Press, 1970; Moede, Gerald F., ed. *Ecumenical Exercise II: The Church of God, The Russian Old Ritualists, and The Church of the Nazarene.* Geneva: World Council of Churches, 1971; "Old Believer Home Page," <http://www.oldbelievers.org/>; Robson, Roy R. *Old Believers in Modern Russia.* De Kalb, IL: Northern Illinois University Press, 1996; "The Russian Church: An Overview," <http://www.sptimes.com/Treasures/TC.5.4.7.html>, 09/29/99; Scheffel, David. *In the Shadow of Antichrist: The Old Believers in Alberta.* Peterborough, Ontario: Broadview Press, 1991.

OLYMPIC TORCH

The Olympic torch is a flame that burns during each session of the Olympic Games, held once every four years in the summer and once every four years again in the winter. The torch was chosen by the International Olympic Committee to represent the purity embodied in the eternal youth of the Olympic philosophy. It is a symbol designed to encourage all competitors to work toward the lasting unity of humankind.

The Olympic games originated in Greece in the year 776 B.C. The games were held every four years on the Peloponnesian peninsula and were attended by athletes from all over Greece. The games were abolished in 394 A.D. with the rise of Christianity, and were not restored until 1896.

The tradition of lighting an Olympic torch goes back to the original games. By tradition, a flame was carried by runners to announce the end of a war. The use of a flame was eventually incorporated in the games themselves as a symbol of the rebirth of the spirit of dead heroes. The torch was allowed to burn throughout the games, and was then extinguished.

The modern tradition of using an Olympic flame was reinstituted in 1928 for the Amsterdam Summer Olympics. The flame was ignited at the location of the games. The torch tradition then died out until the 1936 games held in Berlin. For those games, the flame was lighted on Mount Olympus using a lens to focus the sun's rays on a combustible material. The flame was then carried by a series of runners from Mount Olympus to Berlin.

That tradition has been maintained in nearly every Olympic Games since. The honor of carrying the torch the last distance into the Olympic stadium is accorded to some athlete of special significance from the host country.

Further Reading: Findling, John E., and Kimberly D. Pelle, eds. *Historical Dictionary of the Modern Olympic Movement.* Westport, CT: Greenwood Publishing Company, 1996; Kristy, Davida. *Coubertin's Olympics: How the Games Began.* Minneapolis: Lerner Publications Company, 1995; "The Olympic Flame," <http://www.greece.org/olympics/flame/intro.html>, 07/21/99; Swaddling, Judith. *The Ancient Olympic Games.* Austin: University of Texas Press, 1984.

The tradition of the Olympic torch goes back to the original games in Greece. Shown here is Cathy Freeman at the opening ceremony of the 2000 Sydney Olympics. © 2000, ALLSPORTUSA/Jed Jacobson. All Rights Reserved.

ORDEAL BY FIRE

Ordeal by fire is one form of a practice used virtually throughout human history in many cultures to determine whether or not a person is guilty of a given crime. The principle behind ordeal is that the final judgment about a person's guilt will be made not by human judges and juries, but by supernatural forces, such as spirits or gods.

The ordeal could take many forms. For example, during the Middle Ages, dueling was one form of ordeal. When two knights met on the field of combat, it was assumed that the winner of that contest would not necessarily be the better fighter, but the one on whom God (or the gods) had given his or their blessings.

The ordeal by fire goes back at least to early Indian cultures. We know most about the way it was used by the Christian church during the Middle Ages. The ceremony typically took place at or near a church and began as an iron bar was being heated during the mass. At the end of the mass, the hot iron bar was sprinkled with holy water and placed in the hands of the accused. That person was then required to walk a distance of nine feet while holding the bar. Alternatively, the accused might be required to walk on nine red-hot plowshares.

The actual test involved an examination of the accused's hands or feet three days after the ordeal. If they had healed without the development of infection or festering, the accused was judged innocent. If blisters or infections had appeared, the accused was guilty. Ironically, records indicate that the results of an ordeal might conflict with valid information held by the accusers, the community, or the church, but such information was thought to be irrelevant in light of God's sign during the ordeal.

Ordeals are still practiced today by some cultures in some parts of the world. For example, anthropologists have reported on a tribe in modern-day Liberia that tests a person's guilt or innocence by rubbing a red-hot knife on the individual's leg. If the leg is burned, the individual is thought to be guilty. *See also* RELIGIOUS ALLUSIONS TO FIRE.

Further Reading: Bartlett, Robert. *Trial by Fire and Water: The Medieval Judicial Ordeal.* Oxford: Clarendon Press, 1986; "Catholic Encyclopedia: ORDEALS," <http://www.knight.org/advent/cathen/11276b.htm>, 07/09/99; Farmer, Sharon, and Barbara H. Rosenwein. *Monks and Nuns, Saints and Outcasts: Religion in Medieval Society.* Ithaca, NY: Cornell University Press, 2000; Frazer, James George, Sir. *Myths of the Origin of Fire.* New York: Hacker Art Books, 1974; Lang, Andrew. *Magic and Religion.* New York: Greenwood Press, 1969; "Medieval Sourcebook: Ordeal Formulas," <http://www.fordham.edu/halsall/source/ordeals1.html>, 07/09/99.

ORIGINS OF FIRE

Stephen J. Pyne, author of five books on fire known as the *Cycle of Fire* series, has called Earth a "fire planet." That term reflects the fact that fire is, and probably always has been, a constant factor in the natural world. Given the common occurrence of natural fire caused by lightning, volcanoes, and other phenomena, there has probably been no time in human history when humans were not aware of fire and capable at least of capturing and preserving some form of natural fire. During this early period, however, humans probably had little or no knowledge as to how to make fire on their own. They were entirely dependent on natural sources for fire.

The Origin of Fire: An Anthropological Question

An important question for many anthropologists is when and how humans first learned to control fire. Given the antiquity of this discovery, it seems unlikely that we will ever have precise knowledge as to the origins of fire practices, such as systems for cooking and heating. However, some clues have been discovered that provide hints as to the earliest use of fire for human activities.

For example, excavations in caves have yielded evidence of fire use as far back as 1.5 million years ago. This evidence consists of scorched walls and charred remnants of animal bones and teeth, and was once considered sufficient to set the beginning of fire use by humans. Later studies indicate that this evidence, found primarily in the Zhoukoudian cave in China, can be explained by factors other than human fire use. For example, spontaneous combustion of owl droppings could have produced fire in the cave.

Many scientists now believe that the earliest date at which fire use can positively be credited is about 300,000 years ago. The evidence for this date comes primarily from an ancient beach location in southern France called Terra Amata. There appears to be clearcut evidence of the existence of a stone hearth containing charred bones at this site. The combination of a structure apparently built for the purpose of cooking (the stone hearth) and the scorched bones seems to provide reasonably sound evidence that humans were using fire in this location at this date.

The question as to when humans first developed the ability to make and use their own fire may very well never be answered. One problem is that experts find it difficult to agree on the criteria on which this question can be answered. A second problem is that excavations typically yield equivocal findings that may or may not be acceptable to the anthropological community as definitive proof of the beginning of the "Age of Fire Used."

Mythological Answers

Whatever scientists may eventually decide about the origins of fire, many cultures long ago came up with answers for themselves. These answers appear in the form of folktales and stories that "explain" for people how their ancestors first discovered fire and learned how to use it. The existence of such tales is easy to understand. Fire plays such a crucial role in every human society, that it seems essential for people to have an explanation as to how humans received this marvelous gift.

Sir James G. Frazer's book *Myths of the Origin of Fire,* is a marvelous summary of dozens of fire-origin myths taken from every part of the world. He begins with the oldest society known at the time he was writing (Tasmania) and concludes with some of the most recent fire-origin myths (ancient Greece and India).

One interesting characteristic of many of these myths is that they begin with a very recent time when fire was not available. The members of a Tasmanian tribe from Oyster Bay, for example, believe that their grandfathers, and even their own fathers, did not have fire. For these individuals, then, the discovery of fire was a very recent phenomenon.

A common theme in fire-origin myths is that fire was once the exclusive property of gods and goddesses, who often existed in the form of animals. Alternatively, it may have been just the animals themselves who knew about fire and kept it for themselves. For example, one tribe of Australian aborigines once believed that fire belonged exclusively to a flock of crows living in the Grampian Mountains. Humans gained access to fire when a bird known as *Yuuloin keear* ("fire-tailed wren") stole a firestick from the crows and set the whole country on fire. From that point on, humans had access to fire.

The Menri tribe who lived on the Malay Peninsula believed that their ancestors received fire from a woodpecker. Their story begins with an encounter between the Menri and the Malays, who already knew how to make fire. On one occasion, a stag stole a firestick from the Malays and placed it high in a hut that was his home to prevent it from being stolen. But a woodpecker found the burning stick and carried it to the Menri. Because of this tradition, woodpeckers are revered by the Menri and may not be killed.

Many myths claim that fire was first obtained from the Sun. For example, the Loango tribe in Africa tell that a spider once made a very long thread stretching from the Earth to the Sun. A woodpecker climbed up the thread and pecked holes in the sky, making the stars we see today. Later, humans also climbed up the thread and took some fire from the Sun.

The Apache tribe in North America has a long and quite beautiful story explaining how their ancestors obtained fire. At one time, only certain animals knew about fire, but humans had not yet learned about it. The fox decided that he would like to find out where fire comes from and convinced a flock of geese to show him where fire was kept. The fox eventually fell into the middle of a settlement occupied by fireflies, where a fire was kept burning in the center of the village. At first the fireflies would not allow the fox to take fire from their village. But the fox convinced the fireflies to hold a big celebration around the central fire. While they were occupied with the celebration, the fox stole fire and brought it back to the Apache tribe. As punishment for stealing the fire, however, the fox has never itself learned how to use fire.

A Finnish story about the origin of fire is contained in the nation's classic epic, *The Kalevala*, and was the basis for a composition by Jean Sibelius for choir and orchestra called *Tulen Synty* ("The Origin of Fire").

In summary, the range of explanations for the origin of fire is very broad. At one extreme, these explanations may have connections with some very real sources for fire. For example, many tribes explain that fire came to an area first when a lightning bolt carried fire from the sky to the Earth, an explanation, of course, that is not far wrong. Other traditions claim that fire was first produced when the branches of trees rubbed against each other, presaging one of the earliest common methods by which humans made fire. At the other extreme, fanciful folktales describe how every sort of animal—from a coyote, eagle, flycatcher, iguana, muskrat, robin, or toad—have aided humans by stealing fire from other animals or deities and brought it back to Earth. *See also* RELIGIOUS ALLUSIONS TO FIRE.

Further Reading: "Beyond Vegetarianism," <http://www.beyondveg.com/nicholson-w/hb/hb-interview2c.shtml>, 07/27/99; Campbell, Joseph. *The Masks of God: Primitive Mythology.* New York: The Viking Press, 1959, Chapter 9; Frazer, Sir James G. *Folklore in the Old Testament.* New York: Tudor Publishing Company, 1923, passim; Frazer, Sir James George. *Myths of the Origin of Fire.* London: Macmillan and Company, 1930; Pyne, Stephen J. *Vestal Fire: An Environmental History, Told through Fire, of Europe and Europe's Encounter with the World.* Seattle: University of Washington Press, 1997, pp. 25–48; Sauer, Carl. "Fire and Early Man." In John Leighly, ed. *Land and Life.* Berkeley: University of California Press, 1963, pp. 288–299.

OXIDATION. *See* COMBUSTION

OXIDES OF NITROGEN. *See* AIR POLLUTION

OXIDES OF SULFUR. *See* AIR POLLUTION

OXYACETYLENE TORCH

An oxyacetylene torch is a device for producing very high temperatures by burning acetylene (ethyne; C_2H_2) gas in pure oxygen. The torch is fed from two separate containers, one that holds oxygen gas under high pressure and the second that holds acetylene gas, usually dissolved in acetone. The oxyacetylene torch produces the highest temperature available for cutting and welding of metals; a temperature of about 3,300–3,500°C (5,900–6,300°F). It was invented in 1903 and for many purposes replaced the oxyhydrogen torch that had been widely used before it. The highest temperature available from an oxyhydrogen torch is about 2,700°C (4,900°F).

The torch usually consists of a mixing chamber to which the two gases are fed from their separate containers. The gases are then ignited with a flint and steel sparker, and the flame is adjusted by means of valves that control the flow of each gas to the mixing chamber. In general, the torch operates on a principle similar to that of the Bunsen burner.

Oxyacetylene torches are a specific example of a general type of device known as oxygen-fuel torches. Other oxygen-fuel torches are named for the fuel used for combustion with oxygen and include oxybutane torches (maximum temperature 2,800°C [5,000°F]), natural gas torches (maximum temperature 2,500°C [4,600°F]); oxypropane torches (maximum temperature 2,600°C [4,800°F]); and oxygasoline torches (3,000°C [5,300°F]). These torches differ from each other not only on the basis of the temperatures they produce, but also on the basis of the stability of their flames, how cleanly they cut or weld, and how much they cost.

Further Reading: Baird, Ronald J. *Oxyacetylene Welding: Basic Fundamentals.* Tinley Park, IL: Goodheart Willcox Publisher, 1991; Bowditch, Kevin E., Ronald Baird, and Mark A. Bowditch. *Oxyfuel Gas Welding.* Tinley Park, IL: Goodheart Willcox Publisher, 1999; Finch, Richard. *Welder's Handbook: A Complete Guide to MIG, TIG, ARC, and Oxyacetylene Welding.* Berkeley, CA: Berkeley Publishing Group, 1997; Geary, Don. *The Welder's Bible.* New York: McGraw-Hill Professional Publishing, 1992; Griffin, Ivan H., Edward M. Roden, and Charles W. Briggs. *Basic Oxyacetylene Welding.* Albany, NY: Delmar Publishers, 1984.

OXYGEN

Oxygen constitutes one of the three requirements for any fire, often referred to as the "Fire Triangle." Those requirements include: (1) a fuel, something that will burn; (2) an oxidizing agent (such as oxygen), something that will cause the fuel to burn; and (3) enough heat to raise the fuel to its kindling temperature, the temperature at which it will begin to burn.

Oxygen is one of about 100 chemical elements, substances that cannot be broken down into any simpler form. It has an atomic number of 8, an atomic mass of 15.9994, and a chemical symbol of O. It was discovered in 1771 by the Swedish chemist Carl Wilhelm Scheele (1742–1786) and again three years later by the English chemist Joseph Priestley (1733–1804). The element was named by the great French chemist Antoine-Laurent Lavoisier (1743–1794). Lavoisier suggested the name *oxygen* from two Greek words meaning "acid" (oxy-) and "forming" (-gen) because he thought oxygen occurred in all acids. (A notion about which he was incorrect.)

Physical Properties

Oxygen is a colorless, odorless, tasteless gas with a density of 1.429 grams per liter and a specific gravity of 1.108 (compared to air). Its boiling point is -182.96°C, and its freezing point is -218.4°C. Oxygen exists in three allotropic forms: (1) nascent oxygen, consisting of one atom of oxygen per molecule; (2) dioxygen, consisting of two atoms of oxygen per molecule; and (3) ozone, consisting of three atoms of oxygen per molecule. Both nascent oxygen and ozone are unstable and revert to dioxygen under normal circumstances.

Chemical Properties

At room temperature, oxygen is relatively inactive. It reacts slowly with many metals to form a metal oxide, the best known example of which is iron oxide, or *rust*. Oxygen also reacts slowly with many types of organic matter to produce *decay*. The usual products of decay are carbon dioxide and water.

At higher temperatures, oxygen becomes much more active. For example, it may begin to attack wood, paper, coal, oil, natural gas, and many other materials in a process known as *oxidation*. During the oxidation process, the carbon present in the fuel is converted to carbon dioxide and, usually, carbon monoxide, while the hydrogen present in the fuel is converted to water vapor. The oxidation of methane (the primary component of natural

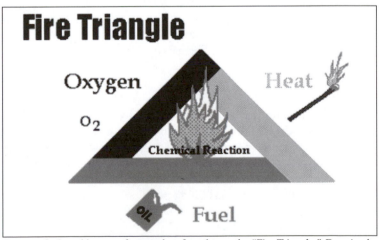

Oxygen, fuel, and heat are frequently referred to as the "Fire Triangle." *Drawing by Leslie Anne Miller.*

gas), for example, can be represented by the following chemical equation:

$$CH_4 + 2O_2 = CO_2 + 2H_2O$$

Occurrence

Oxygen is the most abundant element in the Earth's crust. It makes up about 46.6 percent of all atoms present in the crust. It occurs in the form of oxides, carbonates, phosphates, sulfates, silicates, and other more complex compounds.

Oxygen is also present in the Earth's atmosphere, where it is the second most abundant gas after nitrogen. About one-fifth of the Earth's atmosphere consists of oxygen. Oxygen is also present in the Earth's hydrosphere, its oceans, lakes, rivers, and other bodies of water. It constitutes about 89 percent of the mass of water.

The presence of oxygen in the atmosphere means that fire is an inherent and natural part of the Earth's environment. The Earth's surface is covered with a plethora of potential fuels, ranging from open seams of coal and pools of oil to trees, bushes, grasses, and other forms of plant life. All that is needed for a fire to begin, then, is the presence of the third factor: some form of heat that will cause oxygen and fuel to react with each other. This heat can be provided by any number of natural events, such as the eruption of a volcano, the friction of two rocks rubbing against each other (as during a rock slide), or, most important of all, lightning. Given the ready availability of oxygen in the atmosphere, it is hardly surprising that Stephen J. Pyne and other observers have referred to the Earth as a "fire planet."

Preparation

Oxygen is prepared for commercial use by the fractional distillation of air. Air is first cooled to the point where all of the major component gases present in it liquefy. The liquid air is then allowed to warm up. As each gas in the mixture reaches its boiling point, it evaporates and is collected. Oxygen boils off a liquid air mixture at -182.96°C.

Uses

Many uses of oxygen are well known to the average citizen. For example, pure oxygen is often provided to people with lung disorders or those who have been injured in an accident. The pure oxygen helps them to breathe more easily. People with long-term lung disorders may sometimes be confined to an oxygen tent, in which the life-giving gas is provided either in the form of air with a higher-than-normal fraction of oxygen or as pure oxygen.

Oxygen is also used in high-temperature torches, such as oxyacetylene and oxyhydrogen torches. Torches of this type produce temperatures in the thousands of degrees and are used for welding, metal cutting, and other specialized jobs. It is also used as a rocket propellant. Many different combinations of a fuel and oxidizer are used for rocket launches, but one of the most common is a mixture of liquid oxygen and liquid hydrogen. In this mixture, oxygen is the oxidizer and hydrogen is the fuel. The mixture produces one of the largest propellant forces per kilogram of any fuel/oxidizer combination.

By far the most important commercial use of oxygen gas, however, is in the manufacture of metals. Oxygen is used to burn off the impurities that are usually present in ores of a metal. For example, iron ore may be heated in a blast furnace, a Bessemer convertor, or some other device to obtain pure iron. Pure oxygen or normal air is blown through the liquid iron ore so that it can remove impurities such as carbon, silicon, and sulfur. The type of reactions that typically take place are as follows:

$$Fe(C, Si, S \text{ as impurities}) + O_2 = Fe \\ + CO_2 + SiO_2 + SO_2$$

The gaseous products of this reaction (such as CO_2 and SO_2) escape into the air, while the solid products (such as SiO_2) float to the top of the liquid mixture in the form of *slag* and are skimmed off. *See also* COMBUSTION; FUELS.

Further Reading: Connor, J.A., J.J. Turner, and E.A.V. Ebsworth. *The Chemistry of Oxygen*. New York: Pergamon Press, 1973; Farber, Edward. *Oxygen and Oxidation Theories and Techniques in the 19th Century and the First Part of the 20th Century*. Washington, DC: Washington Academy of Sciences, 1967; Grayson, Martin, ed. *Kirk-Othmer Encyclopedia of Chemical Technology,* 3rd edition. New York: John Wiley, 1978, Volume 16, pp. 653–673; Joesten, Melvin D. et al. *World of Chemistry*. Philadelphia: Saunders College Publishing, 1991, Chapter 11; "Oxygen" in Bridget Travers, ed. *The Gale Encyclopedia of Science*. Detroit: Gale Research, 1996, Volume 4, pp. 2650–2654; Sawyer, Donald T. *Oxygen Chemistry*. New York: Oxford University Press, 1991.

P

PESHTIGO FIRE

The Peshtigo Fire is generally regarded as the most costly fire, in terms of human lives lost, in American history. In one of the great ironies of history, the fire occurred on the same day as the more famous fire in Chicago, Illinois, on 8 October 1871.

Setting

At the time, Peshtigo was a town of about 2,000 residents located along the Peshtigo River in northeastern Wisconsin. By far the most important industry in town was lumbering, with nearly half of all residents employed in logging or sawmilling in one way or another. Prospects for the town in early 1871 were very good. An announcement had been made that the town would soon be joined to Milwaukee and Chicago by a railroad line. The line would provide easy transport for the town's products—raw lumber and wood products—to its much larger cousins in southern Wisconsin and Illinois.

The town was also located in an ideal spot for lumbering. It was surrounded by heavy forests, and local sawmills daily produced an average of 150,000 board feet of lumber. In addition, the town had a factory that produced a variety of wood products ranging from pails, tubs, broom handles, clothespins, and shingles. Although lumbering was by far the major commerce in the town, a few small farms had sprung up, and fishing in the Peshtigo River was a contributing factor.

The Disaster

Life in Peshtigo in 1871 was not, however, free of trouble. As the year developed, nor-

The Peshtigo, Wisconsin fire of 1871 was the most costly, in terms of lives, in U.S. history. *Courtesy of the State Historical Society of Wisconsin.*

mal spring and summer rains failed to appear. As fall approached, forests and grasslands had become extremely dry. With each passing day, residents became more and more aware of the potential danger of fire. As soon as small fires broke out and were discovered, volunteers from the town effectively extinguished them.

Some time during the day on 8 October, indications of a forest fire outside the town began to appear. Smoke began to drift into the town, and the sky became abnormally black. By evening, the fire had obviously gotten out of control. Residents claimed that they could hear the sound of the approaching firestorm, roaring towards them like a tornado or speeding freight train.

By 9:00 P.M., the fire had reached the town and was sweeping everything in its path. Reports claim that people attempting to escape the fire were burned to cinders as they tried to run away. Those who could do so jumped into the river. Even the river was not sufficient protection, however, as heads were scalded as soon as people surfaced to catch their breath.

The Results

By the time the fire had burned out, more than 1,200 people, nearly two-thirds of the town's population, had died. The fire had destroyed more than 1.25 million acres (about 2,000 square miles) of forest. It also swept through the countryside destroying other nearby communities.

Most Americans heard nothing about the Peshtigo Fire. Because of Chicago's larger population and greater fame, it was "Mrs. O'Leary's Fire" that earned headlines in newspapers around the world. It was not until a week later when a newly established local paper, the *Marinette and Peshtigo Eagle*, published an extra edition describing the fire, that the world heard the details of the horrible catastrophe.

It took three years for the survivors in Peshtigo to rebuild their community. When they did so, they no longer had the lumbering industry on which to rely: There were no more trees left in the area to harvest. Instead, the region turned to dairying as its new source

of employment. *See also* FIRESTORM; FOREST FIRES.

Further Reading: Christianson, Rich, "Scorched Earth: The Great Peshtigo Fire Remembered," <http://www.iswonline.com/archives/eclectic/peshtigo.shtml>, 12/22/98; Pernin, Peter. *The Great Peshtigo Fire: An Eyewitness Account*. Madison: University of Wisconsin Press, 1999; Wells, Robert W. *Fire at Peshtigo*. Upper Saddle River, NJ: Prentice-Hall, 1968.

PETROLEUM REFINING. *See* FURNACES

PHLOGISTON THEORY

The phlogiston theory was a theory that attempted to explain two of the most fundamental problems in the early history of chemistry: combustion and calcination of metals. Combustion is the scientific name for burning, while calcination is the change that takes place when a metal is heated in air. Calcination is comparable to the process known today as the oxidation of metal.

Origin of the Phlogiston Theory

The seeds of the phlogiston theory can be found in the writings of the German scholar Johann Joachim Becher (1635–1682). In 1669, Becher wrote a book called *Physicae Subterraneae* in which he says that all matter is composed of five fundamental substances: air, water, and three types of earth. The three types of earth were inflammable earth (*terra pinguis*), mercurial earth, and fusible or "vitreous" earth.

Becher's ideas were later elaborated on by the German physician and chemist Georg Ernst Stahl (1660–1734). Stahl wrote a textbook on chemistry entitled *Fundamenta Chymiae* in 1723 in which he attempted to solve the problems of combustion and calcination. In order to do so, he introduced the concept of *phlogiston*, which he associated with Becher's *terra pinguis* and which he represented by the Greek letter Φ (phi). Stahl wrote that phlogiston was a fundamental component of any combustible material and all metals. He described it as "the matter and principle of fire, not fire itself" (as cited in Partington, p. 87).

Phlogiston: Combustion and Calcination

Stahl believed that combustion was the process by which phlogiston was released from a burning material. For example, suppose that coal is burned. Stahl would explain that process by saying the phlogiston is released from coal in the burning process, leaving ash behind:

$$coal —(heated)— = \Phi + ash$$

Materials such as wood, coal, and paper were, therefore, rich in phlogiston, while noncombustible materials such as water, rock, and air contained little or no phlogiston.

Calcination was explained as the process by which a metal gives up its phlogiston upon heating. For example, when zinc metal is heated, it changes into a white powder. Stahl's explanation was that the heating process drives phlogiston out of the metal, leaving behind the "dephlogisticated" remnant, zinc calx:

$$zinc —(heated)— = \Phi + zinc\ calx$$

Support for the phlogiston theory came from a simple experiment in which zinc calx is heated with coal. According to the theory, coal is rich in phlogiston, which it should be able to give back to the calx, turning it back into a metal:

$$zinc\ calx + \Phi\ (from\ coal) —$$
$$(heated)— = zinc + ash$$

In fact, this reaction had been well known for centuries and was the process by which zinc and other metals could be recovered with their ores. When heated with coal or charcoal, the ore is converted to the pure metal.

Downfall of the Phlogiston Theory

The phlogiston theory may seem like a quaint misunderstanding today. In the later seventeenth and early eighteenth century, however, it explained the vast majority of phenomena involving combustion and calcination. The success of the phlogiston theory was due to at least two conditions present in the early years of chemical sciences. First, scientists knew very little about the nature of gases and were largely unaware that distinct types of gases exist. Second, good quantitative measurements were not yet possible, so the importance of weighing reactants and products in a chemical reaction had not yet been appreciated.

As these two circumstances began to change, questions were raised about the phlogiston theory. For example, quantitative studies of calcination reactions showed that the calx of a metal generally weighs more than the metal itself. That fact alone would suggest that the heating of a metal must result in the *addition* of something to that metal in order to account for its increase in weight. But the phlogiston theory required that something *escape from* the metal, namely phlogiston.

One attempt to explain this contradictory discovery was to claim that phlogiston had a negative weight. Thus, the loss of phlogiston from a metal could account for its gain in weight. The problem with this explanation, however, was that combustion often resulted in a loss of weight, as would make sense if phlogiston had (positive) weight. For example, when wood is burned, the ashes that remain behind generally weigh less than the wood from which they came. That result makes sense if phlogiston with positive weight is lost during combustion.

But how can phlogiston have both positive weight in combustion reactions and negative weight in calcination reactions?

In essence, one important reason for the continued popularity of the phlogiston theory into the late eighteenth century was that no alternative explanations for combustion and calcination were available. Thus, many of the most prominent scientists of the time continued to accept the theory even in light of some fairly substantial contrary experimental evidence. But in addition, the phlogiston theory retained its dominance simply because scientists in general tend to be conservative scholars. They tend to hold on to concepts that have served them well in the past and still work reasonably well in the present.

The phlogiston theory finally met its downfall late in the eighteenth century, however, when the French chemist Antoine Laurent Lavoisier (1743–1794) proposed the theory of oxidation as an explanation for both combustion and calcination. This theory instantly explained all the characteristics of these phenomena as well as the phlogiston theory, but without its attendant difficulties. Still, it was not received throughout the entire chemical community for a number of years after it was first proposed. *See also* BECHER, JOHANN JOACHIM; COMBUSTION; LAVOISIER, ANTOINE LAURENT; STAHL, GEORG ERNST.

Further Reading: Conant, James B. *The Overthrow of the Phlogiston Theory: The Chemical Revolution of 1775–1789*. Cambridge: Harvard University Press, 1950; Lagemann, Robert T. *Physical Science: Origins and Principles*, 2nd edition. Boston: Little, Brown and Company, 1969, Chapter 13; Partington, J.R., and D. McKie. *Historical Studies on the Phlogiston Theory*. New York: Arno Press, 1981; White, John Henry. *The History of the Phlogiston Theory*. New York: AMS Press, 1973; Wightman, William P.D. *The Growth of Scientific Ideas*. New Haven, CT: Yale University Press, 1951, Chapter 16.

PHOENIX

The phoenix is a mythological bird that is reborn after being consumed by fire. It appears in a variety of forms in many different cultures, including the Chinese, Arabic, Greek, and Native American traditions.

Perhaps the best known stories about the phoenix come from Greek mythology. In these stories, the phoenix was described as a bird that lives near a well in Arabia. It sings a song each morning that is so lovely that the Sun stops in its path across the sky to listen. After a certain set period of time, usually given as 500 years, the phoenix feels that death is approaching. At that point, it builds a nest, which it then sets on fire. The bird is consumed by the fire, but then is reborn from the ashes of the fire. The reincarnated bird encloses the ashes of its predecessor in an egg made of myrrh, which it carries to the sun god.

Greek stories of the phoenix were often based on earlier Egyptian legends, in which the phoenix was called the bennu. The bennu was said to have lived on top of the sacred Persea tree in the city of Heliopolis. The egg that the bennu bird carries to the sun god is said to be a replica of the sacred Geb's egg from which the sun was originally born.

A legend similar to the Greek and Egyptian tales was also told in ancient China. The Chinese bird appeared in two forms, male and female, known respectively as the hoang and the phuong. It could be found only during peaceful times and, when trouble spread across the land, it went into hiding.

The phoenix is described in various ways by different cultures. Among the Chinese, for example, it was thought to have a large bill, the neck of a snake, the back of a tortoise, and the tail of a fish. To the Egyptians and Greeks, the phoenix variously took the form of a heron, a peacock, or an eagle, often with fantastic shapes and plumage. Some authorities believe that the phoenix is the same bird mentioned in mythological tales from around the world, including the roe, discussed in the Arabian legend of *The Thousand and One Nights* and the thunderbird spirit found in many Native American legends.

Further Reading: Broek, R. van der. *The Myth of the Phoenix, According to Classical and Early Christian Traditions*. Translated by I. Seeger. Leiden: E.J. Brill, 1972; "Encyclopedia Mythica," <http://www.pantheon.org/mythica/articles/>, 07/13/99; "Greece: Myth and Logic." In Pierre Grimal, ed., *Larousse World Mythology*. New York: G.P. Putnam's Sons, 1965, pp. 96–175; Nigg, Joe. *The Book of Fabulous Beasts: A Treasury of Writings from Ancient Times to the Present*. New York: Oxford University Press, 1999; Nigg, Joe. *Wonder Beasts: Tales and Lore of the Phoenix, the Griffin, the Unicorn, and the Dragon*. Englewood, CO: Teacher Ideas Press, 1995; "Phoenix on Eliki," <http://www.eliki.com/ancient/myth/phoenix/content.htm>, 08/07/99.

PRAIRIE FIRE

Prairies are open lands whose primary vegetation is grass. Prairies are a distinctive ecosystem of North America, probably formed after the last Ice Age, about 8,000 years ago. The name *prairie* comes from the French word *prérie*, meaning "meadow." When the first European explorers arrived in the midwestern regions of the United States and Canada, they encountered an ecosystem that they had never seen before. The ecosystem consists of thousands of square miles of open space, dominated by grass and lacking, to a large extent, any large bushes or trees. Without any other term fitting such an ecosystem, they adopted the French term for meadow.

Fire is a critical element in maintaining prairies. When fires strike a prairie, they tend to burn down nearly all plant life. In many cases, fires are carried over very great distances by prevailing winds, burning away bushes, small trees, and other forms of brush. Grasses are destroyed by the fires also. But grasses are usually able to survive in such conditions, because they tend to have very deep root systems, often extending down as far as fifteen feet or more.

The regrowth of grasses often benefits from fires. Flames clear away dead plant life, opening up the soil to the warmth of the sun and encouraging the germination of new grass seed. Nutrients are also released from burned plant matter, providing those needed for the new growth of grass.

Historically, prairie fires have been started by lightning as well as by humans. Native American Indians, for example, are thought to have used fire extensively to restore and renew grasslands when natural fires failed to do that job. Indians apparently observed the effects of natural fires on the health of prairies, and introduced fire themselves when lightning fires were inadequate for the purpose.

Further Reading: Boyd, Robert, ed. *Indians, Fire, and the Land in the Pacific Northwest.* Corvallis: Oregon State University Press, 1999; Collins, Scott L., and Linda L. Wallace. *Fire in North American Tallgrass Prairies.* Norman: University of Oklahoma Press, 1990; Coupland, Robert T., ed. *Natural Grasslands: Introduction and Western Hemisphere.* New York: Elsevier, 1992; "Fire and the Prairie," <http://www.elnet.com/~prairie/fire.html>, 09/16/99; "Grassfires: Fuel, Weather, and Fire Behaviour. Contents," <http://www.ffp.csiro.au/nfm/fbm/gr_conts.html>, 08/24/99; Madison, John. *Tall Grass Prairie.* Hel-

ena, MT: Falcon Press, 1993; "Tallgrass Prairies in Illinois," <http://tqjunior.advanced.org/3568/prairie.html>, 09/16/99.

PRESCRIBED BURN

A prescribed burn is a fire that is set intentionally in a forest, on a prairie, or at some other location. The primary purpose of a prescribed burn is to prevent the outbreak of a much larger and more disastrous natural fire in the same area. Prescribed burns are also known as *controlled burns* and, more commonly in the past, were also known as *light burns*. The practice of prescribed burns is also known by other terms, reflecting the nations in which they are used, such as *petit feu* in France and *queimada* in Portugal.

Changing Views about Natural Fire

The United States and other nations have traditionally held mixed views about natural fires. In the early history of this nation, such fires were not regarded as a terribly serious issue. The continent appeared to hold far more trees, grassland, and other natural plant resources than could ever be harmed by natural fires. So the loss of a few acres or even a few thousand acres of forests or prairie was not considered a policy issue of much consequence.

That attitude changed, however, as the frontier moved steadily westward. It became apparent that there was not, in fact, an endless supply of trees, brush, and other natural plant resources. One consequence of this dawning realization was the creation of the National Park Service, the U.S. Forest Service, and other conservationist agencies. Some of these agencies took very seriously the need to protect our natural forests, grasslands, prairies, and other biomes. To that end, efforts to prevent and put out natural fires received a high priority.

There can be little doubt that these efforts have had some very positive effects. Many forests, grasslands, prairies and other natural plant resources would have been destroyed had humans been less aggressive about fighting fire in its natural settings. But, by the 1960s, the dominant philosophy about pre-

Training officer Mike Gagarin cuts a swath of fire across a meadow. At times, the flames from this controlled burn reached 12 feet in the air. *Courtesy of the University of California, Santa Cruz.*

prevented from taking their natural course. Unfortunately, the enthusiasm for making use of prescribed burns was greater than the knowledge as to how and where such burns should be conducted and greater than the financial resources available to develop such knowledge. As a result, the federal government's policy toward prescribed burns went through an awkward, relatively uncontrolled, and not entirely productive period in the 1970s and 1980s.

The use of prescribed burns today is, however, still widespread. In 1998, for example, there were 1,107 prescribed burns conducted by the federal government, covering a total of 259,353 acres. By comparison, there were only 428 natural wildfires covering an area of 57,679 acres of federal lands. These numbers were remarkably similar to the annual averages for the preceding five-year period (896 prescribed burns on 224,575 acres and 495 natural fires on 258,962 acres).

venting and fighting natural fires began to change.

Benefits of Natural Fire

One reason for a change in attitudes was a growing understanding of the role of fire in natural systems. Rather than thinking of fire as a natural enemy, ecologists began to realize that most, if not all, ecosystems need to be exposed to fire at some point in their evolution.

An important function of fire is to clear away dead vegetation and live plants that would otherwise smother the growth of other plants. Fire also results in the formation of ash, which becomes part of soil. Ash contains nitrogen, phosphorus, potassium, calcium, and other minerals from dead plants that are needed to replenish the fertility of soils. Fire is also needed by some kinds of plants in order for reproduction to occur. Some seeds will not germinate, for example, unless they are first broken apart by the heat of fires. An area swept by fire may also be more congenial to certain kinds of animal life.

By the 1960s, many government officials had enthusiastically embraced the principle that natural fires are, or can be, a good thing. In fact, programs were developed to initiate burns in areas where natural fires had been

Objections to Prescribed Burns

Not everyone is enthusiastic about the use of prescribed burns as a method of land management. For example, lumber companies suggest that they should have the opportunity to remove dead trees rather than having them burned off. People who live in the vicinity of prescribed burns may object to the smoke produced by such fires. People with asthma, allergies, and respiratory disorders are especially likely to be inconvenienced by prescribed burns. Other observers worry that intentional fires may actually threaten the habitat of some animals and will almost certainly kill some of them.

Finally, many Americans have a strong mind-set against the intentional setting of fires in the wilderness. Since 1945, Smokey Bear has been cautioning Americans about the dangers of forest fires, and with excellent results. People are quite aware of the harm natural fires can cause and take special precautions to make sure they are not responsible for fires on grasslands and prairies or in forests. It should not be surprising, then, that many of us are still hesitant to accept the fact that some natural fires are of value and that prescribed burns can be a contribution to the natural evolution of an area.

On rare cases, the issue of prescribed burns becomes front-page news when a specific burn gets out of control. In July 1999, for example, a prescribed burn by the U.S. Bureau of Land Management in northern California got out of control, swept across 2,000 acres of land, and destroyed twenty-four homes. Total cost of the disaster was estimated at $1.7 million ("Controlled Burn Debate Flares in West," *Ashland Daily Tidings,* 8 July 1999, p.1). A study of the accident by the U.S. General Accounting Office raised questions not only about the loss of personal property by such fires, but also about the added air pollution and threat to wildlife they posed.

About a year later, this scenario was repeated in New Mexico. A prescribed burn was set on 4 May 2000 in Bandelier National Monument, northeast of Albuquerque. The burn was supposed to be confined to a 968-acre region, but it soon got out of control. Winds that reached more than fifty miles per hour swept the fire in a northeasterly direction, towards the city of Los Alamos. Concern focused on the potential damage not only to structures and human life, but also to the Los Alamos National Laboratory (LANL), home of a large amount of nuclear-based research. The fire raged for more than a week before it was brought under control. Overall, 47,000 acres of land were burned, 380 structures were destroyed, and about 25,000 people were forced to evacuate their homes. There was relatively little damage at LANL, at least partly because some of the most dangerous materials at the laboratory were stored in specially designed fireproof areas.

Ironically, one of the experiments being conducted at LANL involved the Multispectral Thermal Imager (MTI) satellite, a satellite for the study of temperature patterns on the planet's surface. Data from the satellite were being received and processed at the laboratory automatically, even after buildings at LANL were being evacuated because of the conflagration.

An investigation of causes of the fire was conducted by a twenty-member Interagency Advisory Committee. The committee determined that a number of related errors were made in deciding to go ahead with the burn even after contrary data had been received. Among the most serious problems were inadequate training in prescribed burns by the officer in charge of the exercise and an unreasonably long delay by him in calling for assistance in fighting the fire. The National Weather Service was also criticized for not providing adequate information about the risk of high winds in the area. *See also* FIRE MANAGEMENT; FOREST FIRES; YELLOWSTONE FIRES OF 1988.

Note

The U.S. Forest Service has published a number of leaflets, brochures, and other materials dealing with prescribed burns on various types of land and in various parts of the United States. Contact a local library or the Forest Service for further information.

Further Reading: Baumgartner, David M. et al., eds. *Prescribed Fire in the Intermountain Region.* Pullman: Washington State University, 1989; Biswell, Harold H. *Prescribed Burning in California Wildlands Vegetation Management.* Berkeley: University of California Press, 1989; "Conflicting opinions on Los Alamos fire heard on Capitol Hill," <http://www.cnn.com/2000/US/06/07/control.fires.02>; Higgins, Kenneth F., Arnold D. Kruse, and James L. Piehl. *Prescribed Burning Guidelines in the Northern Great Plains.* U.S. Fish and Wildlife Service, Cooperative Extension Service, South Dakota State University, U.S. Department of Agriculture EC 760, 1989; Freedman, Bill. "Prescribed Burn." In Bridget Travers, ed. *The Gale Encyclopedia of Science.* Detroit: Gale Research, 1996, pp. 2921–2923; Koehler, John T., "Comprehensive Training: The Use of Prescribed Burning as a Wildfire Prevention Tool." *Fire Report Newsletter,* 21 July 1997, 1; also at <http://www.firefighting.com/FRN/fr2-12-1.htm>, 07/05/98; Martin, Robert E. *Pre-*

scribed *Burning: Decisions, Prescriptions, Strategies*. Bend, OR: Pacific Northwest Forest and Range Experimental Station, Silviculture Laboratory, 1978; Matthews, Mark. "In Western Woodlands, Forest Service Begins Setting Fires to Prevent Fires." *The Washington Post*, 19 April 1997; also at <http://forests.org/gopher/america/sintburn.txt>, 07/05/98; Pyne, Stephen J. *World Fire: The Culture of Fire on Earth*. Seattle: University of Washington Press, 1995, passim; Wright, Henry A., and Arthur W. Bailey. *Fire Ecology: United States and Southern Canada*. New York: Wiley-Interscience, 1982.

PROMETHEUS

Prometheus was a Titan who, in Greek mythology, stole fire from Mount Olympus and gave it to humans. The Titans, in Greek mythology, were the children and grandchildren of Uranus and Gaea, who ruled the Universe until they were dethroned by Zeus and the Olympians. Prometheus and his brother Epimetheus had refused to join their fellow Titans in the war against the Olympians. As a reward, Zeus spared Prometheus and Epimetheus from banishment to Tatarus, the fate of the other defeated Titans. He gave the brothers the task of repopulating the Earth, which had been devastated in the war.

Epimetheus attacked his job with vigor, creating all kinds of animals and showering on them all the good gifts the gods had to give. Prometheus worked more slowly and made man out of clay. The goddess Athena then blew life into a lump of clay, creating the first human. But Prometheus had no gift left to give humans. His brother had given the animals the gifts of sharp sight, keen sense of smell, speed, better hearing, and other attributes. Prometheus decided that he would ask Zeus for the gift of fire to give humans. Zeus, however, declined this request. Fire, he explained, belonged only to the gods. It could not be given to mere mortals. In despair, Prometheus decided to steal fire from Mount Olympus. He brought it to Earth and showed humans how to use it to cook foods, warm their homes, and scare off wild animals.

Zeus was infuriated when he discovered what Prometheus had done. But Prometheus found a way to appease Zeus' anger. He had an ox butchered and divided its carcass into two parts. One part contained the best parts of the animal, wrapped in bones and entrails. The other part consisted of more bones and waste products wrapped in fat. He then offered Zeus his pick of the two packages. Because it looked and smelled so much tastier, Zeus chose the second package. When he opened the package, he found that he had been cheated out of the best part of the ox. He was even more angry than before. Prometheus had committed a second sin in teaching humans how to deceive the gods.

As punishment for Prometheus' sins, Zeus ordered the Titan to be chained to a high peak in the Caucasus Mountains. Each day, an eagle flew down out of the sky and pecked away at Prometheus' liver. Prometheus was sentenced to stay on the mountain until two conditions were met. First, another god must agree to die for him. Second, a mortal must capture and kill the eagle and then unchain Prometheus. Prometheus was finally freed when Chiron, the Centaur, agreed to die in his place and when the mortal Heracles killed the eagle and set Prometheus free.

The story of Prometheus is told in a number of Greek tales, each with a slightly different emphasis and story line. The most famous of these stories are Hesiod's *Theogony*, Hesiod's *Works and Days*, and Aeschylus' play *Prometheus Bound*. The tale of Prometheus was taken up again many centuries later by the poet Percy Bysshe Shelley in his epic poem *Prometheus Unbound*. Shelley uses the Prometheus myth to advance his belief in the perfectibility of humanity. He argues that evil is an "accident" of human existence that can be erased if individuals have enough faith in their own goodness and their willingness to work for their own perfectibility. *See also* GODS AND GODDESSES.

Further Reading: Andrews, Tamra. *Legends of the Earth, Sea, and Sky*. Santa Barbara, CA: ABC-CLIO, 1998, pp. 80–82; "Encyclopedia Mythica," <http://www.pantheon.org/mythica/articles/>, 07/13/99; Erikson, Joan M. *Legacies: Prometheus, Orpheus, Socrates*. New York: W.W. Norton, 1993; "Greece: Myth and Logic." In Pierre Grimal, ed. *Larousse World Mythology*.

New York: G.P. Putnam's Sons, 1965, pp. 96–175; Kerenyk, Karl. *Prometheus: Archetypal Image of Human Existence.* Translated by Ralph Manheim. New York: Pantheon Books, 1963; Kreitzer, L. Joseph. *Prometheus and Adam: Enduring Symbols of the Human Situation.* Lanham, MD: University Press of America, 1994; Richardson, I.M. *Prometheus and the Story of Fire.* Mahwah, NJ: Troll Associates, 1983; Text of *Prometheus Bound* on a number of Web sites, including <http://www.princeton.edu/~rhwebb/prometheus.html>, 09/30/99. Text of *Prometheus Unbound* is also available at a number of sites, including <http://www.bartleby.com/139/shel1160.html>, 09/30/99.

PROTEST AND DEFIANCE BURNINGS

Fire has long been used as a way of expressing disapproval of certain laws or ideas and condemnation of certain individuals and groups. In many cases, rioters set fires to structures as part of their protest against some policy or practice to which they object. One of the most famous examples of riot-related fires was the fires set during riots that took place in the Watts neighborhood of downtown Los Angeles between 11 and 16 August 1965. The riots began when Los Angeles police officers made a routine traffic stop of a black driver in the South Central section of the city. Residents of the area, already in a state of outrage over state legislators' reluctance to pass civil rights legislation, soon began to riot. One form of action they took was the torching of hundreds of buildings in the Watts area. One of the memorable results of those riots was a phrase that was to become famous in the American language: "Burn, baby, burn!" It is reported that many of the rioters, pleased with the fires they had set, stood, watched, and expressed their anger and frustration with this phrase.

The following are some additional examples of the use of fire for dissent and rebellion.

Book Burning

Book burning is the act of destroying by fire books that are objectionable to some person or group of people in authority. In some cases, the action is symbolic because the burners understand that they cannot destroy every copy of a book with which they disagree. But some episodes of book burning have been carried out with the expectation that the objectionable pieces of literature will be forever eliminated from Earth. Book-burning-type demonstrations have also been carried out with musical recordings and other forms of art and expression to which people have objected.

Scholars have found references to book burnings as far back as the second century B.C. The Chinese emperor Shi Tuang Ti is said to have burned all the books in his kingdom except for a single copy of each, which he retained for his own library. Those books were burned after he died. The emperor is also said to have buried alive 460 Confucian scholars to prevent them from writing any additional books.

In one of the most famous examples of book burning in history, the Caliph Omar (ca. 581–644) burned the entire collection of 200,000 books in the library of Alexandria. Inspired by the Christian teachings that the end of the world was at hand, the caliph explained that books that opposed the teachings of God should be destroyed, while those that confirmed those teachings were not needed. Many of the greatest books of antiquity were destroyed.

In 1497 and 1498, the Dominican monk Girolamo Savonarola goaded the citizens of Florence to renounce their worldly pleasures and to destroy their most treasured clothes, jewelry, books, and art in giant bonfires, later called the "bonfire of the vanities." Savonarola's campaign was a short-lived success and he, himself, was burned as a heretic only a year later.

A debate over the translation of the Bible into English came to a head in 1525, when 6,000 copies of William Tyndale's version of the holy scriptures were ordered burned by the Church of England. The books had been printed in English in Cologne and smuggled into England. When they were found, they were destroyed because of the Church's position that the Bible should be available only in Latin, not in a language that the general public could read.

Book burning has sometimes been carried out by governmental agencies, in some cases, by aroused mobs, and in some instances, by ardent individuals. In 1932, for example, some unknown individual bought up every copy of James Joyce's newly published *Dubliners* in order to burn them all.

Arguably the most famous instance of book burning in recent history occurred in the early 1930s, after the Nazi party had come to power in Germany. Chancellor Adolf Hitler called upon the German people to collect and set fire to the works of more than 200 authors whose political views were unacceptable to the new regime. Most of those authors were of Jewish heritage, and they included scientists, philosophers, political writers, poets, and novelists, such as Bertolt Brecht, Thomas Mann, and Albert Einstein. A long list of foreign writers, including Helen Keller, Jack London, Margaret Sanger, H.G. Wells, Sigmund Freud, and Marcel Proust were included. At some of the largest book burnings, the attendance was estimated at more than 40,000 people.

Book burning has by no means disappeared. In 1998, for example, a minor storm was created in the Russian city of Yekaterinburg when the local bishop of the Russian Orthodox Church ordered the burning of books by three authors whose theological views he found objectionable. The authors, theologians with worldwide reputations, were Alexander Schmemann, John Meyendorff, and Alexander Men.

Bra-Burning Feminists

The term *bra burning* is based on an oft-repeated story that in 1968 a group of women at the Miss America Pageant in Atlantic City, New Jersey publicly burned their brassieres in protest against the contest, claiming that it denigrated women. It is not at all clear that bra burning ever took place as described. It appears that one of the organizers of the protest, Robin Morgan, consulted with Atlantic City officials to determine their attitudes about a public demonstration that might in-

volve the burning of brassieres. Although there seemed to be no objection from those officials, it appears that no bra burning ever occurred. A report in the *New York Times* of the protest makes no mention of such actions. Interestingly enough, however, the *Times* did make mention of bra burnings at the Miss America protest a few weeks after the contest was concluded.

In fact, one historian suggests that stories about bra burning may actually have been written *before* the Atlantic City protest. She writes that columnist Art Buchwald humorously and rhetorically wondered whether women would try to emulate the burning of draft cards by men by burning their own bras at the Miss America protest ("Feminist Myths"). It appears that very few verifiable reports of bra burning appeared during the 1960s. One of the best known may have been an event staged by a disc jockey in Chicago in which female models were paid to burn brassieres in public as a publicity stunt for the radio station.

Draft-Card Burning

In the United States, the burning of one's draft card has sometimes been an expression of protest toward some aspect of the nation's foreign policy, such as its conduct of a war. A *draft card* is an identification card issued to all men when they become eligible for selective service registration, usually at the age of eighteen. The draft card is a legal document which, under a 1965 amendment to the Universal Military Training and Service Act (the Selective Service Act), must be in one's possession at all times.

The burning of draft cards dates back to World War II. In 1947, about 500 veterans of the war and conscientious objectors held rallies in front of the White House and at the Labor Temple in New York City. At these rallies, those eligible for the draft burned their registration cards in protest against the nation's policy of conscripting young men to fight in the war.

Probably the most famous draft-card burnings, however, took place in the early 1960s in protest against the war in Vietnam.

Young men throughout the United States expressed their disagreement with government policy about the war by burning their draft cards at rallies. It is not clear in how many cases actual draft cards were burned and in how many cases representations of cards were burned in their place.

The legality of draft-card burning was decided in a 1968 U.S. Supreme Court case, *United States v. O'Brien* (391 U.S. 367). The issue before the Court was whether draft-card burning was a legitimate form of protest protected under the free-speech provisions of the First Amendment to the Constitution. The Court ruled that the federal government had the right to limit the rights of a person's free speech when national security was a consideration. O'Brien knowingly and intentionally attempted to thwart the national interest by burning his draft card and was, therefore, guilty under the provisions of the 1965 amendment to the Selective Service Act.

Self-Immolation

Immolation is the process of offering a sacrifice, often, a burnt sacrifice. Self-immolation is the process by which a person sacrifices his or her life, usually by being set on fire.

Some of the most memorable acts of self-immolation in recent history occurred in Vietnam in the early 1960s. The first took place on 11 June 1963 when Thich Quang Duc, a 73-year-old Buddhist monk, committed suicide in front of a large group of Vietnamese and American newspaper reporters, photographers, and others. The monk sat down in the middle of a busy intersection in downtown Saigon (now Ho Chi Minh City). Two fellow monks poured gasoline over Duc and then set him on fire.

Duc was protesting the South Vietnamese regime led by premier Ngo Dinh Diem. Among the policies to which Duc was objecting was the intrusion of Diem's government into the autonomy of the nation's powerful Buddhist monasteries. Reports of Duc's suicide appeared in newspapers worldwide the next day, but many papers declined to publish the grisly pictures taken of the monk's fiery death.

In following months, three more Buddhist monks followed Duc's example and committed self-immolation in public view. Slowly, these actions gained greater attention in American newspapers and began to draw attention to the policies of the American-supported Diem. A headline in the *New York Times* on 15 June, for example, announced that "U.S. Warns South Vietnam on Demands of Buddhists." The headline was remarkable in that the original story of Duc's suicide appeared only in a short article on page three of the June 12th edition of the paper.

The acts of Duc and his fellow monks are not unique. Many other individuals have used self-immolation to make strong statements about their views on social and political issues. Some other examples of self-immolation in recent years are the following:

A Buddhist monk commits ritual suicide in protest of anti-Buddhist policies in Vietnam in 1963. © *Bettman/ CORBIS.*

- In 1995, Sabine Kratze, a twenty-five-year-old German Buddhist monk in Hanoi, committed suicide by self-immolation in protest over the persecution of the United Buddhist Church of Vietnam by the national government. She had come to Vietnam in 1991 to study Vietnamese language, medicine, and culture.

- In 1996, three Kurdish prisoners being held in Turkish prisons set fire to themselves to protest the deaths of eleven of their comrades in the jail. The two men and one woman poured paint thinner over themselves and then set themselves on fire. All survived.

- In 1998, a member of the Tibetan Youth Congress, Thupten Ngodup, attempted self-immolation in New Delhi during a visit by General Fu Quanyou, chief of China's People's Liberation Army. Ngodup was protesting China's occupation of his homeland and Fu's visit to India. He suffered burns over 90 percent of his body, but survived.

Further Reading: Bradbury, Ray. *Fahrenheit 451*. New York: Simon & Schuster, 1993; Campbell, Joseph. *The Masks of God: Oriental Mythology*. New York: The Viking Press, 1962, Chapter 7; Drogin, Marc. *Biblioclasm: The Mythical Origins, Magic Powers, and Perishability of the Written Word*. Totowa, NJ: Rowan & Littlefield, 1989; Farrer, James A. *Books Condemned to be Burnt*. London: Elliot Stark, 1892; "Feminist Myths: Bra-Burning Discussion (June 1998)," <http://h-net2.msu.edu/~women/threads/disc-braburn.html>, 10/03/99; Hole, Judith, and Ellen Levine. *Rebirth of Feminism*. Chicago: Quadrangle Books, 1971; Martin, Joanna Foley, "Confessions of a Non-Bra Burner," *Columbia Journalism Review* (July, 1971): 11–15; McCutcheon, Russell T. *Manufacturing Religion*. New York: Oxford University Press, 1997; Mutahhari, Murtaza. *The Burning of Libraries in Iran and Alexandria*. Teheran: Islamic Propagation Organization, 1983; "The Self-Immolation of Thich Quang Duc," <http://www.smsu.edu/RelSt/thiqngdc.html>, 10/03/99; Thomas, Cal. *Book Burning*. Westchester, IL: Crossway Books, 1983; *United States v. O'Brien* (1968), <http://www.anarchytv.com/speech/obrien/htm>, 09/11/99; *United States v. O'Brien*, <http://www.bc.edu/bc_org/avp/cas/comm/free_speech/obrien.html>, 09/11/99.

PSYCHOANALYSIS OF FIRE

Psychoanalysis of Fire is a book written by French philosopher Gaston Bachelard in 1938. Bachelard argues that fire is perhaps the most important natural phenomenon through which humans express their innermost thoughts and feelings. He points out that fire itself is a physically neutral occurrence on which humans overlay their own experiences, hopes, and expectations. He opens the book with this thought:

> We have only to speak of an object [such as fire] to think that we are being objective. But, because we chose it in the first place, the object reveals more about us than we do about it. (Bachelard 1964, 1)

Bachelard goes on to remind the reader of the enormous variety of ways in which fire has been viewed and interpreted throughout human history:

> Among all phenomena, it is really the only one to which there can be so definitely attributed the opposing values of good and evil. It shines in Paradise. It burns in Hell. It is gentleness and torture. It is cookery and it is apocalypse.... It is well-being and it is respect. It is a tutelary and a terrible divinity, both good and bad. It can contradict itself; thus it is one of the principles of universal explanation. (Bachelard 1964, 7)

In the seven chapters of *Psychoanalysis of Fire*, Bachelard reviews some of the ways in which humans have talked about fire, and what these interpretations say about the people who developed the interpretations. He talks in Chapter Three, "The Novalis Complex," for example, of what seem to him some deep and profoundly significant sexual meanings for many of the festival fire legends discussed by Sir James Frazer in his classic study, *The Golden Bough: A Study in Magic and Religion*.

Bachelard's book is still important more than sixty years after he wrote it because of the fresh and challenging new views he provides of some important classical interpretations of fire.

Further Reading: Bachelard, Gaston. *The Psychoanalysis of Fire.* Translated by Alan C.M. Ross. Boston: Beacon Press, 1964.

PYNE, STEPHEN J.

Stephen J. Pyne is one of the world's authorities on fire and its cultural significance. He is the author of an important series of books published in paperback by the University of Washington Press (Seattle) called *Cycle of Fire.* The books in that series, with their original publishers are *World Fire: The Culture of Fire on Earth* (New York: Henry Holt, 1995), *Burning Bush: A Fire History of Australia* (New York: Henry Holt, 1991), *Fire in America: A Cultural History of Wildland and Rural Fire* (Princeton, NJ: Princeton University Press, 1982), *The Ice: A Journey to Antarctica* (Iowa City: University of Iowa Press, 1986). and *Vestal Fire: An Environmental History, Told through Fire, of Europe and Europe's Encounter with the World* (Seattle: University of Washington Press, 1997).

Pyne was born in San Francisco on 6 March 1949. He earned his B.A. in English at Stanford University in 1971, and his M.A. and Ph.D. in American Civilization at the University of Texas in Austin in 1974 and 1976, respectively. His first teaching position was in the Department of History at the University of Iowa from 1981 to 1986. He then became Associate Professor and later Professor of American Studies at Arizona State University West between 1986 and 1996. From 1996 to 1999 he was Professor of History at Arizona State and, in 1999, became Professor of Biology at the university. For eighteen summers, Pyne also worked as a firefighter for the National Park Service on the North Rim of the Grand Canyon, at Rocky Mountain National Park, and at Yellowstone National Park.

In addition to *Cycle of Fire*, Pyne has written four books on the Grand Canyon, known collectively as *Grand Canyon Suite*, and four books and reports on fire science. Among his many honors include a Fulbright Fellowship to Sweden in 1992, an Antarctic Fellowship from the National Endowment for the Humanities in 1981–1982, and a MacArthur Foundation Fellowship for 1988–1993. *See also WORLD FIRE.*

PYROMANCY

Pyromancy is divination by fire. Divination is the practice of predicting the future by discovering hidden messages in certain omens. Historians believe that divination was practiced in China as early as the third millennium. By the time of the Shang dynasty (about 1500–1100 B.C.), the use of animal bones to predict the future was a highly developed art. The practice first appeared in the Western World among the Etruscans in the first millennium B.C.

Practitioners of divination used a host of materials in making their predictions. The specific type of divination practiced was named after the material used. Thus, aeromancy involved predictions based on weather phenomena; ceromancy, on the behavior of wax dropped on water; haruspicy, on the appearance of an animal's entrails; pegomancy, on the appearance and behavior of water in a fountain; and tephromancy, on the appearance of sacrificial ashes.

In one form of pyromancy, the priest or diviner examined the flame produced in a sacrificial fire. If the flame was strong, clear, steady, and bright, the signs were regarded as favorable. The presence or absence of smoke, the color of the flame, the sounds produced during burning, the direction in which the flame turned, how long the flame lasted, and the shape of the flame itself were also thought to provide information about the future.

Other types of flames could also be used in making predictions. The way a torch burned in a home, for example, would foretell the family's future or the health of those living in the home. A flame with a single peak, for example, was regarded as a favorable sign, while one with a double peak was an unfavorable omen. A flame with three peaks, on the other hand, was the most favorable indication of all. Flames might also be studied after a flammable material, such as sulfur or pitch, was thrown into the fire. *See also* RELIGIOUS ALLUSIONS TO FIRE.

Further Reading: Gibson, Walter B., and Litzka R. Gibson. *The Complete Illustrated Book of Divination and Prophecy*. Garden City, NY: Doubleday, 1973; Manas, John H. *Divination, Ancient and Modern*. New York: Pythagorean Society, 1947; Morgan, Keith. *Simple Candle Magick: The Ancient Art of Pyromancy Available to All*. Edmonds, WA: Holmes Publishing Group, 1995; Shaw, Eva. *Divining the Future: Prognostication from Astrology to Zoomancy*. New York: Facts on File, 1995.

PYROMANIA

Pyromania is an irresistible desire to start fires. This behavior pattern was first recognized and named in the early 1840s, at a time when the science of mental health was just beginning. For many years, mental health experts were uncertain about the precise nature of pyromania and often regarded it as a form of moral insanity or some other mental malfunction.

Over the years, professionals have attempted to develop a psychological and/or psychoanalytic explanation for the pyromaniac's behavior. Some analysts have attempted to find sexual explanations for the pyromaniac's behavior, while others have classified the condition as a neurosis or obsession.

At the present time, the American Psychiatric Association's *Diagnostic and Statistical Manual of Mental Disorders*, 4th edition (1994), lists a number of symptoms used in identifying pyromania. These include:

- The intentional setting of a fire on more than a single occasion.
- Feelings of arousal prior to committing the act.
- Fascination with the materials and settings associated with a fire.
- Feelings of pleasure, gratification, and/or relief when witnessing a fire.
- Fire setting as an act performed for its own purpose, with no intent to gain monetary reward, revenge, or some other motive.

Fire safety workers point out the necessity of distinguishing between arsonists and pyromaniacs, the primary difference being motive. Arsonists usually intend to cause harm, to profit from a fire, to conceal a crime, or to achieve some other external goal. Pyromaniacs, by contrast, are interested almost entirely in satisfying their own internal needs and desires.

Individuals of all ages can be pyromaniacs, although the condition occurs much more often among males than females. A substantial body of literature, however, is devoted to the study of pyromania in juveniles. Various researchers have claimed to find a correlation between pyromania and a number of other conditions, such as poor school work performance, truancy, disruptive behavior and/or hyperactivity, inability to form close friendships, bedwetting, and cruelty to animals. *See also* ARSON.

Further Reading: Garry, Eileen M., "Juvenile Firesetting and Arson." Office of Juvenile Delinquency and Prevention, U.S. Department of Justice, Fact Sheet #51, January 1997; also at <http://www.ncjrs.org/txtfiles/fs9751.txt>, 08/08/99; Gaylord, Jessica. *The Psychology of Child Firesetting: Detection and Intervention*. New York: Brunner/Mazel, 1987; Lewis, Nolan D., and Helen Yarnell. *Pathological Firesetting (Pyromania)*. New York: Mental Disease Monographs, 1951; Little, Peggy. "Juveniles & Arson," <http://www.paralegals.org/Reporter/Summer98/arson.htm>, 08/08/99; "Review of Literature on Child Firesetters," <http://www.wa-ic.org/wic/Jfsref.htm>, 08/08/99; Schwartman, Paul, Hollis Stambaugh, and John Kimball. "Arson and Juveniles: Responding to Violence." United States Fire Administration, Report 065, Major Fires Investigation Project; also at <http://www.usfa.fema.gov/usfa/techreps/tr095.htm>, 08/08/99; Stadolnik, Robert F. *Drawn to the Flame: Assessment and Treatment of Juvenile Firesetting Behavior*. Sarasota, FL: Professional Resource Exchange, 2000.

PYROPHORIC MATERIALS

Pyrophoric materials are generally defined as any material that will ignite spontaneously in air at temperatures above about 54°C (130°F). The term has also been defined in a number of other closely related ways, however. For example, some sources define a pyrophoric material as anything that gives off sparks when rubbed, scratched, or struck. Some authorities restrict the term to finely divided powders, but many other kinds of matter, including pieces of metal, metal turnings, chips, and pellets can also be pyrophoric.

Common examples of pyrophoric materials include sodium metal, phosphorus, titanium dichloride, lithium hydride, trialkylaluminum compounds, and alkylboranes. Carbon monoxide gas reacts with a number of metals to form compounds known as carbonyls. The compound iron pentacarbonyl ($Fe(CO)_5$) is an example. The carbonyls are oily liquids that are very pyrophoric and that may explode spontaneously at relatively low temperatures.

The primary application of pyrophoric materials has, for a long time, been in cigarette lighters and similar devices. These devices consist of two parts: one a pyrophoric material, and the second a material used to produce friction. When the pyrophoric material is scratched or rubbed, it gives off a spark or small flame that can be used to ignite some other object, such as a cigarette.

A more recent use of pyrophoric materials is as decoys against heat-seeking missiles.

A heat-seeking missile is a weapon used to track down and destroy military aircraft. The missile is designed to recognize very high temperatures, such as those produced by a jet engine. When fired, it "locks in" on the heat source, follows the hot object (the aircraft) through the air, and eventually strikes and blows it up.

Canisters containing pyrophoric materials can be used to confuse heat-seeking missiles. They can be fired by aircraft in a variety of directions at right angles to its own flight path. Once the pyrophoric materials ignite, they will tend to attract any heat-seeking missiles in the area, providing protection for the aircraft itself. *See also* IGNITION.

Further Reading: McIntyre, R., and C.P. Dillon. *Pyrophoric Behavior and Combustion of Reactive Metals*. Huntington, WV: MTI Publications, 1988; Schmitt, Jeff, ed. *Pyrophoric Materials Handbook*. Available only on the Internet at <http://saber.towson.edu/~schmitt/pyro/>, 08/27/99.

PYROTECHNICS. *See* FIREWORKS; FLARES

R

RELIGIOUS ALLUSIONS TO FIRE

Fire plays a central, and often extensive, role in most religions. In many instances, the role of fire in modern religions can be traced to mythic practices and beliefs from much earlier times.

As an example, many Christian customs were adopted from preexisting practices carried out by pagans in countries where Christianity was adopted. Many of the customs associated with Christmas, Lent, Easter, and other holy days can be traced to midwinter fire festivals practiced by early inhabitants of Europe. Understanding the role of fire in Christianity (and other major religions) requires, therefore, some appreciation of the way in which earlier cultures thought about and used fire festivals.

An interesting confirmation of this fact is found in efforts made by the early Christian church to discourage heathen practices among people who were new converts to the faith. For example, in 734, a synod under the leadership of Saint Boniface promulgated an "Index of Superstitions and Heathenish Observances" that listed activities in which Christians were not allowed to participate. Included on the Index was the ceremony of need-fire. Evidence suggests, however, that most people either ignored the restrictions on such practices or tended to incorporate the practices into their new religion.

In general, fire tends to have one or more of the following functions in a religion:

1. A source of light.
2. An instrument of punishment and retribution.
3. A means of purification.
4. A way of sending messages to a god or gods.
5. A representation of the god figure herself or himself.
6. An explanation of the circumstances under which the world began and will end.

A Source of Light

In some cases, fire and flames are depicted as a way by which a god or the gods guide humans through darkness, either real darkness (as at nighttime) or symbolic darkness (separation from a god). For example, Chapter 14 of the Book of Exodus in the Bible tells how God commanded the Israelites to flee from Egypt under the leadership of Moses. When the Israelites saw that they were being pursued by the Egyptian army, they became very concerned. But God sent down an angel to relieve their fear and tell them that He would guide their way into the Promised Land. To make sure that they would know the way, he promised to send "a pillar of cloud" by day and a "pillar of fire" by night not only to direct their paths, but also to show that He was truly God and to "trouble the host of the Egyptians" (Exodus 14).

The pillar-of-fire imagery is repeated many other times in the Bible as, for example, in Exodus (33:9–10); Numbers (14:14); Numbers (16:42); Nehemiah (9:12); Psalms (99:7); and Isaiah (19:1).

An Instrument of Punishment

A god or gods also use fire as a way of punishing evildoers or unbelievers. Psalms 106 of the Bible, for example, tells of God's punishment for Korah and his followers when they rebelled against Moses and Aaron. The writer tells that, "His lips are full of indignation, And His tongue like a devouring fire. A fire was kindled in their company; the flame burned up the wicked." (Psalms 106).

Other holy books also describe the use of fire as an instrument of punishment. In the Koran, for example, Muhammad warns many times of the terrible fate that will befall unbelievers and evildoers. For example, in "The Cow," he says:

> And whoever disbelieves, I will grant him enjoyment for a short while, then I will drive him to the chastisement of the fire; and it is an evil destination. (The Cow, 126)

And later, he repeats that:

> Thus will Allah show them their deeds to be intense regret to them, and they shall not come forth from the fire. (The Cow, 167)

The same message is repeated in other books. For example, in "The Family of Imran," the prophet warns that:

> As for those who disbelieve, surely neither their wealth nor their children shall avail them in the least against Allah, and these it is who are the fuel of the fire. (The Family of Imran, 10)

And again:

> And guard yourselves against the fire which has been prepared for the unbelievers. (The Family of Imran, 131)

This admonition is repeated more than 100 times in various parts of the Koran.

The Mormons are perhaps unique in arguing for a very different view of fire in the afterlife. Prophet Joseph Smith taught that the source of eternal fire after death was heaven, not hell. He wrote that, after death, eternal heavenly fires burned away all flesh and bones and left only a pure spirit to survive forever in heavenly grace. He said that, "All men who are immortal dwell in everlasting burnings." He goes on to explain that, "God Almighty Himself dwells in eternal fire; flesh and blood cannot go there, for all corruption is devoured by the fire" (see Further Reading, "Everlasting Burnings").

Purification

Many religions draw on the way fire is used in everyday life by metalworkers to separate pure metals from dross. In that process, an ore of the metal may be heated with charcoal to obtain the pure metal, leaving behind unusable waste products. Or, an impure metal may be heated so as to drive off as vapors impurities that occur along with the metal.

In much the same way, some religions teach that a supernatural being will expose individual humans or humans in general to a "refining fire." That fire will burn away the evil in an individual's soul or, alternatively, it will burn the evildoers and nonbelievers in a society and leave only the pure of heart. For example, the Shaker Book of Holy and Eternal Wisdom quotes God as saying that:

> I have come as a refiner and purifier of silver; and with the fire of the gospel I will try every son and daughter of Adam's race, that cometh unto me; for I have come to purify and cleanse as with refiner's fire, and fuller's soap; and no one shall escape the purifying of their own souls, by the fire of my coming. . . . (Divine Book of Holy and Eternal Wisdom, "The Female Spirit Placed in Her Order, in Christ, as the Second Eve," 27)

The Bible also alludes to the use of fire as a way of purification. In Luke, Chapter 3, for example, John the Baptist is preaching about the coming of Jesus. He tells that the

Savior will "gather the wheat into his garner, but the chaff he will burn with fire unquenchable" (Luke 3:17).

Messages and Offerings to Gods

In some religions, fire is a way of sending messages and offerings to a god or gods. The principle behind this practice is easy to understand. A physical, substantial, material object can be set on fire and converted to a vaporous, unsubstantial, immaterial spirit that can ascend to the supernatural realm. There it can relay prayers, requests, praise, or other messages from humans to the spirit world.

A central ceremony in the Buddhist religion, for example, is the *puja*, or fire ceremony. In this ritual, candles, incense, ghee (clarified butter), grains, or other materials are burnt while traditional chants are said. These objects are offered as foods to the gods, who do not have mouths with which to feed themselves. They do, however, have the sense of smell and sight with which to take in the burnt offerings.

Representation of the God or Gods

In some religions, fire, in one form or another of its manifestations, takes on the role of the central god figure. In perhaps no religion is this more true than in Zoroastrianism. Zoroastrianism evolved from the teachings of the Iranian prophet Spitaman Zarathushtra (known to the Greeks as Zoroaster) in about 1500 B.C.

Fire has a central role in Zoroastrianism because the prophet saw it to be the physical representation of Asha, which represents order, truth, and righteousness. All ceremonies in the religion are conducted in front of a fire, and the priest conducting a service refers to the fire as *Atash Prashah*, or king. The holiest temples of the religion hold an eternal, consecrated fire known as *Atash Behram* or *Atash Aderan*. Zoroastrian holy writings describe the supreme importance of fire and the careful attention that must be given to it. In the writings of Adar Frobag, for example, fire is described as:

A thing differing from other principal shining bodies (i.e., the Sun, the Moon, the Stars, etc.) which give luminosity to the things on this Earth; and it is the original principle of water, of air, and of every visible matter. To it is due the existence and Sustenance of men. One should abstain from extinguishing it. (http://www.avesta.org/denkard/dk5s.html)

In other religions, fire is not equated with the god or gods themselves, but is taken as a representation of them. In the Roman Catholic Church, for example, candle flames are not taken to constitute any physical representation of God Himself, but are symbols to remind the congregation of His presence at the service. In one rite, for example, the priest lights a three-branched candle and repeats a prayer that asks God to:

> make us enlightened by that light and inflamed with the fire of Thy brightness; and as Thou didst enlighten Moses when he went out of Egypt . . . ("Liturgical Use of Fire," *The Catholic Encyclopedia*)

The Beginning and the End of the World

Many religions teach that the world was created out of fire and/or will end in fire. The Aztecs, for example, believed that the world was created by Xiuhtecuhtli, also known as Huehueteotl or "the old god." He was generally regarded as the "personification of light in the darkness, warmth in coldness, and life in death" (http://www.pantheon.org/mythica/articles/x/xiuhtecuhtli). Priests of the religion were responsible for maintaining sacred fires in the god's honor. Once every fifty-two years, the fires were ceremoniously reconsecrated with the bodies of victims whose hearts had just been torn from their bodies. It was through these acts, the Aztecs believed, that Xiuhtecuhtli would continue to provide the Earth with light, fire, and warmth.

In many religions, fire is also thought to be the ultimate cause of the world's demise. For example, a fiery end to the world is recounted in many parts of the Bible. For example:

Fire that precedes him as he goes,
devouring all enemies around him;
his lightning lights up the world, earth
observes and quakes. The mountains
melt like wax at the coming of the
Master of the world. (Psalms 97:3–
5)

Old Testament prophets also foretold of a fiery conclusion to the physical world. Ezekiel, for example, tells that God:

will collect you in my furious anger
and melt you down; I will collect you
and stoke the fire of my fury for you,
and melt you down inside the city.
(Ezekiel 22:19–22)

See also GODS AND GODDESSES OF FIRE; OLD BELIEVERS; RITUAL FIRE; ROAD OF FLAME.

Further Reading: Boyce, Mary. *Zoroastrians: Their Religious Beliefs and Practices*. London: Routledge & Kegan Paul, 1979; "Divine Book of Holy and Eternal Wisdom: Excerpts," <http://www.libby.org/password/SHAKER-MANUSCRIPTS/Holy-Wisdom/wisdm7x.htm>, 05/28/99; "Everlasting Burnings," *Encyclopedia of Mormonism*, Volume 1, <http://www.mormons.org/basic/afterlife/burnings_eom.htm>, 07/21/99; Goudsblom, Johan. *Fire and Civilization*. London: Allen Lane, 1992, Chapter 5; The Holy Bible, King James Version; The Koran; Levinson, David, vol. ed. *Encyclopedia of World Cultures*. Boston: G.K. Hall & Company, 1996; "Liturgical Use of Fire." *The Catholic Encyclopedia*, <http://www.knight.org/advent/cathen/06079a.htm>, 07/20/99; Morgenstern, Julian. *The Fire upon the Altar*. Chicago: Quadrangle Books, 1963; Staal, Frits. *Agni: The Vedic Ritual of the Fire Altar*, 2 vols. Berkeley: University of California Press, 1983; "Who Are We Zoroastrians?, "<http://coulomb.ecn.purdue.edu/~bulsara/AOROASTRIAN/wawz.html>, 07/20/99.

RITUAL FIRE

Ritual fire is fire used for some ceremonial purpose. A large variety of rituals exist in which fire plays a prominent role. These rituals include religious, mythological, cleansing, punitive, and memorial events.

Religious and Mythological Ceremonies

Almost every religion and mythical system has ceremonies involving fire in either a primary or subsidiary role. One function of the votive candles in Roman Catholic churches, for example, is to carry prayers to heaven. In the Buddhist religion, candles, incense, ghee (clarified butter), grains, or other materials are burnt as offerings to the gods, who have no mouths with which to feed themselves. In Zoroastrianism, fires are the center of every ceremony as the primary focus of that religion's devotions.

One of the most common roles of fire is to symbolize the renewal of life. In many cultures, some central fire is allowed to burn throughout the year as a symbol of the continuity of life. At the end of some given period of time, the fire may be extinguished ceremonially and a new fire ignited. The new fire often represents the unending cycle of life in which even death (the extinction of last year's fire) is followed by rebirth (the kindling of the new year's fire).

For example, the Creek Indians in the southeastern United States long practiced an elaborate ritual each year in which all remnants of an existing central fire were destroyed and a new fire was created by a shaman. Elaborate provisions were made to ensure the purity of the new fire, and it was believed that the fire atoned for all sins (except murder) committed during the preceding year. This ritual began to die out in the eighteenth century as Creek communities were destroyed and scattered by European invaders.

Some scholars believe that certain modern-day customs may derive from these centuries-old traditions. The custom of lighting and blowing out the candles on a birthday cake may be one such custom. Candle-lit dinners, when other sources of light are available for illumination, may be remnants of some long-forgotten traditions held sacred by our ancestors.

Eternal Flames

One use of fire that appears in many religious and mythical traditions is the eternal or perpetual flame. Such flames seem to provide people with a constant reminder of the presence of a god or goddess and a focus for their worship of that deity. The eternal flame also acts as a way of unifying a community.

Among the Lithuanians, for example, an eternal flame was traditionally the most important religious symbol. It long burned at the center of Vilnius, the nation's capital, as a way of unifying the living community and providing a link with ancestors who have died and gone to live with the gods.

Eternal flames may also represent secular power. The perpetual fire maintained by the Vestal Virgins at Rome, for example, may at one time have been dedicated to mythological gods, but it eventually came to represent the power of the Roman government itself.

Whatever they represent, perpetual fires are always maintained according to very strict rules. They may not be defiled by having unclean objects placed in them, they must be tended by only certain anointed individuals, and they could be renewed only according to very specific rituals.

Purification Rituals

In most cultures, fire has been thought to provide cleansing actions. This belief probably grows out of the physical reality that fire can burn up all types of matter, producing only a very pure flame as the final product. Many cultures have ceremonies, for example, in which people are exposed to the smoke of a fire or to the fire itself to purify and cleanse their bodies of disease and their souls of sin. Even in modern days, some Chinese physicians use the smoke of burning incense sticks to treat some forms of bodily disorders. The practice of jumping over a burning fire has been noted in a great many cultures, with the practice believed to purify the jumpers and guarantee that they will lead a happy, healthy, and productive new year.

One modern-day practitioner of fire purification ceremonies in New Jersey explains that fire "cleanses and rebalances. It helps you retain harmony," she says, "and once you carry the sacred fire with you, it's there forever" (McCloud, 1999).

Punishment

Fire was long used as an instrument of punishment and a way of killing individuals found guilty of violating religious or social mores. Perhaps the best known example of this practice was the custom of burning witches, heretics, and others who stood outside mainstream society through much of the Middle Ages in Christian nations. A horrible punishment such as burning could sometimes be justified, at least to those who used it, since it provided a method by which the sinful body of the accused could be destroyed in order to free his or her spirit to go on to another level.

Cremation

Cremation is another method by which a person's eternal soul, by whatever name it is called, is freed from its Earthly body to join the supernatural realm in which a particular culture believes. In the Hindu practice of *sati*, for example, a wife joins the body of her dead husband on the ceremonial fire with which he is cremated so that she can join him on his journey to the next world.

Sacrificial Fires

Fire has been used for the purpose of offering sacrifices to gods and goddesses through much of human history. A fundamental principle on which sacrificial fires are based appears to be that an effective way of sending gifts and messages to deities is by setting them on fire. The gifts and messages are then converted to smoke and vapors which travel upward into the realm of gods and goddesses. These gifts and messages have a variety of purposes. In many instances, sacrificial fires are clearly meant to show respect or to honor a deity. Of the more than four dozen references to burnt offerings in the Old Testament, for example, the intent of such offerings is almost always to show honor to Jehovah. In fact, much of the language dealing with burnt offerings involves specific instructions as to what kinds of offerings are acceptable and how they are to be offered.

In Chapter 1 of Leviticus, for example, God gives Moses very detailed and specific instructions as to how burnt offerings of cattle and smaller animals are to be offered. He directs that the animal shall be an "unblemished male" that is "slaughtered before

God" by Aaron's sons, the priests of the religion, who must then "arrange the cut pieces, the head, and the fatty intestinal membrane on top of the wood that is on the altar fire." The animal is also to be slaughtered only "on the north side of the altar," and its internal organs and feet "shall [first] be washed with water." Such highly specific instructions for the presentation of a burnt offering are very common, not only in Christianity, but also in most other religions and mythic systems.

Sacrificial fires are also conducted to carry messages to gods and goddesses and to ask for their help and assistance. In some Polynesian cultures, for example, priests would entrust particular commissions to the human victim of a sacrificial fire, knowing that the god or goddess to whom the victim was going would eat the soul of the victim and receive the entreaty. Similarly, many Native American tribes believed that tobacco smoke was a special form of sacrificial fire certain to reach the deities. They often prayed to their gods as they smoked for help in defeating an enemy or having good crops.

Many types of plants and animals, including humans, have been consigned to sacrificial fires. The Bible refers to many instances in which human victims were used for such fires. Two early kings, Manasseh and Ahaz, are mentioned specifically in 2 Chronicles as sending their children to sacrificial fires. And the prophet Jeremiah complains that the Israelites have ignored his warnings to stop sending their children to the sacrificial fires, as had been the custom of the heathens who were their neighbors.

Human sacrifices by fire were also common more recently in many parts of the world. In Ireland, for example, Druids sacrificed to their gods and goddesses by placing victims in wickerwork cages and setting them on fire. In one part of the island, the festival of Samain (1 November) was celebrated by the sacrifice of one-third of all children by fire.

Most burnt offerings conducted today involve the use of plants, oils, or other inanimate objects. Among the Hindus, for example, sacrificial fires are an important part of many ceremonies and are conducted using ghee, or melted butter. In China, some people use fake money in sacrificial fires because they believe that their dead ancestors will be able to use the currency in the afterlife. In Japan, prayers are printed on rice paper before being burned so that their messages can be carried to the deities in their smoke.

Transformation

Fire rituals are sometimes performed to bring about important changes in a person's life. Such rituals are called *transformational rituals* and are used to help a person deal with physical, emotional, or other problems in their lives. The steps to be followed in such rituals are generally very clearly defined and involve participants' interacting with a central fire in some way or another. For example, participants may be asked to bring with them to the ceremony some object, such as a piece of cloth, to offer to the fire as a representation of the issue with which they are trying to deal. As the fire burns up and cleanses the cloth, it is thought to perform a similar cleansing action on the person who made the offering.

Festival Fires

Nearly every human culture celebrates one or more important events in its history by means of a festival fire. One of the best known examples is the Guy Fawkes bonfires that are set every year on 5 November across England. Although this celebration is based specifically on an aborted effort by Fawkes to blow up the Parliament buildings in 1605, the history of festival fires goes much farther back in English history and in the traditions of nearly every other nation. *See also* BONFIRES; CREMATION; ETERNAL FLAMES; NEW-FIRE; RELIGIOUS ALLUSIONS TO FIRE; VESTA; WITCH BURNING.

Further Reading: "Agni Yoga," <http://www.agniyoga.net/AgniYoga.htm>, 08/17/99; "Fire Ceremony," <http://www.triquetra.com/firecere.htm>, 10/26/98; Frazer, J.G. *Balder the Beautiful: The Fire-Festivals of Europe and the Doctrine of the External Soul*, 2 vols. London: Macmillan and Company, 1913, Chapter 4; Frazer, Sir James George. *The Golden Bough: A*

Soldiers discover conspirator Guy Fawkes (1570–1606) attempting to blow up the House of Parliament. © *Hulton Getty/Archive Photos.*

Study in Magic and Religion. New York: The Macmillan Company, 1958, Chapter 64, passim; Goudsblom, Johan. *Fire and Civilization.* London: Allen Lane, 1992, pp. 74–80; McCloud, Minx, "Therapist Borrows from Incas," <http://www.pacpub.com/new/news/6-4-98/incas.html>, 03/02/99; Pyne, Stephen J. *Vestal Fire: An Environmental History, Told through Fire, of Europe and Europe's Encounter with the World.* Seattle: University of Washington Press, 1997, pp. 68–78; Staal, Fritz. *Agni, the Vedi Ritual of the Fire Altar.* Berkeley, CA: Asian Humanities Press, 1983; Trinkunas, Jonas. "Ritual of Fire," <http://www.romuva.ot/fire_e.htm>, 07/20/99; Tylor, Edward Burnett. *Religion in Primitive Culture.* New York: Harper Torchbooks, 1958, passim.

RIVERSIDE FIRE LABORATORY

The Riverside Fire Laboratory was created on 29 March 1960 by the U.S. Congress largely in response to devastating wildfires that had taken place in southern California in 1953 and 1958. The laboratory was situ-ated near the University of California at Riverside, partly because of its proximity to other fire research facilities. The initial legislation envisioned a facility that would include an instrument shop, electronics laboratory, staging area, forest meteorology facility, humidity chambers, forest fuels laboratory, and chemistry laboratory. The original structure consisted of three buildings, an administrative office, a technical center, and a laboratory, covering a total of 31,000 square feet overall. The buildings were dedicated on 11 September 1963.

Over the years, the mission of the Riverside Fire Laboratory has changed to some extent. Today it focuses on a variety of forest fire problems, including research on prescribed fire and fire effects, the effects of various meteorological factors on fire forecasting, atmospheric deposition effects in the Western United States, and fire management at the wildland/urban interface.

For further information, contact:
Riverside Fire Laboratory
4955 Canyon Crest Drive
Riverside, CA 92507
(909) 680–1500
http://www.rfl.psw.fs.fed.us/
See also FIRE MANAGEMENT; FOREST FIRES.

ROAD OF FLAME

The Road of Flame is a doctrine first developed in ancient India that describes one of the things that can happen to a person after he dies. The doctrine appears to have applied exclusively to men since women were regarded as too impure to pass through this course after death.

The Road of Flame was contrasted to the Road of Smoke, a second pathway through which the dead might pass. The road one traveled depended on the life that one had lived. A virtuous man would, after cremation, travel along the Road of Flame, first to the Moon, then to the Sun, and then into lightning. Present in the lightning was a nonhuman spirit who took the person to the gods, where he remained forever.

In contrast, a person whose life was not virtuous was doomed to travel the Road of Smoke that led first to the Moon and then to the Sun. At that point, however, they travel to the world of their ancestors, from which they return again to the Earth. They are transformed first into a wind and are then converted to a cloud, after which they fall back to Earth in the form of rain. They are reborn again in the form of rice, barley, an herb, a tree, or some other kind of plant.

In this form, the person has little chance of being reborn as a human, but is more likely to appear again as a pig, dog, or some other type of animal. Only if the person (in the form of a plant) is eaten by another person and becomes part of that person's semen can the dead be reborn as a human. *See also* RELIGIOUS ALLUSIONS TO FIRE.

Further Reading: Campbell, Joseph. *The Masks of God: Oriental Mythology*. New York: The Viking Press, 1962, Chapter 4.

ROCKET. *See* JET ENGINES

ROMAN BATHS

The Roman public baths were facilities designed for the pleasure of all Romans, including men, women, and children, both free and slaves. They are an illustration of the way in which heat produced by a fire can be used to maximum advantage in a closed structure.

The baths consisted of five primary areas: apodyterium, tepidarium, caldarium, frigidarium, and unctorium. The apodyterium was the first room in which the bather changed from street clothes to bathing apparel. The next room visited, the tepidarium, was a warm room in which the bather first began to perspire. He or she then moved on to the caldarium, the hottest room in the baths.

The heating system was connected directly to the caldarium. Fire was used to boil water, which then passed through pipes beneath the caldarium. As heat was lost from the hot water into the caldarium, the water continued through the piping system to the tepidarium. What little heat it retained was then given off in the "warming room."

Finally, the water passed beneath the frigidarium, the cooling room. The water, now cool, removed heat from the frigidarium, keeping it at the lower temperature needed in this space. The water was then piped back into the furnace, where it was reheated by the fire. Meanwhile, the bather continued onward, out of the cool frigidarium, into the unctorium. The unctorium was the room in which he or she was given a massage to complete the bathing ritual.

The baths were popular not only as a place of relaxation for bathers, but also as a meeting place where gossip and business were conducted. In more elaborate buildings, the baths also included other facilities, such as swimming pools, game rooms, and lounging areas. *See also* HEAT.

Further Reading: DeLaine, Janet. *The Baths of Caracalla: A Study in the Design, Construction, and Economics of Large-Scale Building Projects in Imperial Rome*. Portsmouth, RI: Journal of

Roman Archaeology, 1997; Nielsen, Inge. *Thermae et Balnea: The Architecture and Cultural History of Roman Public Baths.* Aarhus, Netherlands: Aarhus University Press, 1990.

ROME. *See* BURNING OF ROME

RUMFORD, COUNT, BENJAMIN THOMPSON

Count Rumford was one of the most extraordinary scientists and politicians of the late eighteenth and early nineteenth centuries. The experiments he conducted on the nature of heat were instrumental in the downfall of the caloric theory and seminal in the development of the science of thermodynamics.

His Life

Thompson was born in Woburn, Massachusetts, on 26 March 1753. At the age of nineteen, he married a rich widow, considerably older than himself. As troubles between the colonies and England developed in the 1770s, Thompson aligned himself with the mother country. When war broke out, he left his homeland (and his wife and young child) and spent the next few years serving in various minor positions in the English government. When the war ended with the revolutionaries victorious, Thompson knew that he could not return to the United States.

By 1783, Thompson had worn out his welcome in England and moved to Bavaria, where he became an administrator in the government of Elector Karl Theodor. His work so pleased Theodor that the Elector named him Count Rumford, the title coming from the town in New Hampshire where Thompson had lived briefly with his wife.

In 1799, Rumford decided it was time to leave Bavaria and return to England. The Elector had died, and Rumford's machinations had lost him support at court. Shortly after returning to England, he was persuaded to establish the Royal Institution, a center for research and education in the sciences. Five years later, Rumford moved to France where he met and married another rich widow, the former wife of Antoine Laurent Lavoisier. The marriage proved to be an unhappy one and lasted only four years. Rumford's daughter came from America to care for him for the last few years of his life. He died in Auteuil, near Paris, on 21 August 1814.

Scientific Accomplishments

In spite of his relatively unpleasant personality, Rumford was a serious and dedicated scientist. He became interested in the question of the nature of heat in the late 1790s while supervising the boring of cannons in Munich. He observed that an endless amount of heat appeared to be produced when cannons were being bored, with virtually no loss of metal. This observation appeared to be in direct contradiction to the then-popular caloric theory of heat, which held that heat was a form of matter. Rumford went one step further and attempted to calculate the amount of heat produced when a given amount of energy was used in the boring process. This research was one of the earliest attempts to determine the mechanical equivalent of heat.

In later research, Rumford very carefully weighed a sample of water in both liquid and solid form. He found no mass difference whatsoever in the two cases, a result he took to provide further confirmation of the falsity of the caloric theory. *See also* HEAT.

Further Reading: Asimov, Isaac. *Asimov's Biographical Encyclopedia of Science & Technology.* 2nd revised edition. Garden City, NY: Doubleday & Company, 1982, pp. 242–244; Brown, G.I. *Scientist, Soldier, Statesman, Spy: Count Rumford: The Extraordinary Life of a Scientific Genius.* Stroud, Gloucester: Sutton Publishing, 1999; Brown, Sanborn C. *Count Rumford: Physicist Extraordinary.* Garden City, NY: Anchor Books, 1962; Brown, Sanborn Conner. *Benjamin Thompson, Count Rumford.* Cambridge, MA: MIT Press, 1979; Holton, Gerald, and Duane H.D. Roller. *Foundations of Modern Physical Science.* Reading, MA: Addison-Wesley Publishing Company, 1958, Chapter 19; Larsen, Egon. *An American in Europe: The Life of Benjamin Thompson, Count Rumford.* London: Rider Publishers, 1953; Orton, Vrest. *Observations on the Forgotten Art* *of Building a Good Fireplace: The Story of Sir Benjamin Thompson, Count Rumford, an American Genius and His Principle*. Dublin, NH: Yankee Press, 1974; Porter, Roy, ed. *The Biographical Dictionary of Scientists*, 2nd edition. New York: Oxford University Press, 1994, pp. 591–592.

S

SAFETY LAMP. *See* DAVY SAFETY LAMP

ST. ELMO'S FIRE

St. Elmo's fire is the name given to a naturally occurring phenomenon that takes on the form of a fireball in the sky. The phenomenon was first observed by sailors who thought they saw balls of flame suspended at the top of their ships' masts. They gave the name *St. Elmo's fire* to these phenomena in honor of the patron saint of sailors, St. Elmo.

St. Elmo's fire does not involve flames or any other kind of combustion-related event. Instead, it is an electrical phenomenon caused when static electricity builds up on a surface and is suddenly released into the air.

Air normally consists of neutral molecules, containing equal numbers of positively charged atomic nuclei and negatively charged electrons. In some circumstances, an excess of electrical charge of one kind or another can build up on a surface. The charge is not distributed equally across the surface, but tends to concentrate on a point or sharply curved surface.

When the electrical charge on the point or the surface exceeds some value, the charge escapes from the point or surface into the surrounding air. It causes air molecules to break apart into positively and negatively charged particles. This effect is known as a *corona discharge,* an *aurora*, an *electric corona*, or simply a *corona*.

A corona discharge gives off a glow as electrons that had been torn from their nuclei return to them a fraction of a second later. As they do so, they give off energy in the form of visible light. With air molecules, the light produced is a bluish-purple, the color of St. Elmo's fire.

Sailors saw St. Elmo's fire at the top of a mast because that was the most rounded part of the mast. It was the region in which electrical charges were able to collect to a high enough density to cause a corona discharge. In today's world, St. Elmo's fire is also observed commonly during aircraft flights. The tips of a plane's wings and tail have the shapes that permit the accumulation of electrical charges that lead to a corona discharge and the formation of St. Elmo's fire.

Further Reading: Forrester, Frank. *1001 Questions Answered about the Weather*. New York: Dover Publications, 1981, pp. 182–183; Schaefer, Vincent, and John A. Day. *A Field Guide to the Atmosphere*. New York: Houghton Mifflin, 1983, p. 196; "What Causes the Strange Glow Known as St. Elmo's Fire?" <http://www.sciam.com/askexpert/physics/physics35.html>, 12/22/98.

ST. FLORIAN (ca. 250 A.D.–304)

St. Florian is regarded as the patron saint of firefighters. He was thought to have been born in about 250 A.D. in the region then known as Cetium, now part of modern Austria. He became a soldier in the Roman army and rose to become an administrator in the region of Noricum, also in modern-day Austria.

During his term of office, the Emperor Diocletian pronounced an edict against the Christian faith, declaring that anyone who practiced the religion would be rounded up and persecuted. Florian declined to carry out this order, however. Instead, he made his own confession of faith to the governor of the province, one Lorch.

As punishment for his disobedience, Florian was scourged, flayed, and threatened with death by burning. As he was tied to the stack of wood on which he was to be burned, however, he gave a warning to his fellow soldiers. "If you burn me," he is reputed to have said, "I will ascend into heaven on the flames."

The threat terrified his executioners. They decided to drown him rather than burn him. He was thrown into the river Enns with a stone around his neck.

Florian was later beatified and then raised to sainthood. In 900, a monastery was erected near the spot where Florian had been buried six centuries earlier. A small town also grew up around the monastery.

In 1138, King Casimir of Poland asked Pope Lucius III to send St. Florian's relics to his nation. Shortly thereafter, Florian was adopted as the patron saint of Poland.

Florian's connection with firefighting comes from an event that occurred shortly after his relics were sent to Poland. The story is told that a person was saved from a fire by calling out the saint's name. Since that time, Florian has been regarded as the patron saint of firefighting.

Further Reading: Thurston, Herbert, and Donald Attwater, eds. *Butler's Lives of the Saints*. New York: P.J. Kennedy & Sons, 1926, Volume 22, pp. 230–231.

SANTA ANA WINDS

Some of the most predictable fires to occur annually are those caused by the Santa Ana winds. The Santa Ana winds are warm, dry winds that blow out of the Great Basin region in the U.S. West. They flow through a gap between the San Gabriel and San Bernadino Mountain ranges east of Los Angeles. They occur every year, with greater or less severity, usually producing damaging fires in their wake.

Santa Ana winds typically occur between October and February each year. They have their beginning in the Great Basin where high-pressure regions develop and are driven in a southwesterly direction by a clockwise circulation of air. As they move toward the Pacific Coast, the winds are driven through a unique topographical region consisting of narrow canyons. Some of the strongest gusts occur in the Santa Ana Canyon, from which the winds get their name.

The winds become drier and pick up speed as they pass through the canyons, reaching speeds in excess of forty miles per hour with gusts of up to 100 miles per hour. Meteorologists generally reserve the term *Santa Ana wind* for winds with speeds greater than twenty-five knots (about thirty miles per hour).

When the winds reach the western slopes of the San Gabriels and San Bernadinos, they sweep across the land. As they travel, they become hotter, drying out nearly all vegetation in their path. The risk of fire quickly becomes very great, and a single spark or flame can ignite a wildfire. It is not unusual for fires to destroy thousands of acres of vegetation and residential areas.

One of the worst of these fires occurred in the Malibu area of California in 1996. More than 2,500 firefighters aided by tanker planes and helicopters attempted to bring the fire under control. Eventually, about 35,000 acres were burned and more than 100 homes and other structures were destroyed. Fortunately, there were no deaths and only about a dozen injuries, primarily to firefighters. *See also* Forest Fires; Firestorm.

SAVONAROLA, GIROLAMO. *See* Bonfire of the Vanities

SCALDS. *See* Burns and Scalds

SCORCHED EARTH POLICY

A scorched earth policy is a military tactic that attempts to destroy fertile land by fire and leave an attacking army with no natural resources or other assets with which to continue its battle. The policy appears to have been practiced since time immemorial by military units. One of the first reports of such a policy relates to the battle of Granicus River in 334 A.D., between the Persian army and the forces of Alexander the Great. Historical records say that Memnon, a Greek advisor to the Persian army, recommended that the Persians withdraw to the east and pursue a scorched earth policy in order to leave nothing for Alexander's army on which to subsist. The Persians distrusted Memnon, however, and decided not to lay waste to their own lands. In the end, Alexander was able to cross the Hellespont, engage the Persians on the shores of the Granicus, and gain a victory. The victory was a crucial step in opening the Orient to Alexander's forces.

Military history is replete with example after example of the use of scorched earth policies by retreating, conquering, and vengeful armies. In order to pacify the Philippine Islands that had been ceded to it in 1898, the U.S. Army undertook a fierce conquest of the young Philippine Republic. It used many of the same policies against the islands that would be used seventy years later in the Viet Nam War.

One of the most extensive scorched earth policies prior to the twentieth century was pursued during the U.S. Civil War. General William Tecumseh Sherman's "March to the Sea" in 1865 was little other than a scorched earth policy in which rebel Confederate cities, towns, and lands were ravaged by the advancing Union Army.

Scorched earth policies are often accompanied by inhuman attacks by advancing or retreating military forces on opposing forces and, even more commonly, on the civilian populace. In their terrible attack on Nanking, China, on 12 December 1937, Japanese troops were said to have robbed and then burned more than 12,000 stores and houses. They then attacked, raped, and killed every female they could find between the ages of 10 and 70.

As armies became larger, wars and battles more extensive, and technology more advanced, scorched earth policies correspondingly became more severe. Some of the most dramatic examples of scorched earth policies arose during World War II. For example, the Soviet Union practiced such policies as the German army moved into the Ukraine in the early days of the war. Joseph Stalin decided that every effort be made to deprive the advancing Germans from having access to any resource whatsoever. He released a directive on 3 July 1941 that ordered:

> In case of a forced retreat . . . all rolling stock must be evacuated, the enemy must not be left a single engine, a single railway car, not a single pound of grain or gallon of fuel. . . . In all areas occupied by the enemy, guerilla units . . . must set fire to forests, stores, and transports. ("World War II in Ukraine")

Later in the war, conditions were reversed. By December 1941, it was the Germans who were in retreat. It was then Adolf Hitler's time to announce a scorched earth policy. On the 22nd of that month, Hitler ordered:

> Each area that has to be abandoned to the enemy must be made completely unfit for his use. Regardless of its inhabitants every locality must be burned down and destroyed to deprive the enemy of accommodation facilities . . . the localities left intact have to be subsequently ruined by the air force. ("World War II")

The policy proved to be a huge success, at an enormous loss to the people of the Ukraine. By one account, more than 28,000 villages and 714 cities and towns were totally destroyed. More than 10,000,000 people were left without shelter. In addition, 16,000 businesses, 27,910 collective farms, 872 state farms, and 32,930 schools and other educational institutions were destroyed.

Scorched earth policies remain popular even in modern day warfare. During the Gulf War of 1991, for example, the Iraqi army set fire to oil wells in parts of Kuwait from which they were fleeing in order to provide cover and to deprive the Kuwaitis and their allies of a valuable natural resource.

Further Reading: Carell, Paul. *Scorched Earth: The Russian-German War, 1943–1944*, London: Schiffer Publishing, 1994; Editors of Time-Life Books. *Scorched Earth*. Alexandria, VA: Time-Life Books, 1991; Pyne, Stephen J. *Vestal Fire: An Environmental History, Told through Fire, of Europe and Europe's Encounter with the World*. Seattle: University of Washington Press, 1997, pp. 291–295; Thomas, William. *Scorched Earth: The Military's Assault on the Environment*. Philadelphia: New Society Publishers, 1995; "World War II in Ukraine," <http://www.infoukes.com/history/ww2.html>, 12/27/98.

SHIFTING CULTIVATION

Shifting cultivation is a form of agriculture also known as *slash-and-burn*. The practice has been used for centuries by peoples who live in areas with rich expanses of natural vegetation, such as the Amazon, Indonesia, and India. Some other terms used to describe shifting cultivation or very similar practices include *kaski* (in Finland); *ladang* (in Malaysia); *kyttlandsbruk* and *svedjebruk* (in Sweden); *landnam* (in Denmark); *reutberge* or *waldfeldbau* (in Germany); paring and burning, or *devonshiring* and *denshiring* (in England); *swidden* (in Scandinavia); *tavy* (in Madagascar); and *zalezhé* (in Russia).

The Method

The first step in the slash-and-burn system is, as the name suggests, cutting down trees in order to clear land for agricultural crops. The trees are burned, and the ashes produced are spread on the land as fertilizer. The use of ashes as fertilizer is important because it helps restore to the land nutrients that are removed from the soil by the growing trees.

When treated in this way, cleared lands can usually be employed for the growing of crops for up to ten years. Farming is done most efficiently, however, if different crops are grown during the cycle. That is, certain crops will exploit some nutrients in the soil, and other crops will exploit others. Over many centuries, peoples who lived in forested re-

gions and practiced shifting cultivation learned which crops to grow at which times of the cycle.

When the soil in the cleared land was exhausted of nutrients, people simply left the area and went elsewhere. They then repeated the cycle of cutting and burning trees and growing crops on the cleared land. The land that had been cleared and cultivated in the first cycle was allowed to remain fallow, permitting natural vegetation to regrow and nutrients to collect once more in the soil. At some future time, that same plot of land might be once more ready to be cut and cultivated.

Problems with the System

A successful program of shifting cultivation depends primarily on the availability of very large stretches of land. People have to be able to move on to new areas as soon as the land on which they are working has been depleted of nutrients. In addition, shifting cultivation is inherently not a very efficient form of agriculture because the land on which it is practiced tends to be low in nutrients under the best of circumstances.

Whatever value shifting cultivation may have had in some cultures in the past, it has now become much less successful as a form of agriculture. The primary reason for this change is that various forms of development have vastly diminished the amount of land available to nomadic peoples for the practice. The drilling for oil, search for minerals, construction of new cities and towns, and rapid deforestation of many regions formerly used for shifting cultivation has now made it virtually impossible to practice this form of farming.

Other Applications

The principles of slash-and-burn agriculture have been adopted in recent decades by large corporations for purposes quite different from those described above. For example, lumber companies have been practicing the slash-and-burn technique in the forests of Borneo and Sumatra since 1983. The companies would go into an area, strip it of trees, and then spray the area with gasoline. When the gasoline was ignited, huge fires would burn away waste trees, underbrush and other unwanted vegetation. Foresters could then go in and replant young trees in anticipation of the next harvest.

The success of this technique, however, depends on the regular appearance of the monsoon rains that typically arrive in East Asia in July and August. In 1997, primarily because of the *El Niño* weather patterns then in place, the rains never came. As a result, the fires set by logging companies grew out of control. Huge clouds of smoke spread across Indonesia and into other parts of Southeast Asia. Businesses and schools closed for days at a time, and people were often able to travel out of doors only while wearing masks.

Some experts placed the cost of the incident at tens of billions of dollars in lost economic production, timber, and tourism. They saw the disaster as a "wake-up call" to Indonesia and other Southeast Asian governments about their insensitivity to land use and other environmental practices. *See also* CLIMATE CHANGE; FOREST FIRES.

Further Reading: Conklin, Harold C. *The Study of Shifting Cultivation*. Studies and Monographs VI. Washington, DC: Panamerican Union, 1963; International Workshop on Shifting Cultivation. *The Future of Shifting Cultivation in Africa and the Task of Universities*. Rome: Food and Agriculture Organization, 1984; Peters, William J., and Leon F. Neuenschwander. *Slash-and-Burn: Farming in the Third World Forest*. Moscow: University of Idaho Press, 1988; Ramakrishnan, P.S. *Shifting Agriculture and Sustainable Development*. Paris: UNESCO; New York: Parthenon Publishing Group, 1992; Robison, Daniel M. *Shifting Cultivation and Alternatives: An Annotated Bibliography, 1972–1989*. Wallingford: C.A.B. International; "Shifting Cultivation," <http://hammock.ifas.ufl.edu/txt/fairs/12046>, 12/18/98; "Shifting Cultivation," <http://www.sru.edu/depts/artsci/ges/lamerica/d-5-13.htm>, 12/18/98; Warner, Katherine. *Shifting Cultivators: Local Technical Knowledge and Natural Resource Management in the Humid Tropics*. Rome: Food and Agriculture Organization of the United Nations, 1991.

SLASH-AND-BURN. *See* SHIFTING CULTIVATION

SMELTING. *See* FURNACES

SMOKE

Smoke is a colloidal dispersion of a solid in a gas. A colloidal dispersion is a physical state consisting of two components, one of which consists of finely divided particles suspended in a gas, solid, or liquid. As an example, fog is a colloidal dispersion in which droplets of water are dispersed (suspended) in air.

Types of Smoke

To most people, the term *smoke* is a familiar term. It applies to the white to gray to black cloud that forms when paper, wood, coal, or other combustible materials are burned. *Wood smoke* or *coal smoke* actually consists of tiny particles of unburned carbon produced during combustion of the fuel (wood or coal, in this case). The color of smoke depends largely on the size and composition of the dispersed particles. Those in wood smoke are made up of ash and are smaller than those in coal smoke. The particles in coal smoke are made of unburned carbon and are larger than those from wood smoke. Thus, coal smoke tends to reflect light less efficiently and is darker than wood smoke.

Metallic smoke is another type of smoke. It is formed when metals or metallic ores are heated. A metallic smoke consists largely of unburned particles of the metal or the metal ore. These particles tend to reflect light efficiently and may have any one of a variety of colors.

Chemical smoke is smoke produced by the combustion of a particular element or compound, such as the element phosphorus or the compound titanium tetrachloride ($TiCl_4$). In many instances, a chemical smoke is produced intentionally for military purposes (such as providing a smoke screen for a potential target for aerial bombing) or for entertainment purposes (as in a magician's stage show).

Cigarette smoke is a highly complex mixture in which the dispersed phase consists of more than 100 distinct compounds, including nicotine, cresol, pyridene, benzopyrene, and polonium (a radioactive element).

Health Issues

A variety of health and environmental issues are associated with all forms of smoke. The issues surrounding cigarette smoke have been perhaps more intensely studied than those relating to any other kind of smoke. At this point, medical researchers have demonstrated that the substances dispersed in cigarette smoke are responsible for a number of medical problems, including emphysema, chronic bronchitis, lung cancer, other respiratory disorders, and heart problems.

Metallic and chemical smoke also pose health problems because the particles of which they consist are often toxic, carcinogenic, or teratogenic to anyone who inhales them. The effect of such smokes depends on the size and chemical composition of the particles they contain. The larger the particles, the more likely they are to be trapped in the human respiratory system, giving them a longer period of time to exert their harmful effects on the body.

The primary health effects of wood and coal smoke also result from the size of the particles of which they consist. Particles larger than about two microns (micrometers, or millionths of a meter) tend to become lodged in the bronchi of the respiratory system. The bronchi are two tubes that lead from the windpipe to the lungs. As they become clogged with smoke particles, moving air through the respiratory system into the lungs becomes more and more difficult. In such cases, emphysema, bronchitis, and other respiratory disorders may arise.

Environmental Issues

The kinds of health issues created by inhalation of smoke by humans are reproduced in some regards in other organisms. For example, plants that grow in an area where wood and coal smoke are common may become unhealthy. The particles that make up this smoke can clog the stomata through which plants exchange gases with the atmosphere. The toxic nature of certain metallic and chemical smoke also exert their effects on plants and other animals. For example, the chemical smoke produced near a smelter often carries sufficient toxic substances to cause the death of all living organisms in the area around the smelter.

Wood and coal smoke also create aesthetic problems in areas where they are produced in abundance. One characteristic of any colloidal dispersal is that the dispersed particles may eventually settle out of the material in which they are suspended. Thus, the tiny coal particles from coal smoke tends eventually to settle out of the air onto cars, buildings, people's clothing, and other exposed surfaces. Although they may not create a health problem for any living organism, they may make it necessary to spend large amounts of money to clean the surfaces that have become covered with coal dust.

Prevention and Treatment

A number of methods are available for reducing the amount of smoke, of all kinds, released to the air. In the case of cigarette smoke, it is possible, for example, to attach a filter to a cigarette to cut down on the amount of smoke released to the air. The filter may capture particles of some of the harmful chemicals found in smoke. The use of filters in cigarettes is somewhat problematic, however. They tend to reduce the flavor of the cigarette, so that manufacturers may find it necessary to increase the amount of tobacco used in such cigarettes. In such cases, use of the filter achieves only a portion of its intended objective.

At one time, industrialists and others responsible for production of large quantities of smoke were not particularly concerned about the health or environmental effects of smoke. For example, people living in England during the early decades of the Industrial Revolution probably gave little thought to the health and environmental effects of the vast quantities of smoke being produced by factories and other wood- and coal-burning operations. In many cases, the new prosperity created by coal-burning technology was a fair tradeoff for any inconvenience or medical discomforts they may have felt.

Today, most individuals and corporations are far more sensitive to the need to control the production of smoke. One general method for doing so is to increase the efficiency of combustion devices. For example, furnaces that operate at very high temperatures are more inclined to bring about more complete combustion of fuel, resulting in the release of smaller quantities of unburned carbon.

Methods for removing smoke particles from effluent gases have also been developed. For example, many smokestacks today are equipped with devices that remove particles of unburned carbon and other smoke dispersants before they leave the stack. One method makes use of charged electric fields on the inside of the stack. The electrical charges on the stack attract tiny particles of carbon and other dispersants, preventing them from escaping into the air. Another method involves the passing of effluent gases through chemical solutions suspended in the smoke stack. Pollutant gases react with the chemical solutions to form precipitates, thus preventing the gases from escaping into the air. *See also* COMBUSTION.

Further Reading: Green, H.L., and W.R. Lane. *Particulate Clouds: Dusts, Smokes, and Mists: Their Physics and Physical Chemistry and Industrial and Environmental Aspects*, 2nd edition. London: E.&F.N. Spon, 1964; Harrison, R.M., ed. *Pollution: Causes, Effects, and Control*, 2nd edition. Cambridge: Royal Society of Chemistry, 1990, pp. 135–143; Hasegawa, Harry K., ed. *Characterization and Toxicity of Smoke*. Philadelphia: ASTM, 1990; Hilado, Carlos J., ed. *Smoke and Products of Combustion*. Westport, CT: Technomic Publishing Company, 1973; Lewis, Richard J., Sr. *Hawley's Condensed Chemical Dictionary*, 12th edition. New York: Van Nostrand Reinhold Company, 1993, p. 1044; Rossotti, Hazel. *Fire*. Oxford: Oxford University Press, 1993, Chapter 14; Turk, Jonathan, Amos Turk, and Karen Arms. *Environmental Science*, 3rd edition. Philadelphia: Saunders College Publishing, 1984, Chapter 21.

SMOKE BOMBS. *See* SMOKE GENERATION

SMOKE DETECTOR

A smoke detector is a fire safety device that is able to detect the presence of smoke in a room. The presence of smoke typically indicates that a fire exists in the room. When smoke is detected, the device sends out a piercing sound that warns occupants to leave the room.

Construction

Two types of smoke detectors are made: photoelectric detectors and ionization detectors. In a photoelectric detector, a beam of light is shined from one side of the detector to the opposite side. The receiving side of the detector is coated with a light-sensitive material.

In the absence of smoke, the beam of light travels across the device without interruption. If smoke enters the detector, however, the amount of light transmitted decreases. The detector measures the decrease in light intensity and trips a signal mechanism, such as a sound. For hearing-impaired individuals, smoke detectors can also be modified so that they turn on a light, rather than producing a sound, when the light beam has been interrupted.

Today, ionization detectors tend to be more common than photoelectric detectors. The core of an ionization detector is a very small piece of a radioactive material. The synthetic element americium (americium-241) is usually used for this purpose.

Americium-241 decays radioactively (breaks down with the release of radiation) producing alpha and gamma rays. Alpha rays ionize air molecules within the detector. Ionization is the process by which neutral molecules are bro-

A smoke detector sends out a piercing sound in the presence of smoke. *Courtesy of Gentex Corporation.*

ken apart to produce electrically charged particles. The presence of these electrically charged air molecules permits an electric current to flow from one side of the detector to the opposite side. A very small amount of electricity flows, therefore, through an ionization detector in which smoke is absent.

If smoke particles do get into the detector, however, they tend to attach themselves to ionized air molecules. In so doing, they reduce the electrical charge on the molecules and, in turn, reduce the flow of electrical current through the device. The reduction in electrical flow is detected within the device, and a signal (sound or light) is triggered.

Use

Smoke detectors are widely regarded as one of the most effective fire protection devices one can install in a home, office, or other structure. They are inexpensive and, when properly maintained, highly reliable. The fact that the American public has become convinced of the value of smoke detectors is reflected in the fact that thirteen out of every fourteen homes in the United States has at least one smoke detector.

Two problems remain, however. First, the small fraction of homes that do *not* have a detector are at high risk for serious fires. In the late 1990s, nearly half of all residential fires occurred in those homes without a detector. More than 60 percent of all fire-related deaths occurred in those same homes. Second, too many homeowners do not bother to keep their smoke detectors in proper working order. "Proper working order" means essentially that batteries are replaced on a regular basis. Studies have shown that the number of homes with nonworking detectors is actually greater than the number of homes with no detector at all. About one-third of the homes with nonworking detectors have fires each year, fires in which hundreds of people are killed.

Fire safety experts now suggest that homeowners check their smoke detectors twice a year when they adjust their clocks because of time changes.

Further Reading: Bukowski, Richard W. *Smoke Detector Design and Smoke Properties*. Washington, DC: Center for Fire Research, 1978; "Smoke Detector Information," <http://www.cfdonline.org/detector.htm>, 12/22/98; Yu, Jessen, "A:Eeeeeeeee! There are Two Different Types of Smoke Detectors—Photoelectric Detectors and Ionization Detectors." *The Stanford Daily*, 7 May 1996 and at <http://daily.stanford.org/5-7-96/NEWS/index.html>, 12/22/98.

SMOKE GENERATION

Smoke generation is the process by which smoke is created intentionally for some specific purpose. Historically, one of the earliest and most widely forms of smoke generation was that used for military purposes: the smoke screen. Today, smoke generation is used in a large variety of applications, such as training of firefighters and other personnel for emergency situations, leak testing, air tunnel and other air-flow visualizations, home and office security, and entertainment programs.

Smoke Screens

It seems likely that military strategists throughout human history must have understood the potential value of smoke in hiding certain types of operations. Indeed, there are records of officers attempting to hide their movements or to befuddle their enemies by laying down small smoke screens.

However, it was not until World War I that the value of *large area smoke screens* (LASS) became obvious. With the introduction of air warfare, factories, oil refineries, and other potential targets required a new form of protection. LASS seemed to be a promising way of hiding such facilities from enemy bombers. But the problem turned out to be somewhat more complex than was originally thought. To begin with, the LASS really had to be *large*. The smoke covering had to be spread out over a sufficiently large area that bombers were confused by it rather than having the target pinpointed by the smoke. In addition, the science of smoke production had not been developed to a point where large quantities of relatively safe and persistent smoke could be produced and controlled.

There was, in general, relatively little progress in LASS technology between the two world wars. Evidence for this fact comes from the techniques used by Great Britain during the early years of World War II. As soon as the Germans had conquered France and the Low Countries, their air force had bases from which they could easily launch attacks on England. The LASS system used by the British to protect against these attacks was very primitive.

The British first canceled all smoke abatement regulations that had earlier been imposed on factories. Instead of trying to cut down on the amount of smoke produced by industry, the government encouraged an increase in smoke. Second, several thousand smudge pots were installed around the countryside. These smudge pots were very low-tech devices, consisting of a large pot in which petroleum was burned. The huge volumes of black smoke produced by these pots was moderately effective in disguising potential Luftwaffe targets and confusing pilots.

By the time the United States entered World War II, military strategists had become convinced of the need for some form of LASS. One primary area of concern was the Panama Canal. As an important link in the nation's military transportation system, the Canal had to be defended from enemy bombing attacks at all costs. Experiments conducted in 1940 at the Canal produced conflicting results. The LASS produced was effective in hiding the Canal, but two problems arose. First, the technology needed for an effective LASS was still not well developed. Second, officers complained that the smoke produced in the experiment seriously interfered with their antiaircraft efforts from the ground.

Building on this experience, the Chemical Warfare Service of the U.S. Army moved forward on its LASS research. By early 1942, thirty-four chemical smoke-generating companies had been created within the Service. The first eleven of these companies were stationed in Panama, at the Sault Ste. Marie Locks in Michigan, at the Edgewood Arsenal, and at the Aberdeen Proving Grounds in Maryland.

These companies still relied on a relative primitive form of smoke generation—the commercially available smudge pot used in agricultural fields. Eventually they were able to use a more sophisticated model of smoke generator that released a spray of superheated oil into the air. The oil condensed upon hitting cooler air, forming a dark, persistent LASS. The generator was known officially as the M1 mechanical smoke generator, but its nickname, the "Esso," came from the company that developed it for the army.

The need for LASS technology by U.S. forces diminished rapidly. After the Battle of Midway in June 1942, the Japanese threat to the continental United States was vastly reduced. Research and development on LASS technology continued for other battle locations, but it never again assumed the importance it had shortly after the attack on Pearl Harbor.

Smoke Generators Today

The role of LASS in modern military planning is greatly diminished for a number of reasons. For example, aircraft no longer depend on visual sightings to identify their targets. New technology has made it possible to see through clouds more easily with instrumentation than by human eyesight. Also, no one really thinks that laying down a cloud of smoke will protect any area from most kinds of weapons available to the military today.

Nonetheless, smoke generation is still an important field of technology, with applications in many different fields. One obvious use is in the training of firefighters. One constant problem which firefighters have to deal with is smoke produced during a fire. Part of their training, then, involves the production of clouds of artificial smoke that allow them to have practice in navigating their way through such an environment.

Other workers who face the possibility of working in smoke-filled conditions benefit from similar training. For example, airline stewards and stewardesses must be trained to deal with sudden emergencies in which an airplane cabin may fill with smoke. Commercial smoke generators are now available for such training sessions.

Smoke generators are also used to study airflows, as in a wind tunnel. It is not possible, of course, to watch normal airflow over the wings and/or body of an experimental aircraft. However, the addition of a small amount of smoke to the airflow makes its shape and pattern clearly visible.

Smoke can be used similarly in testing devices for air leaks. For example, a leak in an airplane luggage compartment might be too small to be visible to the human eye, but large enough to pose a risk to the plane when it is in flight. By forcing smoke through the luggage compartment, such leaks can be easily identified.

Finally, smoke generators are widely used in the entertainment business. At times, their purpose is to produce smoke that mimics some real disaster, as, for example, war scenes in which large amounts of smoke are produced. Smoke is also used purely for aesthetic reasons, as when a magician heightens the effects of his or her acts in front of an audience.

Smoke Technology

A variety of technologies are currently available for producing smoke on demand. One popular technique continues the tradition used in the earliest LASS devices: petroleum products are heated to high temperatures and then sprayed into the air. As these products burn, cool, and condense, they form a thick smoke. Commercial oils, such as mineral and vegetable oils, have also been used in place of petroleum products in smoke generators.

Some petroleum products can cause respiratory discomfort and, in high concentrations, health problems, however. As a consequence, other techniques have also been developed. Certain chemicals, for example, produce thick clouds of white smoke when heated. Three widely used materials are zinc chloride ($ZnCl_2$), titanium tetrachloride ($TiCl_4$) and ammonium chloride (NH_4Cl). While relatively safe when used with care, each of these compounds has potentially serious health effects.

One of the newest forms of smoke technology involves the use of glycols. Glycols are a class of organic alcohols that includes ethylene glycol ("glycol") and glycerine. Glycols are not without their medical hazards, but overall they seem to be the safest of all materials used in smoke-generating devices.

Smoke Bombs

A smoke bomb is, in some regards, a miniaturized version of large-area smoke screens. It is generally used for a very specific purpose in a very limited area. For example, police officers might choose to use a smoke bomb to force out criminals hiding within a restricted area, such as a private home or small building. Observers have also pointed out that smoke bombs have the potential for use by terrorists or criminals themselves.

An airman moves away after setting off a smoke bomb during a training exercise in Zweibrucken Air Base in Germany, 1990. *Courtesy of DOD Defense Visual Center.*

Smoke bombs could be used to force the evacuation of buildings that terrorists or criminals wanted to take over.

Smoke bombs usually consist of two kinds of combustible materials: one type that burns very well at a high temperature and a second type that burns poorly and produces large quantities of smoke. For example, the first type of material could consist of white phosphorus or black powder that would ignite the bomb. The second type of material might be rubber, plastic, titanium compounds, or other materials that burn slowly with the production of large amounts of smoke.

Further Reading: "Chapter 7. Special Effects," <http://artsnet.heinz.cmu.edu:70/0/csa/books/film/film7.txt>, 01/05/99; "Corona Smoke Generators," <http://www.smokemachines.com/Colt.html>, 01/05/99; Kleber, Brooks E., and Dale Birdsell. *The Chemical Warfare Service: Chemicals in Combat*. Washington, DC: Office of the Chief of Military History, United States Army, 1966, Chapter VII.

SMOKE JUMPERS

Smoke jumpers are firefighters who parachute from an airplane in order to fight forest and brush fires in areas that are inaccessible by road or other means. The first smoke-jumping experiments were carried out in 1939, but the concept of using aircraft in firefighting can be traced to the late 1910s. At that time, the Chief Forester of the U.S. Forest Service, Henry S. Graves, became interested in the use of aircraft to watch for fires in remote regions of public lands. Graves contacted the U.S. Army Air Corps to find out if they could provide aircraft and personnel for such purposes.

By May of 1919, a program of this design had been established in the Angeles National Forest region, north and northwest of Los Angeles, California. Over time, the use of aircraft for fire detection was expanded to most of the Western region under the responsibility of the Forest Service.

By the early 1930s, some fire officials began to consider the possibility of using aircraft to drop equipment to firefighters working in areas that could not be reached by conventional means. At first, the idea seemed too dangerous and impractical, but some authorities in the Forest Service refused to abandon hope. By the end of 1935, an experimental program of this design had been authorized in certain parts of California. The program was designed to drop equipment to firefighters as well as to drop chemicals and water on fires.

The experimental project was eventually deemed a failure, and it was discontinued in the summer of 1939. Forest Service authorities decided to use unexpended funds from the project, however, to pursue another approach to the fighting of fires in remote regions: smoke jumping. The purpose of the smoke jumping program was to use aircraft to bring firefighters to isolated fires, where they could be dropped by parachute. The first test drops of this kind were carried out in the Chelan National Forest in Washington State between 5 October and 15 November, 1939. Eventually, about sixty jumps were completed without injury to any firefighter.

The results of the Chelan experiment were considered successful enough to establish a full-fledged program. The program was first initiated in Region 1 of the Forest Service's area of responsibility in 1940. Six firefighters were selected from a group of nearly 100 applicants. The first actual jumps were made by members of this group at a fire in Nez Perce Forest in Idaho on 12 July 1940.

Smoke jumpers eventually became an important part of the all-out war against wildfire that was conducted on public lands over the next four decades. As Stephen Pyne has pointed out, firefighting after World War II took on many of the aspects of the kind of battles that had been used to win that war, especially the use of airpower. The general philosophy seemed to be, he explains, that "if fire was an attacking enemy, firefighters could counterattack with the same methods that proved so successful during the war" (Pyne 1995, p. 200). Included among the tools used in such counterattacks were helicopters, air tankers, patrol planes, and, of course, smoke jumpers.

Today, the dominant philosophy about fire management has changed dramatically since the 1950s and 1960s. However, smoke jumpers continue to be an important component of the team used to fight wildfires, particularly those in remote regions.

Further Reading: Freeman, Paul. *Smoke Jumpers*. Indianapolis: Baskerville Publishers, 1992; Joregensen, Nels. *Smoke Jumpers*. New York: Bouregy & Curl, 1954; Maclean, Norman. *Young Men & Fire*. Chicago: University of Chicago Press, 1992; Maclean, John N. *Fire on the Mountain: The True Story of the South Canyon Fire*. New York: William Morrow & Company, 1999; The National Smokejumpers Association maintains a website at <http://www.smokejumpers.com/Page/Pages/>, 09/30/99.

SMOKE SCREEN. *See* SMOKE GENERATOR

SMOKE SIGNALS

Smoke signals were a method of communication used by Native American tribes in parts of the United States and Canada. The system was usable only in relatively flat areas, like the prairies of the Midwest. In such regions, the smoke puffs that carried messages could be seen for miles in every direction.

Smoke signals are produced when wet grasses and other materials are piled on top of a smoldering fire. The smoke produced by such a fire can then be controlled by placing a blanket or some other type of cloth above the fire. Smoke can be released in larger or smaller puffs, similar in concept to the dots and dashes used in Morse code messages.

Smoke signals were a simple and effective method by which individuals and groups separated from each other by relatively large distances could communicate with each other. They made possible the transmission of information about the location of game, pending battles, peace agreements, or other matters.

There appears to be no readily available systematic classification of the actual codes used for smoke signal messages by various tribes.

SMOKEY BEAR

Smokey Bear (not "Smokey *the* Bear") has been the official symbol of efforts to reduce forest fires in the United States for more than half a century.

History

Early in World War II the U.S. Forest Service became aware of the potential threat of enemy bombing on the nation's forests. This concern arose not only out of the desire to protect an important national resource, but also to insure an abundant supply of an important war materiel: wood.

As a result of this concern, the Forest Service developed a program known as the Co-operative Forest Fire Prevention Program in 1942. The purpose of the program was to alert citizens of the need to prevent forest fires. The Forest Service also asked the War Advertising Council to assist in developing and promoting this program. The War Advertising Council consisted of advertising professionals who donated their time and energies to develop marketing programs to promote essential wartime programs.

Smokey Bear is the well-known symbol of efforts to reduce forest fires in the United States. *Courtesy of USDA Forest Service and the Advertising Council, Inc.*

The Council's first forest fire prevention campaign featured Bambi the deer from the Walt Disney movie of the same name. Bambi turned out to be a highly effective representative of the campaign. The deer gave fire prevention a "warm and fuzzy" feel that appealed to many people. Before long, however, the Council decided to replace Bambi with another forest creature: the bear. The plan was to design a new campaign centered on a stylized brown or black bear, with an "intelligent, appealing, and slightly quizzical" look. The bear would be depicted as wearing a characteristic fire prevention hat.

The result of this plan was *Smokey*, who first appeared in 1945. The bear was shown on posters along with the slogan: "Smokey says: Care will prevent 9 out of 10 forest fires." Over the years, details of Smokey's image have changed although he remains familiar to the general public. The slogans he promotes have also changed, one of the most familiar being: "Only you can prevent forest fires."

The "Real" Smokey

For the first five years of his "existence" on campaign posters, Smokey was nothing other than a figment of the imagination of the Advertising Council and its artists. An important event occurred in 1950, however, to change that fact. In that year, a black bear cub was found in a tree following a catastrophic forest fire in the Capitan Mountains of New Mexico. The cub was badly burned, but he was nursed back to health by the daughter of a New Mexico Fish and Game warden.

The Forest Service was not unaware of the potential promotional value of this bear cub that had survived a serious forest fire. It decided to christen the cub "Smokey" and to transfer him to the National Zoo in Washington, DC. Before long, a live counterpart of the symbolic Smokey began appearing in forest fire prevention programs. He was also popular among visitors to the zoo.

For some time, there was hope that Smokey and his mate Goldie might produce another generation of Smokeys. However, that was not to be. Zookeepers eventually found an adopted son for Smokey, Smokey II, in 1975. The original Smokey died a year later and was returned to New Mexico for burial. The Smokey Bear Historical State Park was created there in his honor.

Smokey II carried on his adopted father's legacy until 1990 when he, too, died. Since that time, no living representative of the Smokey legend has existed. He continues to deliver his message of fire prevention, but only in the form of an artist's creation.

Smokey Today

Smokey Bear has become one of the best known and most widely utilized symbols in American culture. In addition to Smokey Bear Historical State Park in New Mexico, there is a "Smokey the Bear" song (somewhat incorrectly titled), a book about Smokey (*Guardian of the Forest*, by Ellen Earnhardt), an official birthday (9 August 1944), a Smokey Bear balloon in the Macy Thanksgiving Day Parade, a Smokey float in the Rose Parade on New Year's Day, a Smokey Bear cartoon show, a Smokey comic book, a 1984 20-cent Smokey stamp, and a host of Smokey Bear paraphernalia available from a variety of commercial sources. The highest honors awarded by the Forest Service for wildfire prevention are also known as the Smokey Bear awards. *See also* U.S. FOREST SERVICE.

Further Reading: Morrison, Ellen Earnhardt. *Guardian of the Forest: A History of the Smokey Bear Program*. New York: Vantage Press, 1976; *Smokey Bear*. USDA Forest Service Publication FS-551; "Smokey Bear State Historical Park," <http://www.emnrd.state.nm.us/forestry/smokey.htm>, 01/10/99; "Smokey Bear—The First Fifty Years," <http://www.odf.state.or.us/sound/SMOKEY/bearfaq2.htm>, 01/10/99.

SMOKING (TOBACCO)

Tobacco is the common name for the plant *Nicotania tabacum*, a member of the botanical family that also includes the potato, tomato, eggplant, many peppers, and the deadly nightshade (*Belladonna*).

The use of tobacco products for smoking and chewing is today an essentially worldwide practice. Each year, many billions of cigarettes, cigars, and other tobacco products are sold. The tobacco plant is native to North and South America, and appears not to have been known outside the Western Hemisphere before the exploratory voyages of Christopher Columbus in the 1490s. Archaeological records suggest that the tobacco plant was growing in North and South America as early as 6000 B.C., although human use of the plant can be traced to no earlier than the beginning of the first millennium A.D.

Indian Use of Tobacco

Most Native American cultures have regarded tobacco as a sacred plant. Each tribe has its own explanation as to how the gods made it available to humans. The Huron tribe, for example, believed that the Great Spirit sent a woman to Earth to provide for human needs. She touched the soil in various places, making corn, potatoes, peppers, and other crops grow. At one point, she sat down to rest, and when she arose, tobacco began to grow on the spot.

According to another legend, the first tobacco plant can be traced to the visit of a powerful spirit to the Earth. As the spirit lay sleeping, his hair caught fire. When he arose and ran away, some of his singed hair fell off, where it sprung up from the ground in the form of a tobacco plant.

Tobacco served two quite different functions for Native Americans. First, it was an essential part of many ceremonies and rituals. Smoked in a calumet (peace pipe), it was thought to help a person relax and meditate. Tobacco smoking also served as a bond among members of the community and a way of memorializing important events and agreements. In his ethnographic study of Native American culture written between 1851 and 1857, Henry R. Schoolcraft claimed that a Native American's tobacco pipe was "his dearest companion, and in death is inseparable; for whatever else may be forgotten at his funeral obsequies, his pipe is laid in the grave with him to solace him on his journey to the *happy hunting-ground*."

Tobacco was also valued for its supposed medical benefits. It was used to treat pain, to dress wounds, and to cure all manner of diseases. A common cure recommended for toothache, for example, was the chewing of tobacco.

Introduction to Europe

The introduction of tobacco by members of the Columbus expeditions was met at first with indifference and mild concern. The first Europeans to have taken up the habit of smoking regularly so frightened their neighbors that Columbus was imprisoned by the Inquisition for seven years.

The habit eventually took hold, however, and by the mid-sixteenth century, tobacco was being imported to Europe from Brazil, Cuba, and other parts of South, Central, and North America. Between 1556 and 1559, the first tobacco crops were being planted in Spain, Portugal, and France.

The practice of smoking was met with mixed reactions, however. For some people, it became an enjoyable and possibly healthful practice. For others, it earned the reputation of being a dirty, offensive, and a possibly dangerous habit. The Roman Catholic Church, for example, banned smoking in 1575 in any of its churches in the Spanish colonies.

Smoking in the Twentieth Century

Smoking grew rapidly in popularity during the twentieth century. Among the many reasons for this growth was a brilliant and extensive advertising campaign in which smoking was glorified as an admirable and attractive habit. Most major public figures smoked, and hardly a motion picture was made in the United States that did not include its stars' smoking. By the 1950s, advertisements for cigarettes, cigars, pipe tobacco, and other tobacco products could be found in nearly every arena of life.

World Wars I and II were also a factor in the growth of smoking's popularity. Bored military men and women often turned to smoking as a leisure-time activity, adopting a habit that, in many cases, they retained for the rest of their lives.

Health Issues

It seems likely that few smokers thought about the health effects of smoking before the 1950s. In fact, many advertisements for tobacco products in the first half of the twentieth century depicted white-coated doctors touting the pleasures and, by inference, the safety of smoking.

By the 1950s, however, that situation had begun to change. Medical studies began to show a correlation between smoking and a variety of health problems, such as lung cancer, emphysema, chronic bronchitis, and heart disease. Beginning in the 1960s, governmental agencies began to limit the conditions under which tobacco products could be advertised and, in some cases, sold. For example, the Surgeon General of the United States in 1966 ordered the inclusion of warning labels on cigarette packages and, in 1971, all television advertisements for tobacco products were banned in the United States.

In the 1990s, the backlash against the use of tobacco products became more severe as it became known that tobacco companies had apparently known for many decades about the health effects of smoking and had withheld that information from its customers.

The Chemical Composition of Tobacco and Tobacco Smoke

Tobacco smoke is a complex mixture of more than 10,000 compounds, of which about 4,000 have been identified. The most important components are carbon dioxide, carbon monoxide, hydrogen cyanide, nitrous oxide, aldehydes, phenols, alkaloids (including nicotine), arsenic, certain radioactive materials, and a large variety of hydrocarbons.

Two of the constituents of tobacco smoke that cause the greatest concern among health officials are nicotine and "tar." Nicotine is an extremely dangerous chemical in large doses. It exerts powerful effects on the central nervous system (brain and spinal cord), the peripheral nervous system, the heart, and other body structures. In smaller doses, such

as those received in tobacco smoke, it produces similar, although much more moderate effects. By some estimates, a smoker's life is shortened by 14 minutes for each cigarette smoked, with nicotine being a major factor in that effect.

Nicotine also seems to be moderately addictive. It is this property of cigarette smoke that makes it so difficult for a person to give up the habit of smoking. Although tobacco products are legal for adults in nearly every part of the world, they produce many of the same effects experienced with other addictive drugs, such as cocaine and heroin.

An interesting part of the controversy over tobacco smoking in the United States in the 1990s was the revelation that some tobacco companies may actually have increased the concentration of nicotine in their products artificially. By so doing, they may have been able to increase the chemical dependence of smokers on their products.

The other component of cigarette smoke with serious health effects is "tar." This term is used to describe a complex mixture of tarlike substances formed when tobacco burns. When inhaled, minute amounts of "tar" are drawn into the respiratory system, where they tend to be deposited on the lungs and its ancillary structures. Over time, the accumulation of tar on these structures tends to prevent them from transmitting air into and out of the respiratory system, leading to respiratory disorders such as emphysema and chronic bronchitis. Since some of the components of "tar" are also carcinogenic, the deposition of this material can also lead to lung cancer, now the number one preventable cause of death among adult males in the United States.

Further Reading: "Chemical Constitution of Tobacco Smoke," <http://www.piave.net/salute/eiptdsch.html>, 09/20/99; "The Control and Manipulation of Nicotine and Cigarettes." Testimony before the Subcommittee on Health and the Environment, Committee on Energy and Commerce, U.S. House of Representatives, 21 June 1994; also at <wysiwyg//16http://www.oncolink.com/cancer_news/1994/kessler_nicotine.html>, 09/20/99; Deangelis, Gina. *Nicotine and Cigarettes*. Broomall, PA: Chelsea House Publishers, 1999; Eysenck, Hans J. *Smoking, Health, and Personality*. Piscataway,

NJ: Transaction Publishers, 1999; Gabb, Sean. *Smoking and Its Enemies: A Short History of 500 Years of the Use and Prohibition of Tobacco*. London: Freedom Organisation for the Right to Enjoy Smoking Tobacco (FOREST), 1990. Also at <http://freespace.virgin.net/old/whig/faghist.htm>, 09/19/99; Hirschfelder, Arlene B. *Encyclopedia of Smoking and Tobacco*. Phoenix, AZ: Oryx Press, 1999; Julien, Robert M. *A Primer of Drug Action*, 2nd edition. San Francisco: W.H. Freeman and Company, 1978, pp. 93–98; Kranz, Rachel. *Straight Talk about Smoking*. New York: Facts on File, 1999; "Schoolcraft on the Origin of Tobacco (1851–1857)," http://www.tobacco.org/History/sacred.html>; Sullum, Jacob. *For Your Own Good: The Anti-Smoking Crusade and the Tyranny of Public Health*. New York: The Free Press, 1998; "Tobacco History," <http://www.tobacco.org/History/history.html>, 09/19/99; WRAMC Patient Education: History of Tobacco, <http://www.wramc.amedd.army.mil/education/tobaccohistory.htm>, 09/19/99.

SPACE EXPLORATION FIRES

The United States and Russia (formerly, the Soviet Union) have been conducting space research for nearly half a century. During that time, the possibility of a fire breaking out in a piloted ("manned") spacecraft has often been an issue of serious concern. In only two cases, have serious fires occurred that caused the death of space travelers or serious risk to spacecraft and their occupants. The first of these events occurred during a test of the *Apollo-204* spacecraft in 1967. The second took place aboard the Russian *Mir* spacecraft on 23 February 1997.

Apollo-204 (Apollo-1) *Fire*

The worst disaster in the U.S. space program occurred on 27 January 1967. During a preflight test of what was to be the first manned flight to the Moon, then designated as *Apollo-204*, a fire broke out in the spacecraft, killing astronauts Virgil Grissom, Edward White, and Roger Chaffee.

The test had begun as usual at 8:00 A.M. in the Manned Spacecraft Operations building at Cape Kennedy's Launch Complex 34. The astronauts did not themselves enter the spacecraft until about 1 P.M. The first indication of possible problems occurred almost immediately when Grissom reported that he detected a "sour smell" when he entered the spacecraft. After checking and finding no source of the odor, officials decided to continue with the test.

Before long, a second problem developed. Oxygen flow meters inside the spacecraft began to register unacceptably high readings, setting off alarms in the command center. Technicians were consulted and expressed the opinion that the readings were caused by normal movements of the astronauts within the spacecraft.

A third interruption in the test occurred when problems developed in the communication systems connecting various members of the test team. This problem caused a delay in the test until 6:31 P.M. At that point, test supervisors were about ready to restart the test when ground instruments detected a sharp and unexplained rise in oxygen concentration in the spacecraft. Four seconds later, Chaffee announced over his intercom: "Fire, I smell fire." Two seconds later, White affirmed this fact when he reported "Fire in the cockpit."

The *Apollo-204* spacecraft had been designed to permit rapid evacuation of the cabin in case of an emergency. Target time for evacuation was supposed to be no more than ninety seconds, although that goal had seldom been achieved in earlier tests. Before the astronauts were able to open their hatch, fire had spread through the cabin, and the three men killed.

Members of the test team made heroic efforts to break open the spacecraft hatch, but the intense heat and smoke from the fire made their task almost impossible. Nonetheless, they were able to release the hatch only five and one-half minutes after the first "high oxygen" alert had sounded. The delay had been too long, however, and all three astronauts were already dead. Later investigations showed that the astronauts had died of suffocation and from carbon monoxide poisoning. The burns they received probably occurred primarily after they were dead and were probably not sufficient in and of themselves to cause death.

In addition to the loss of life, the *Apollo-204* fire dealt a terrible blow to the U.S. space program. The National Aeronautics and Space Administration (NASA) spent nearly a year investigating the cause of the accident and redesigning the *Apollo* spacecraft to prevent a recurrence of the event. The final report on the tragedy came to nearly 3,000 pages contained in fourteen books. Investigators were never able to identify a single cause for the fire, but they did identify a number of conditions that made the fire possible. These conditions included:

1. The use of pure oxygen inside the cabin, insuring that any fire that might develop would burn very quickly and at a high temperature.
2. The extensive use of highly combustible materials inside the spacecraft cabin.
3. The incorrect electrical wiring systems in the cabin.
4. The use of a plumbing fluid consisting of a combustible and corrosive liquid.
5. The lack of a fast and efficient method of escape for the astronauts.
6. The lack of provision for rescue and medical assistance during testing.

The *Apollo-204* project was later renamed *Apollo-1*, in honor of the astronauts who died in the fire.

Mir *Fire*

The 1997 *Mir* fire began in a supplementary oxygen-generating system used when more than three people are aboard the Russian spacecraft. Oxygen is produced in the generator by heating solid lithium perchlorate ($LiClO_4$). The reaction is similar to one familiar to many high-school chemistry students. In high-school laboratories, oxygen gas is often generated by heating potassium chlorate ($KClO_3$).

For reasons still unknown, the oxygen-generating reaction got out of control. The generator itself caught fire, exploded, and quickly filled the *Kvant 1* command module with flames and smoke. The six crew members—four Russians, a German, and U.S. astronaut Jerry Linenger—donned oxygen masks and began to fight the fire. They eventually brought it under control and the mission continued as planned.

Reports released at the time of the incident differed significantly from later accounts provided by Linenger after he returned to Earth. In a press briefing on 28 February, 1997, NASA *Shuttle-Mir* Program Director Frank Culbertson described the fire as "quite exciting." He indicated that it lasted only ninety seconds, however, and was brought under control rather quickly. Culbertson seemed to suggest that the fire was of only minor importance and that it created some damage to insulation, panels, support brackets, and "things like that." The tone of the briefing seemed to suggest that NASA was not very worried about the event.

When Linenger returned to Earth on 25 June, however, he gave a different description of the fire. In the first place, the fire raged not for ninety seconds, but for fourteen minutes. During that time, flames and burning metal shot out from the oxygen-generating canister a distance of about two feet. One of the passageways to the *Soyuz* "life boat" escape modules was blocked. Linenger portrayed the fire as "serious stuff," an occurrence that "almost looked like SRBs (solid-fuel rocket boosters)—you almost can't look at them they're so bright. It was a hot fire." (Dunn, *Seattle Times*, p. 2).

The possibility of fire on *Mir* was not news to NASA. In the early 1980s, the U.S. space agency had learned that the Soviets were not especially concerned about fire risks during space exploration. They generally did not use fire-resistant materials inside their spacecraft and installed only the simplest firefighting equipment.

The folly of this practice was illustrated time after time in Soviet/Russia spaceflights, however. For example, fire broke out on the *Salyut-6* flight of September 1978 that produced "plenty of smoke and fire" according to cosmonaut Vladimir Kivalyonok. Russian cosmonaut, engineer Georgiy Grechko, also described another fire that was so bad that "I couldn't see the other end of the station—it was in smoke, all in smoke" (Oberg, p. 4).

Much of the information about fires aboard Soviet/Russian spacecraft was never released to the general public, however. It became known only through press interviews with cosmonauts at much later dates or during briefings with NASA officials during initial planning for joint United States/Russian space experiments in the 1990s.

Further Reading: "*Apollo* Accident Report." *Aviation Week and Space Technology*, (6 February 1967): 29–36; Brooks, Courtney, G., James M. Grimwood, and Lloyd S. Swenson Jr. *Chariots for Apollo: A History of Manned Lunar Spacecraft*. Washington, DC: NASA SP-4205, 1979. Also on the Internet at <http://www.hq.nasa.gov/office/pao/History/Apollo204/chariot.html>; Dunn, Marcia. "Close-up: *Mir* Fire Was Worse than Initially Reported," *Seattle Times*, 26 June 1997. Also on the Internet at <http://www.seattletimes.com/extra/browse/html97/fire_062697.html>; Oberg, James. "Fire (And Smokescreens) On *Mir*," <http://www.jamesoberg.com/russian/fire.html>; Oberg, James. "Shuttle-*Mir's* Lessons for the International Space Station." *Spectrum*, (June 1998): 28–37; "Project Apollo Annotated Bibliography, Ch. 5". <http://www.hq.nasa.gov/office/pao/History/Apollo204/biblio.html>; Report of the *Apollo 204* Review Board, <http://www.hq.nasa.gov/office/pao/History/Apollo204/as204report.html>; "Small Fire Extinguished on Mir." NASA News, 24 February 1997, Press Release 97–30. Also on the Internet at <http://www.jsc.nasa.gov/pao/media/rel/1996/97-30.html>.

SPARKY™ THE FIRE DOG

Sparky™ the Fire Dog is the official "spokesdog" for the National Fire Protection Association (NFPA). He is featured in countless brochures, coloring books, and other fire safety products, and is the centerpiece of National Fire Prevention Week messages for children.

Sparky was "born" on 18 March 1951, when he was chosen as a symbol of fire safety prevention by the NFPA. He is described as an adult Dalmatian with "three freckles on each side of his muzzle and five spots on each ear." He is always shown dressed in full firefighter protective clothing. Sparky was chosen by NFPA for the purpose of commu-

Sparky is the official symbol for fire safety at the National Fire Protection Association. *Sparky™ and Sparky the Fire Dog™ are registered trademarks of the National Fire Protection Association, Quincy, MA.*

nicating fire safety information with a positive approach.

Since 1996, Sparky has also appeared in a series of television public service announcements. The dog's voice has been supplied by actor Dick Van Dyke. Sparky's story can be found on his home page at http://www.sparky.org/story/story.htm. *See also* NATIONAL FIRE PROTECTION ASSOCIATION.

SPONTANEOUS COMBUSTION

Spontaneous combustion is a process by which slow oxidation produces enough heat to raise a fuel to its kindling temperature, thus making it possible for a fire to begin.

Normally, combustion occurs when heat is applied to fuel intentionally in order to raise its temperature to the kindling point. For example, a group of logs can be ignited in a fireplace by touching a lit match to some papers. The heat released by the burning papers raises the logs to their kindling temperature, setting them on fire.

But fuels can catch on fire unintentionally also. One such example is the case when oily rags are packed tightly together in a container. Oils react slowly with oxygen of the air to form carbon dioxide, water, heat, and other products. This reaction takes place slowly and under most circumstances, the heat produced will escape into the surrounding air. If the oily rags are packed tightly together, however, the heat may be trapped. As the oxidation of oils proceeds over time, more and more heat is produced and trapped within the rags. Eventually, the temperature of the rags may reach their kindling temperature, and they will catch on fire spontaneously.

Barns that contain hay that is insufficiently dry are also prime candidates for spontaneous combustion. Moist hay tends to oxidize slowly, but much more rapidly than perfectly dry hay. As the moist hay oxidizes, it may release enough heat to dry out and ignite the hay, setting the barn on fire.

Dust explosions of various types may also be examples of spontaneous combustion. For example, the dust formed in a coal mine contains very fine particles of coal. Those coal particles may react with oxygen of the air, producing small amounts of heat that would normally be dissipated in an open environment. In a mine, however, the heat may have less opportunity to escape and may raise the coal dust particles to their kindling temperature. The added factor in this case is the very large surface area provided by coal dust particles in contrast to chunks of coal. The heat released by the oxidation of coal particles may be sufficient to cause a sudden combustion of the particles, causing an explosion in the mine. *See also* COMBUSTION; FUELS.

Further Reading: Bacharach, J.P.L. et al. *A Review of Spontaneous Combustion Problems and Controls with Applications to U.S. Coal Mines: Final Technical Report*. Washington, DC: Department of Energy, Division of Fossil Fuel Extraction, Mining Research and Development, 1978; Smith, Alex C. *Large-Scale Studies of Spontaneous Combustion of Coal*. Washington, DC: U.S. Department of the Interior, Bureau of Mines, 1991; Smith, Alex C. *Spontaneous Combustion Studies of U.S. Coals*. Pittsburgh, PA: U.S. Department of the Interior, Bureau of Mines, 1987; "Spontaneous Combustion Explained," <http://www.inspect-ny.com/arson/spontcom.htm>, 08/27/99.

SPRINKLER SYSTEMS

A fire sprinkler system is a method used in homes, offices, hotels and motels, schools, and other structures to control and stop the spread of a fire. The first fire sprinkler system was invented in 1874 by an American, Henry S. Parmalee. Parmalee developed his system in order to protect the factory in which he built pianos.

Until the 1940s, sprinkler systems were used primarily in large warehouses and factories. Many people felt that they were far too expensive to install in other types of buildings, such as hotels and office buildings. In warehouses and factories, the cost of a sprinkler system was usually recovered in only a few years because of the savings achieved in insurance premiums.

The attitude about sprinkler systems began to change in the 1940s as the result of a series of terrible fires that resulted in the loss of hundreds of human lives. The most dramatic of these was the Coconut Grove Nightclub fire in Boston in 1942, in which 492 people were killed. Four years later, two other large fires in public buildings resulted in more deaths: 61 in the LaSalle Hotel fire in Chicago and 119 in the Winecoff Hotel fire in Atlanta.

Today, fire safety experts and most business people have become convinced of the value of sprinkler systems in protecting loss of property and lives. Most cities require the installation of sprinkler systems in large structures or buildings where the risk to human life by fire is especially serious. For example, New York City requires sprinkler systems in all high-rise buildings. Chicago requires sprinklers in all nursing homes. Some communities even require sprinkler systems in private homes. San Clemente, California, is one such community.

How Sprinklers Work

The core of a fire sprinkler system is a small, fan-shaped metallic device that is usually suspended from the ceiling of a room or hallway. The device is connected by pipes to a source of water under pressure. The device contains a heat-sensitive plug or a small glass

Fire sprinkler systems stop and control the spread of fire in buildings. *Courtesy of the Reliable Automatic Sprinkler Co., Inc.*

bulb. When the temperature in a room rises (as it would when a fire has started), the plug melts or the glass breaks. In either case, water in the systems is released through the device and sprayed over an extended area in the room.

The heat-sensitive plug in a sprinkler device is made of a material known as a *fusible alloy*. An alloy is a mixture of two or more metals that has properties different from each of the metals themselves. A fusible alloy is one in which the combination of metals melts at a relatively low temperature.

For example, the alloy known as Wood's metal is a mixture of 50 percent bismuth, 10 percent cadmium, 13.3 percent tin, and 26.7 percent lead. Wood's metal melts at a temperature of 70°C (158°F). The melting point of a fusible alloy can be adjusted very precisely by altering the proportion and types of metals used in the alloy. The alloys used in sprinkler system heads are usually designed to melt at about 75–80°C (165–175°F).

Types of Sprinkler Systems

Sprinkler systems are classified into three major types: wet-pipe systems, dry-pipe systems, and pre-action systems.

1. A wet-pipe system is one in which the pipes attached to the sprinkler heads are constantly filled with water. Once a fire breaks out and the fusible alloy melts, water is released immediately to the burning area. Wet-pipe systems are very popular because they are simple, dependable, easy to modify, and relatively inexpensive to install and maintain. They have two major disadvantages, however. First, they are not suitable for use in areas where temperatures drop below the freezing point. Water in the pipes may freeze and expand, causing the pipes to break and release water into an area. Second, they may also develop leaks at any time, presenting a risk to the rooms in which they are installed.

2. A dry-pipe system is one in which pipes are normally filled with pressurized air or nitrogen gas. The pressure of the air or nitrogen pushes against a valve, known as the *dry-pipe valve*, preventing water from entering the pipes. When a fire breaks out, the fusible alloy in the sprinkler heads melts, releases the air or nitrogen gas to the room, reduces pressure on the dry-pipe valve, and allows water to flow through the pipes and the sprinkler heads. Dry-pipe systems tend to be more complicated to install and maintain than wet-pipe systems and there may be a delay in operation of the sprinklers that can increase the damage caused by the fire.

3. A pre-action system is similar to a dry-pipe system except that it has one additional component: a *pre-action valve*. The pre-action valve is a device for detecting heat, smoke, or flames independently from the sprinkler heads themselves. If a fire breaks out in a room, two events must occur before water is released from the sprinkler system. First, the pre-action valve must recognize the event and send an electrical signal that will allow water to flow into pipes. Second, the sprinkler fuse must melt, allowing water to escape from the pipes. A pre-action system is more expensive to install and maintain, but it provides an additional level of protection against the accidental release of water from the system.

Modifications of these systems are sometimes used in settings where the release of water might cause serious damage to the materials present in a room. For example, protecting a room full of computers might not be advisable, since the release of water into the room (even in the case of a small fire) might cause as much damage as the fire itself. For such situations, the pipes used in a sprinkler system can be filled with some fluid other than water. An obvious choice for such systems is carbon dioxide gas because it is noncombustible and relatively inexpensive. At one time, halon gases were widely popular for such systems, although the environmental risks of using such gases has resulted in a ban on their use. An increasingly popular fluid for use in dry-pipe systems with special applications is a mixture of nitrogen (52 percent), argon (40 percent), and carbon dioxide (8 percent) known as *Inergen*. Inergen is relatively inexpensive, environmentally harmless, and totally incombustible.

Value of Sprinkler Systems

In the past half century, sprinkler systems have proved their value many times over. In one study conducted by the National Fire Protection Association, an average of twenty lives were lost annually during the period 1971–1975 in buildings with sprinklers. By comparison, an average of 8,000 lives were lost per year in buildings without sprinklers.

Some of the most impressive statistics about sprinkler effectiveness have come from Australia and New Zealand, where fire reporting is required by law. During the period 1886 through 1968, these nations reported that 99.76 percent of all fires were either extinguished or controlled by sprinkler systems. In New York City, records indicate that the success rate for the control of fires in high-rise buildings is 98.4 percent. In addition, there has never been an instance when more than one human life has been lost in a building with a sprinkler system.

Insurance companies are aware of the advantages offered by buildings with sprinkler systems. In such buildings, property protection rates tend to be 47 percent lower for the structures themselves and 43 percent lower for their contents compared to buildings without sprinkler systems.

Additional information about fire sprinkler systems can be obtained from the National Fire Sprinkler Association:

The National Fire Sprinkler Association
P.O. Box 1000
Patterson, NY 12563
Tel: (845) 878–4200 x 133
http://www.nfsa.org
e-mail: info@nfsa.org

Further Reading: Bryan, John L. *Automatic Sprinkler & Standpipe Systems*. Quincy, MA: National Fire Protection Association, 1976; Coon, J. Walter. *Fire Protection: Design Criteria, Options, Selection*. Kingston, MA: R.S. Means Company, 1991; *Fire Protection Systems*, 2nd edition. Quincy, MA: National Fire Protection Association, 1992; "Fire Sprinkler Facts," at <http://www.waycool.com/southwest/intro.html>, 12/30/98; "Homeowners Guide to Fire Sprinkler Systems," at <http://www.nfsa.org/homeown.html>, 12/30/98; Puchovsky, Milosh T., ed. *Automatic Sprinkler Systems Handbook*. Quincy, MA: National Fire Protection Association, 7th edition, 1996; Tatum, Rita, "Fire Sprinklers Save Lives—and the Bottom Line," at <http://www.facilitiesnet.com/NS/NS3bk5e.html>, 12/30/98.

STAHL, GEORG ERNST (1660–1734)

Stahl's place in the history of science rests on his development of a theory of combustion known as the *phlogiston theory*. Stahl inherited the seeds of this theory from his teacher, Johann Joachim Becher, from Jan van Helmont, and from other earlier thinkers. However, Stahl is responsible for the final formulation of the phlogiston theory, which dominated scientific explanations of burning, rusting, and other phenomena for nearly a century.

His Life

Stahl was born in Ansbach, Bavaria on 21 October 1660. He was the son of a Protestant clergyman and went on to study medicine at the University of Jena. In 1687, he was appointed court physician to the Duke of Sachsen Weimar, a post he held for seven years. He then became the first Professor of Medicine at the University of Halle. He held that post until 1716, when he became personal physician to king Frederick I of Prussia. He remained in this post until he died in Berlin on 14 May 1734.

The Phlogiston Theory

Understanding the nature of combustion (burning) is a fundamental question in chemistry. In the first place, combustion reactions have a virtually unlimited number of practical applications, from metallurgical reactions by which metals are freed from their ores to heat-generating reactions used to power many forms of machinery. Stahl had the insight to observe that other important chemical reactions, such as the rusting of metals, are related to combustion reactions.

In his attempt to develop a theory of combustion, Stahl drew heavily on the ideas put forth by his teacher, Becher. Becher had argued that some forms of matter contain a volatile substance which he called *terra pinguis* ("fatty earth"). When a material burns, Becher had said, its *terra pinguis* is released to the air.

Stahl expanded and modified the outlines of Becher's theory. He renamed the volatile material *phlogiston*, from the Greek meaning "to set on fire." He then explained in detail how a phlogiston theory could explain most of the known observations about combustion, rusting, and other phenomena. He pointed out, for example, that the smelting of an ore involved a reaction in which phlogiston was expelled from charcoal (a flammable material) and passed into a metallic oxide (the ore). As the ore took on the phlogiston, it became a pure metal which, as everyone knew, could then be made to rust (lose its phlogiston once more) to become an ore.

The main problem with Stahl's phlogiston theory was a quantitative issue: Certain reactions explained as occurring because of the *loss* of phlogiston actually gained weight in the process. For example, the rusting of a piece of iron results in iron's *gaining* weight although it has supposedly *lost* phlogiston.

Stahl was not concerned about this discrepancy. At the time he lived, it was still not regarded as important that researchers know anything about quantitative changes that occur during reactions. It was sufficient to be able to develop rational qualitative descriptions of such changes. In addition, the equipment needed to make good measurements, such as accurate balances, were simply not yet available.

As strange as the phlogiston theory may sound today, then, it held sway among scientists for nearly a century after Stahl first announced his ideas. It was only with a better understanding of the nature of gases, oxygen in particular, that Stahl's theory was finally overthrown in the 1770s to be replaced by the modern theory of combustion. *See also* BECHER, JOHANN JOACHIM; COMBUSTION; LAVOISIER, ANTOINE LAURENT; PHLOGISTON THEORY.

Further Reading: Asimov, Issac. *Asimov's Biographical Encyclopedia of Science and Technology.* 2nd revised edition. Garden City, NY: Doubleday & Company, 1982, pp. 241–242; Porter, Roy, ed. *The Biographical Dictionary of Scientists.* New York: Oxford University Press, 1994, p. 637.

STEAM ENGINE

A steam engine is a device by which heat energy is converted into mechanical energy by means of a piston that moves up and down within a cylinder. A steam engine is an example of an external combustion engine since the fuel that operates the device is burned outside the cylinder in which work is done.

Steps Toward the Steam Engine

Humans have known that steam can do work for many hundreds of years. The Greek engineer Hero (born about 20 A.D.; date of death not known) invented a glass device consisting of a sphere to which two bent arms were attached. He filled the sphere with water and heated it with a flame. As water inside the sphere began to boil, steam escaped through the bent arms, pushing the device around in a circle. Some modern-day lawn sprinklers operate on a similar principle using only the force of water, without using steam, to drive the arms.

Most early "steam engines" were really little more than toys designed to amuse people or to demonstrate scientific principles. The first steam engines intended for actual use were not constructed until the late seventeenth century. An awkward beginning on the research was the work of the Dutch physicist Christiaan Huygens (1629–1695), who invented a "gunpowder engine" in 1680. Huygens' device consisted of a metallic cylinder containing a charge of gunpowder and a piston. When the gunpowder was ignited, the force of the explosion moved the piston outward. After a few moments, the cylinder could be recharged with gunpowder and fired again.

There really wasn't any application for Huygens' gunpowder engine, at least partly because it had to be recharged before each use. Also, working with gunpowder was very dangerous. However, Huygens' assistant, Denis Papin (1647–1712) had a glimpse of the way in which the gunpowder engine could be refined and improved. He wrote in 1690 that:

> Since it is a property of water that a small quantity of it turned into vapour by heat has an elastic force like that of air, but upon cold supervening is again resolved into water, so that no trace of the said elastic force remains, I concluded that machines could be constructed wherein water, by the help of no very intense heat, and at little cost, could produce that perfect vacuum which could by no means be obtained by gunpowder. (quoted in Derry and Williams, p. 315)

Papin proceeded to build a device based on this principle. The device consisted of a tube about twelve centimeters (2.5 inches) in diameter with a little water in the bottom and a piston fitted inside it. When the cylinder was heated, the water was converted to steam and pushed the piston upward.

The First Steam Engines

As with all earlier and similar devices, Papin's invention had no practical use. But it established the principle on which all steam engines were later to be based.

The first practical steam engine was built in the English engineer Thomas Savery (ca. 1650–1715) in about 1690. Savery's device did not quite fit the model of the steam engine since it had no piston. However, it was a true steam engine since it used steam (and a vacuum) to perform useful work.

The purpose of Savery's engine, eventually given the name "The Miner's Friend," was simply to remove water from coal mines. The problem of flooding in mines was a serious one in England since it posed a limit as to how deeply and extensively miners would work. Savery's solution to this problem was to construct a large metallic cylinder with a small amount of water in it. A long tube connected the cylinder to the bottom of the mine shaft. When water in the cylinder was heated, steam was produced, pushing out all of the air in the cylinder. When the cylinder was cooled, air was prevented from returning to the cylinder, leaving a partial vacuum inside it. Normal atmospheric pressure outside the cylinder then pushed water from the bottom of the mine up into the cylinder, from which it could be emptied. By alternately heating and cooling the cylinder, the operator was able to continue pumping water out of the mine.

The final stage in the design of the modern steam engine came about twenty years later when the English engineer Thomas Newcomen (1663–1729) reintroduced the piston to what was essentially Savery's device. Newcomen's steam engine occupied a whole section of Dudley Castle in Worcestershire and first began operation in 1712. In his design a brick boiler was installed on the bottom floor of the building in which the steam engine was housed. Attached to the top of the boiler was a long brass cylinder containing the piston. Attached to the top of the piston was a horizontal wooden beam that could rock up and down. To start the engine, heat was applied to the boiler, where water was converted to steam. The pressure applied by the steam was about the same as atmospheric pressure, so that the horizontal beam did not move up or down.

When cold water was sprayed on the cylinder, however, a partial vacuum was produced. Atmospheric pressure pushing down on the top of the piston forced it downward into the cylinder, causing one end of the horizontal beam to move upward. When steam was reintroduced into the cylinder, the piston was forced back upward again, restoring the beam to its original position.

As with Savery's Miner's Friend, the purpose of Newcomen's machine was to remove water from mines, in this case, tin mines. The machine was able to make twelve strokes a minute, lifting about forty liters (ten gallons) of water a height of forty-five meters (150 feet) from the mine to the surface. Newcomen's engine was about five times as efficient as the Miner's Friend. Its primary shortcoming was simply that machinery was not available for making cylinders and pistons that fit tightly to each other.

Perhaps the most important breakthrough in the development of steam engines came from the work by the Scottish engineer James Watt (1736–1819). Called upon to repair one of Newcomen's engines, Watt noted problem areas and vastly improved the efficiency of steam engines. One of his most important changes was to use two cylinders, instead of one. While steam was being introduced into one cylinder, it was being cooled in the other. This development removed the need for a delay while a single cylinder was first cooled and then heated. Watt also invented gears that converted the back-and-forth motion of the horizontal beam into rotary motion and a centrifugal governor that kept the engine running at constant speed.

Applications

The steam engine can really be said to have powered the Industrial Revolution. It quickly replaced many other, more traditional, forms of power. It was far more effective than human or animal labor, and it had the convenience of being portable, capable of being used almost anyplace. By the middle of the eighteenth century, steam engines had begun to take over almost every known manufacturing task, including textile operations, mining, rolling mills, sawmills, pumping plants, and printing presses. They also brought about a revolution in the field of transportation to power boats and ships, trains, and even automobiles. In the 1880s, steam engines were first used to drive turbines, providing both mechanical energy and a way of making electrical energy.

By the early 1900s, however, the role of steam engines in commerce and industry had begun to decline dramatically. The one problem that could never be solved was the very low efficiency of steam engines. In many cases, no more than a quarter of the energy supplied to the engine came out in the form of useful work. As long as no other type of engine was available, the steam engine solved many industrial problems satisfactorily. But with the development of an option—the internal combustion engine—the days of the steam engine came to an end. *See also* INTERNAL COMBUSTION ENGINE.

Further Reading: Bruno, Leonard C. "Steam Engine," in Bridget Travers, ed. *Gale Encyclopedia of Science*. Detroit: Gale Research, 1996, pp. 3475–3477; Derry, T.K., and Tevor I. Williams. *A Short History of Technology from the Earliest Times to A.D. 1900*. New York: Dover Publications, 1993, Chapter 11; Dickinson, H.W. *James Watt and the Steam Engine*. London: Encore Editions, 1981; Hills, Richard. *Power from Steam: A History of the Stationary Steam Engine*. Cambridge, England: Cambridge University Press, 1993; Rossotti, Hazel. *Fire*. Oxford: Oxford University Press, 1993, Chapter 9; Rutland, Jonathan. *The Age of Steam*. New York: Random House, 1987; Tunzelmann, G.N. von. *Steam Power and British Industrialization to 1860*. Oxford: Clarendon Press, 1978; Watkins, George. *The Steam Engine in Industry*. Ashbourne, England: Moorland Publishers, 1978.

STOVES

A stove is defined as a device in which a fuel is burned or electrical current is used to produce heat for the purpose of heating a room or cooking food. At one time, the word had other meanings. For example, during the fifteenth and sixteenth centuries, the word referred to a heated room used as a hothouse, a steam room, or a drying room. In fact, the modern term *stove* derives from a Middle English word for "heated room." Such rooms were generally not meant for human habitation. Over time, stoves were synonymous to fireplaces. It was not until about 300 years ago that the term was used in its present-day sense.

Types of Stoves

Throughout history, stoves have been made from a great variety of materials including metals such as cast and wrought iron, bronze, and steel; logs covered with mud or clay; brick and porcelain; and soapstone. In most early stoves, wood was used as the fuel, although any available combustible material, such as dung, hay, straw, peat, kerosene, oil, and charcoal was also used. After coal became readily available, it was long the preferred fuel in stoves since it burned with a hotter flame and less smoke than wood and other common fuels. Today, most stoves are heated by natural gas or electricity.

History

Human heating systems have evolved from many sources: outdoor campfires; unventilated fires inside caves or tents; similar inside fires provided with holes through which smoke could escape; fireplaces; and stoves of many designs. Most of these heating systems were inefficient and unpleasant. Most fires produced smoke that interfered with the daily life of the people living in a room. And ventilation systems that allowed smoke to escape also carried with it most of the heat produced by the fire.

Humans certainly understood these problems early on and worked to invent devices that could provide heat under more satisfactory conditions. Possibly the earliest such device was a cast iron stove made in China about 100 A.D. The stove was a box-shaped object with rounded corners supported by four legs. It contained four openings on which containers could be heated and a fifth opening for a chimney. The stove is on exhibit at the Field Museum of Natural History in Chicago.

The Chinese stove contained all the elements needed in any stove: a box in which the fuel is to be burned, an opening through which the fuel can be placed into the box, and an outlet for smoke and other exhaust gases produced during combustion. The presence of cooking units in the Chinese stove, as it happens, was not a feature common to most stoves until the eighteenth century. Until that time, stoves were used primarily for heating, and other methods were used for cooking.

There was relatively little development in stove technology until the late fifteenth century. Some historians claim that the first modern stoves, made of cast iron or brick and tile, were first introduced into Alsace in about 1475. A few decades later, similar stoves had begun to appear in Germany and Scandinavia. By contrast, cast iron stoves apparently were not produced in England until 1759 when the Carron Iron Works in Scotland began their manufacture.

In the United States, colonists made their first stoves out of wood covered on the inside and outside with mud or clay. By 1647, however, the iron works in Saugus, Massachusetts had begun to turn out cast iron stoves. These stoves were often of the "five-plate" variety, also known as *jamb* or *German* stoves. The five-plate stove, as its name suggests, consisted of five rectangular iron plates assembled to form a box open at one end. The stove was placed with the open end protruding into the back or side of a fireplace and the bulk of the box into an adjacent room. Heat from the fireplace warmed the box, which then radiated heat into the room.

By the end of the 1760s, the five-plate stove was being replaced by the six-plate stove. The six-plate stove was essentially a rectangular box in which one plate served as a door that could be opened for the addition of the fuel. The door also had a draft hole to permit air to flow into the combustion chamber and a chimney hole to allow the escape of smoke and combustion gases. These stoves were self-contained heating units and had no connection with a fireplace.

The Franklin Stove

One of the most famous advances in stove technology occurred in 1742 when Benjamin Franklin invented a stove that has since carried his name. As it happens, units now known as "Franklin stoves" are quite different from the device originally designed by Franklin.

Franklin's intent was to create a unit in which a greater portion of the heat generated within a fireplace was returned to the room. He also wanted to reduce the amount of smoke that flowed back into a room from the fire. To accomplish these goals, Franklin designed a hood-shaped metallic box that could be placed directly into the opening of any traditional fireplace. The box contained a metal base where fuel was burned. Running below, behind, and around the sides of the box was a hollow opening. Room air flowed into this opening through a passageway at the front of and below the combustion box. It was warmed by the heat of the fire and then returned to the room through openings at the top and sides of the firebox. Franklin also designed a separate opening through which smoke could escape without entering the room. Franklin's stove greatly increased the amount of heat released by means of convection, conduction, and radiation.

Over the last 250 years, a great many changes have been made to Franklin's original design. The most important of these changes appeared about a century later. The hood-shaped metal box was replaced by a cylindrical metal box, moved out of the fireplace, and set directly in the center of a room. The new Franklin stove looked much like a coal- or wood-burning stove in any modern home. Franklin's concept was retained in that air from the room flowed below, up, and around the firebox. But it was no longer necessary to build a special metal chamber to allow this flow of air. Room air was now warmed directly by the hot stove.

The new "Franklin stoves" evolved in many different directions. For example, the cylindrical metal box was replaced by a bulb-shaped metal container to provide a larger surface area on which room air could be warmed. This bulb-shaped container became better known as a pot-bellied stove that was once the primary heating source in most American homes, as well as in many offices and stores.

Franklin, pot-bellied, and other types of stoves were often as much works of art as they were utilitarian devices. One of Franklin's first models, for example, carried a design of the sun with sixteen rays streaming from it and the motto *Alter Idem* (another like me) on its face. Other stoves carried flowers, vines, classical motifs, and other decorations on their faces and sides. Some were even named on the basis of their artistic designs; stoves, such as the "Jews-harp" and "dolphin" stoves being especially popular.

Today, Franklin and pot-bellied stoves are seldom used for utilitarian purposes, although they continue to have enormous aesthetic appeal to many people. Such stoves are usually designed to burn wood, coal, other solid fuels, and a variety of liquid fuels. As in the past, the artistic design and construction of such stoves may be as important as their heating quality.

Further Reading: Brewer, Priscilla J. *From Fireplace to Cookstove: Technology and the Domestic Ideal in America.* Syracuse, NY: Syracuse University Press, 2000; *Fireplaces and Wood Stoves.* Alexandria, VA: Time-Life Books, 1997; *Homeowner's Guide to Wood Stoves.* Menlo Park, CA: Lane Publishing, 1979; Lyle, David. *The Book of Masonry Stove: Rediscovering an Old Way of Warming.* White River Junction, VT: Chelsea Green Publishing Company, 1998; Peirce, Josephine H. *Fire on the Hearth: The Evolution and Romance of the Heating Stove.* Springfield, MA: Pond-Ekberg Company, 1951; Wright, Lawrence. *Home Fires Burning: The History of Domestic Heating and Cooking.* London: Routledge & K. Paul, 1964.

SULFUR OXIDES. *See* AIR POLLUTION

SUN

The Sun is the star nearest the Earth, around which the Earth, the other planets, and other objects in the Solar System revolve.

Humans have long worshipped the Sun in some form or another as being the source of all life on Earth. In a fundamental sense, that belief is true, since energy provided by solar radiation is what makes life on Earth (and anywhere else in the solar system it may exist) possible. Most early religions installed a Sun god or goddess who was either the primary and most powerful deity, or one of the most important deities, in the religion's pantheon of sacred figures.

For example, the Egyptians worshipped the god Ra (or Re) as the most important of their many gods and the one who had created the Earth and everything that lives on it. Over time, Ra became represented by the scarab, a type of beetle who was thought to push the Sun in its orbit around the sky. The pharaohs, leaders of the Egyptian nation, were thought to be mortal embodiments of Ra.

Among the Navajo Indians of North America, Tsohanoai was revered as the Sun god. He was thought to be responsible for daylight by carrying the Sun on his back across the sky. At night, Tsohanoai hung the Sun on a peg in his house.

In the Shinto religion of Japan, Amaterasu was honored as the Sun goddess. She is said to have been a fickle goddess early in her life and often hid in a cave when her brother Susanowo mistreated her. She was finally tricked into coming out of her cave when the other gods and goddesses had a happy party just outside the cave opening. When Amaterasu left the cave to see what was going on, she saw her reflection in a huge mirror that the gods and goddesses had installed at the opening of the cave. She was so pleased with what she saw, that she never returned to her cave again.

The Sun and Fire

Throughout much of early history, humans believed that the Sun was made of fire or was closely related to fire. In many cases, the same deity who was worshipped as the god or goddess of the Sun also became the god or goddess of fire. For some cultures, it is difficult to separate the two types of deities from each other.

A number of rituals and ceremonies practiced by early cultures confirm the association of the Sun with fire. For example, the Sencis of ancient Peru shot burning arrows at the Sun during an eclipse, at least partly in an effort to keep it from burning out. In New Caledonia, priests made burnt offerings of plants, corals, locks of hair from a child, and teeth from the skull of an ancestor in order to keep the Sun ablaze so that it would burn off the clouds.

For most of human history, the Sun was thought to be a gigantic burning object in the sky. As science developed, however, this simplistic model of the Sun was found to be deficient. If one knows the approximate size of the Sun and the amount of energy it produces, one can calculate the amount of fuel needed to keep it burning. Further, it is easy to estimate how long the Sun can continue to burn given the amount of fuel that it must contain. These calculations showed over 200 years ago that the Sun could not be "burning" in the sense that a wood or coal fire on Earth is burning. The fuel present on the Sun for such a scenario would have been used up long, long ago.

Nuclear Reactions in the Sun

Scientists have simply had no explanation for the way energy is produced on the Sun until the early part of the twentieth century. Then, physicists began to consider the possibility that solar energy is produced by a type of reaction that had only recently been discovered, nuclear fusion. In nuclear fusion, two small nuclei combine with each other to produce a single larger nuclei, with the release of very large amounts of energy. In 1929, two Dutch physicists, R. d'E Atkinson and F.G. Houtermans, proposed that such reactions might be responsible for the production of solar energy.

At that early point, however, there was insufficient understanding of the nature of nuclear reactions to pursue this suggestion in any detail. In fact, it was not until ten years later that such a mechanism for the production of solar energy was rigorously worked out by the German-American physicist Hans Bethe (1906–). Bethe suggested that four hydrogen nuclei (protons) combine with each other to form a single helium nucleus, with the release of large amounts of energy:

$$4H^+ = 2e^+ + He^{2+} + \text{energy}$$

This reaction actually occurs in a series of steps in which carbon and nitrogen nuclei are involved as catalysts. Bethe's theory has, therefore, sometimes been called the *carbon cycle* theory of energy formation in the Sun.

The problem with all theories of nuclear fusion as a source of energy in the Sun is that such reactions normally occur only at very high temperatures, of the order of a few million degrees. Until measurements of the temperatures of the interior of stars like the Sun had been obtained, it was difficult to confirm or reject theories such as that of Bethe.

Today, we know that temperatures sufficient to produce nuclear fusion do exist within the cores of stars such as our own Sun. There is, therefore, no longer any doubt that nuclear fusion is a sufficient mechanism for explaining the way in which the Sun produces its energy.

Further Reading: Fleck, Bernhard, and Zdenek Svestka, eds. *The First Results from SoHo.* Dordrecth, the Netherlands: Kluwer Academic Publishers, 1998; Frazer, J.G. *Balder the Beautiful: The Fire-Festivals of Europe and the Doctrine of the External Soul,* 2 vols. London: Macmillan and Company, 1913, Volume 1, pp. 331–341; Kitchin, Chris. *Observing the Sun.* London: Springer-Verlag, 1999; Mechler, Gary, Robert Marcialis, and Melinda Hutson. *National Audubon Society Pocket Guide: The Sun and the Moon.* New York: Alfred Knopf, 1995; Phillips, Kenneth J.H. *Guide to the Sun.* Cambridge: Press Syndicate of the University of Cambridge, 1995.

T

TALL TIMBERS RESEARCH STATION

The Tall Timbers Research Station was established in 1958 when Henry L. Beadel willed 2,800 acres of land for the creation of an ecological research center. The land borders Lake Iamonia in northern Leon County, Florida. The research center has since grown to nearly 4,000 acres, including a small tract of land in nearby Grady County, Georgia.

The importance of Tall Timbers in fire science grows out of a series of annual conferences on fire ecology sponsored by the center between 1962 and 1974. These conferences were, according to Stephen Pyne, essentially the only forum for the discussion of fire management policies and practices outside of the U.S. Forest Service. Given the strong commitment of the Forest Service to a philosophy of fire prevention and fire suppression, Tall Timbers was an important meeting place for people who held other views about management techniques, especially prescribed burning. Pyne argues that the Tall Timbers conferences "brought before the public a cornucopia of counterexamples [to suppression] from elsewhere in the world . . . and compelled official research to factor the biology of fire into its agenda" (Pyne 1997, 210).

Tall Timbers has continued to sponsor conferences on fire ecology and has promoted research in a number of fields related to forest ecology. Its stated goal is "to accomplish a sustainable balance between people and natural systems through long-term studies [of forest flora and fauna]" (Tall Timbers Research Station).

The work of the station falls into three large categories: research, conservation, and historic preservation. It works towards its goals through educational and research publications, seminars and training sessions, and conferences designed for landowners and managers, scholars, research scientists, concerned citizens, and students.

For further information, contact:

Tall Timbers Research Station
13093 Henry Beadel Drive
Tallahassee, FL 32312–0918
(850) 893–4153
http://www.talltimbers.org/
See also FIRE ECOLOGY; FIRE MANAGEMENT; FOREST FIRES.

Further Reading: *A Quest for Ecological Understanding*. Tallahassee, FL: Tall Timbers Research Station, 1977; Pyne, Stephen J. *World Fire: The Culture of Life on Earth*. Seattle: University of Washington Press, 1997.

THERMITE

Thermite is a mixture of coarsely ground aluminum metal and iron oxide that is often used in incendiary weapons and in metallurgical applications. When heated, the mixture undergoes a chemical reaction in which aluminum oxide and iron metal are formed:

$$2Al + Fe_2O_3 = 2Fe + Al_2O_3$$

The reaction is very exothermic with the evolution of about 180 kilocalories of heat per mole. A very high temperature is required to initiate this reaction. That temperature is usually achieved by igniting the reactants with a strip of magnesium metal.

The thermite reaction was studied by the German chemist Johann (Hans) Wilhelm Goldschmidt (1861–1923). Goldschmidt was actually looking for a way of producing certain pure metals that was less expensive then any available. The thermite reaction he worked with proved to be a solution for this problem and was (and is) used for the preparation of iron, chromium, manganese, uranium, and other metals from their oxides.

The thermite reaction proved to have two other important applications. One application was in the manufacture of incendiary bombs. The reactants (aluminum, iron oxide, and magnesium) are packed inside a metal casing. When the magnesium is ignited by an electric spark, the bomb goes off. The materials of which the bomb is made are spread widely and cause the ignition of other materials on which they are deposited.

The second application of the reaction is in welding. The reactants are placed in a ceramic crucible and then ignited. The molten iron that is formed flows out of a hole in the bottom of the cone onto the metals to be welded. The reaction is useful for this purpose because of the very high temperatures produced, in the range of 2,500°C (4,500°F). At this temperature, the molten iron not only serves as a bond between the two metals, but may also actually melt the metals being welded, increasing the strength of the bond between them. *See also* IGNITION.

TRACER BULLETS

Tracers are a special kind of bullet used to estimate the range of a target, to mark the point of impact of an explosion, to guide the direction of fire, and for other purposes. The composition of the most common type of tracer bullet, a red tracer, is shown in the table below. A red color, produced by strontium compounds, is chosen because it is most easily seen in adverse weather conditions.

The composition of tracers can be altered for special needs. For example, a dim igniter may be attached to the tracer to protect a gunner from being blinded by the tracer bullets. A daylight tracer is one that contains a large percentage of magnesium in order to produce a brighter light that can be seen more easily in the daylight. *See also* INCENDIARY WEAPONS.

Compound	Red Tracer	Dim Igniter	Daylight Igniter
strontium nitrate	40%		
strontium peroxide		85%	
strontium oxalate	8%		
calcium resinate	4%	9%	
barium peroxide			83%
magnesium	28%	6%	15%
other compounds	20%		2%

TRIAL BY FIRE. *See* Ordeal by Fire

TRIANGLE SHIRTWAIST FIRE

The Triangle Shirtwaist Fire occurred on 25 March 1911 in New York City. The fire swept through the top three floors of the ten-story Asch Building at the corner of Greene Street and Washington Place. One-hundred-forty-five lives were lost, the great majority of them young women, and property damage was estimated at $500,000.

The Triangle Shirtwaist Company was in the business of producing shirtwaists (women's blouses). It employed young women between the ages of sixteen and twenty-three primarily as seamstresses. A number of these women were immigrants and did not speak English. While perhaps not properly called a "sweatshop," working conditions at the company were scarcely comfortable.

The Asch Building was generally regarded to be fire-safe. It had been built with fire-proof materials and, in fact, newspaper accounts of the fire emphasized the fact that the building itself had been largely undamaged by the fire.

Course of the Fire

The fire broke out at 4:40 P.M. on the 25th. The timing of the fire was important and somewhat ironic. The workday normally ended at 5:00 P.M. In another few minutes, workers would have been on their way out of the building, or gone completely. As it was, all employees were still at work or finishing their chores when the fire broke out. The fire began in a container of waste fabric on the eighth floor. The precise cause of the fire was never discovered.

Conditions were such that the fire spread quickly. Scraps of highly flammable cloth were strewn across the floor everywhere, serving as ideal fuel.

In spite of its fireproof construction, the Asch Building quickly became a death trap. It contained virtually no avenues of escape for workers surrounded by fire. A single elevator holding twelve passengers was able to make one trip down, but was then disabled. The building had a single fire escape, but it was so flimsy that it also broke almost immediately. Terrified workers tried running down the stairways, but exit doors were too narrow and this escape route was also blocked quickly.

The only remaining way out of the building was windows. Observers on the street watched as young men and women had to decide in a moment's time whether to trust their fate to a jump to the street or to face the oncoming flames.

The city fire department responded quickly, but was hampered by inadequate equipment. Its fire ladders reached only to the sixth floor and its hoses only to the seventh. The department's fire nets proved not to be strong enough to catch those who chose to jump from nearly 100 feet above the street.

The bodies of victims of the 1911 Triangle Shirtwaist Fire in New York City are laid out on the sidewalk across the street from the Asch Building. *Courtesy of Brown Brothers, Sterling, PA.*

Early jumpers broke through those nets, leaving later victims to fall directly on the street.

The fire lasted no more than about fifteen minutes before it burned itself out. By that time, however, more than 140 of the company's 600 employees had died.

As with most major fires of this kind, the Triangle Shirtwaist Fire spurred public officials to rethink their fire safety requirements and to pass new laws designed to reduce the possibility of a repetition of such a disaster.

Further Reading: "141 Men and Girls Die in Waist Factory Fire." *New York Times*, 26 March 1911, p. 1. Also available at <http://www.ilr.cornell.edu/trianglefire/original/newspaper/141men.html>, 12/07/98; DeAngelis, Gina. *Triangle Shirtwaist Company Fire of 1911*. Broomall, PA: Chelsea House Publishers, 2000; David Wallechinsky, and Irving Wallace. *The People's Almanac, #2*. New York: Bantam Books, 1978, pp. 533–534.

U

U.S. FIRE ADMINISTRATION

The U.S. Fire Administration (USFA) is the federal government's primary agency for dealing with all aspects of the nation's fire problems. The agency was created in 1974 by the Federal Fire Prevention and Control Act. The act was passed following an exhaustive study of the troubling number of fires and fire casualties being reported in the United States. Concerns about the risk of fire to the nation prompted President Richard Nixon to appoint a National Commission on Fire Prevention and Control in 1972. In its final report, "America Burning," the commission recommended the creation of a federal entity to deal with fire safety issues on a national level. The agency was originally called the National Fire Prevention and Control Administration and is now known as the U.S. Fire Administration.

Today, USFA is a directorate within the Federal Emergency Management Agency. It has four primary functions: data collection, public education and awareness, training, and technology development. The first of these functions is carried out by the National Fire Data Center, which collects, analyzes, and disseminates data and information on fire and other emergency incidents.

Among USFA's public education functions are a series of press releases and other informational materials about fire safety and prevention. Examples include "USFA

Emphasizes Importance of Sprinkler Systems," "Where There's Smoke, There's an Educational Opportunity," and "Have a Fire-Safe Holiday Season." The primary mechanism for the agency's training function is the National Fire Academy (NFA). NFA offers educational opportunities for professional firefighters at all levels of experience. Courses and training sessions are offered through state and local fire marshals.

USFA works with public and private organizations to improve and develop fire safety and firefighting technology. The results of this research are published and made available at no charge to the general public.

Additional information about USFA is available at <http://www.usfa.fema.gov/about/programs.htm>.

U.S. FOREST SERVICE

The U.S. Forest Service is a division of the U.S. Department of Agriculture. It is the government agency responsible for fire management on more than 191 million acres of national forest system land. In addition, it works cooperatively with fire management agencies in all fifty states, with foreign governments around the world, and with many private companies involved with forestry and timber production.

Early Attitudes about Forest Management

For the first century of this nation's existence, the idea of a federal agency with responsibility for fire management on public lands would have been unheard of. Such was the case for two reasons. First, until the late 1800s, the federal government was primarily interested in converting public lands to private lands. The era of national parks, national forests, national wilderness areas, and other natural regions set aside for conservation and preservation protected by the federal government had barely begun to dawn as the nineteenth century drew to a close.

In addition, the control of natural fire was given relatively low priority by almost everyone concerned with forests, grasslands, and other wilderness areas. The general practice among timber companies, for example, was to harvest an area as quickly as possible and then move onward. Everyone knew that forest fires were a common and recurrent phenomenon in the forests, so there was little effort or concern towards saving trees from blazes. Individuals and companies knew that there was an endless expanse of untouched forests farther to the West for their taking.

By the 1860s, the earliest signs of a dawning conservationist philosophy were beginning to develop. Vermont lawyer, businessman, and Congressman George Perkins Marsh wrote his classic *Man and Nature* in 1864, chronicling the devastation that humans had caused on the natural environment. Soon, pioneers, such as John Wesley Powell, John Muir, Gifford Pinchot, Stephen Mather, and Enos Mills were beginning to plead for a new and different view of nature, a view that required the setting aside of large expanses of land for protection and conservation. It took half a century for the U.S. government to really accept this view with any enthusiasm. And when they did, the question of the role that natural fire plays in protected parks and forests took on a new meaning.

In addition, a lumbering philosophy built on "more just over the horizon" came rather abruptly to an end towards the end of the century as the view "over the horizon" was of the Pacific Ocean, not more trees. Timber companies were faced for the first time with having to make use of the resources they had already discovered and to think about the problems of losing significant portions of those resources to fire.

Early Years of the Forest Service

The U.S. government first became involved with forestry issues in a formal way in 1881 when the Congress established the Division of Forestry within the U.S. Department of Agriculture. The impetus for this action was a growing interest in using the nation's forest reserves for foreign commerce. At the time, there was little or no concern about protecting or conserving the nation's forests.

The Division of Forestry was given permanent official status in 1886 (24 Stat. 100, 103) and was then renamed the Bureau of Forestry in 1901 (31 Stat. 922, 929). In 1905, the responsibility of the Bureau was increased dramatically when all forest lands previously under the Department of the Interior were transferred to it and it was renamed again, this time as the U.S. Forest Service (USFS).

The U.S. Forest Service and Fire Management

A debate of moderate proportions about the role of the U.S. Forest Service in fire management had been smoldering in the U.S. Congress throughout the last third of the nineteenth century. No action was taken, however, to have the Service become involved in this aspect of forest management until 1910. In that year, terrible fires of previously unseen magnitude swept through much of the West, especially through Montana and Idaho. Timber producers and the Congress were presented for the first time with vivid evidence of the destruction that fire could do to the nation's forest reserves.

The immediate result of these fires was the passage by the Congress of the Weeks Act of 1911. The Act was a piece of legislation that had bounced around the Congress for half a decade, but had never received much support. The 1910 fires changed that situation, however, and the Act was finally passed a year later. The Act consisted of two major parts. In one part, the Forest Service was authorized to purchase lands to be set aside for public use on which no commercial operations could take place. In the second part, the Forest Service was authorized to develop agreements with individual states to develop fire management programs that, in theory, would reduce the risk of 1910-like conflagrations.

Over the next decade, the debate over Forest Service fire management policies and practices continued to rage. The debate culminated in an article written by Gifford Pinchot for the December 1919 issue of the *Journal of Forestry*. In this article, Pinchot expressed the view of the Society of American Foresters that the federal government had to take a much more ambitious role in the management of the nation's forest reserves, including the development of a comprehensive, aggressive program of fire management.

The Pinchot article marked a turning point in the national debate about forest management, including fire management in national forests. Timber companies were outraged that an effort was being made to give the federal government expanded control over the nation's forest resources. They argued that private industry was still the most efficient force in determining how to use these resources.

But the tide of Congressional opinion had already turned. Confronted with the horror of out-of-control forest fires, Congress began to accept the fact that it had some responsibility in bringing these fires under control and, in general, assuring the survival of one of the nation's greatest natural resources. This sentiment finally took concrete form in 1924 with the passage of the Clarke-McNary Act. This Act expanded the provisions of the Weeks Act in two important ways. First, it permitted the federal government to purchase private land to be set aside for public use in any area, not just along navigable streams as had been the case with the Weeks Act. Second, the Forest Service's authority for working with states to develop fire management programs was greatly extended. In one fell swoop, the Forest Service became the central organizing agency in the United States responsible for the development of fire management programs on virtually every level, from local through state to federal, including both public and private lands.

Over the next eight decades, the Forest Service was to become the center of the ongoing controversy among fire and wildland specialists as to how it could best carry out its mandate to manage fire on the nation's public lands. *See also* FIRE MANAGEMENT.

Further Reading: "Forest Service and Aviation Management," <http://www.fs.fed.us/fire/aboutus.shtml>, 08/22/99; Frome, Michael. *The Forest Service*, 2nd edition. Boulder, CO: Westview Press, 1984; Robinston, Glen O. *The Forest Service: A Study in Public Land Management*. Baltimore: Johns Hopkins University Press, 1975; Steen, Harold K. *The U.S. Forest Service: A History*. Seattle: University of Washington Press, 1976, Chapter 7, passim; Wild, Peter. *Pioneer Conservationists of Western America*. Missoula, MT: Mountain Press Publishing Company, 1979, especially Chapters 4 and 8.

V

VESTA

Vesta was the goddess of fire and hearth in Roman mythology. She was thought to be the daughter of Saturn and Rhea. As with most other gods and goddesses, Vesta had her counterpart in Greek mythology, Hestia.

During the earliest years of the Roman state, Vesta was worshipped primarily in private homes. Over time, however, she became more important as a symbol for the entire state. According to one story, she was finally raised to her highest position by the early Roman king Numa Pompilius (715–673 B.C.). The king ordered that a temple be built in her honor on the Palatine Hill, in the middle of the Roman Forum. Eventually, the temple became one of the holiest spots in the empire.

The Temple of Vesta was home to the eternal sacred flame that had come to represent the sanctity and power of the Roman state. The job of protecting and maintaining the eternal flame fell to the *Vestal Virgins*. The Vestal Virgins were women selected specifically to devote their lives to guarding and preserving the flame. At first, the Vestal Virgins were probably daughters of the Roman kings. Later, however, they were chosen from among members of the aristocratic families of the Republic.

A Vestal Virgin began her duties between the ages of six and ten. She was required to commit the next thirty years of her life to her work at the temple and to remain a virgin throughout that period of time. Some authorities have suggested that a Virgin spent the first ten years of her time in the temple learning her job, the next ten years carrying it out, and the last ten years teaching it to new Virgins. At the end of her thirty-year commitment, a woman was allowed to leave the temple and marry, if she wanted. It seems likely, however, that few women accepted that choice.

Among a Vestal Virgin's duties were making sure the eternal flame did not go out, bringing water to the temple, baking the sacred bread, keeping wills, and taking care of objects sacred to the Roman state. Virgins were accorded special places of honor at state events and were permitted to pardon prisoners.

Failing in one's duties as a Vestal Virgin was among the most serious crimes in the Roman state. A Virgin who broke her vows of chastity was put to death by being buried alive. An elaborate ceremony was devised for anyone who met this fate. The transgressor was wrapped in heavy linens in order to muffle her cries, and she was then carried to an area known as "The Field of Unhappiness" on Quirinal Hill. There she was buried in a grave carefully covered over so that no one would be able to find the victim's body. Historical records indicate that very few Vestal Virgins broke their vow and met this terrible fate. Indeed, many remained in the service of Vesta even after their thirty-year-long commitment had been met.

As part of his campaign to eliminate pagan practices in an increasingly Christian Rome, Emperor Theodosius ordered in A.D. 394 that the sacred Vestal fire be extinguished. The temple fell into neglect, although its remains can still be viewed today. *See also* GODS AND GODDESSES.

Further Reading: Frazer, J.G, "The Prytaneum, the Temple of Vesta, the Vestals, Perpetual Fires," *Journal of Philology* 14 (1885): pp. 145–172; Frazer, Sir James George. *The Golden Bough: A Study in Magic and Religion*. New York: The Macmillan Company, 1958, passim; Pyne, Stephen J. *Vestal Fire: An Environmental History, Told through Fire, of Europe and Europe's Encounter with the World*. Seattle: University of Washington Press, 1997, pp. 76–78, passim.

VISUAL ARTS

Fire appears relatively infrequently as a primary theme in the visual arts. In some paintings, fire is the major focus of the composition. One of the most famous examples is the depiction of the fire in the town of Borgo that occurred in A.D. 847, as portrayed by Raphael (1483–1520) in 1514. The Borgo fire was used as one of the themes of the frescoes that Pope Julius II commissioned of Raphael in 1508. The frescoes cover the four room apartment on the top floor of the Pope's apartments in the Vatican.

Another well-known series of paintings dealing with actual fires are those done of the burning of the Houses of Parliament by the English painter Joseph Mallord William Turner (1775–1851) in 1834. Other famous fires immortalized in paintings are those that destroyed the cities of Sacramento (Hugo Wilhelm Arthur Nahl [1833–1889]), 1852, and San Francisco (Charles Christian Nahl [1818–1878]), 1856, and Francis Samuel Marryat [1826–1855]), 1850. Fire is also the focus of the French painter Claude Lorrain's (1604?–1682) *The Trojan Women Setting Fire to Their Fleet* (1643). The painting depicts the Trojan women destroying their ships as a way of ending their years of wandering after the fall of Troy.

Fire sometimes appears in a painting as background to the work's major theme. For example, in his painting *Alexander the Great Rescued from the River Cydnus* (1650), the Italian artist Pietro Testa (1612–1650) focuses on Alexander's rescue from the river after he had been taken with a sudden chill. In the background, however, Testa shows smoke from the fires that had been set in Tarsus by Darius's retreating troops.

On rare occasions, fire itself is the major theme of a visual work. Art historians believe that the Italian painter Piero di Cosimo (1462–1521?) may have painted a series of panels depicting the growth of civilization through the control of fire for the Florentine nobleman Francesco del Pugliese in about 1505–1507. Since no signed or documented works of Cosimo exist, this conjecture is based on commentaries by a contemporary critic, Giorgio Vasari (1511–1574). Vasari suggests that the Pugliese panels and possibly other paintings were part of a cycle dealing with the discovery of fire.

Another work of art based on the discovery of fire is a page from the illustrated manuscript Shahnama of Shah Tahmasp, by the Persian artist Sultan Muhammad, dated at about 1520–1522. The page is beautifully illustrated in ink, brilliant colors, silk, and gold and is now held by the Metropolitan Museum of Art in New York City.

Just as it is in Christian theology, fire is sometimes a theme in art dealing with Christian subjects. In a stained glass window made in Canterbury, Kent, in about 1175–1180, for example, the martyrdom of Saint Lawrence is depicted. In contrast to the usual story, in which Lawrence is killed by being grilled on a fire, this representation shows the saint at prayer with fire at his feet and head. The fire at his head is divided into three parts, representing the fervor of faith, the love of Christ, and the true knowledge of God. A painting by William Holman Hunt (1827–1910) depicts the *Miracle of the Holy Fire* (begun in 1896), a ceremony celebrated by the Greek Orthodox Church until recent times in which a holy candle is lit spontaneously by heavenly powers. Hunt included along with his painting a keyplate that described and explained the ceremony, one that he thought had no place in modern religious life.

Finally, fire sometimes plays a symbolic or unexplained role in a work. In 1935 and 1936, respectively, Salvador Dali (1904–1989) painted two works whose symbolic meaning is probably not clear to the viewer, *Giraffe on Fire* and *Burning Giraffes and Telephones*. At about the same time (1934–1935), Belgian painter Rene Magritte produced a work entitled, *The Discovery of Fire*, which showed a French horn on fire. Magritte provided no explanation for the meaning of his painting.

Further Reading: Hartt, Frederick. *Art: A History of Painting, Sculpture, Architecture*, 4th edition. Englewood Cliffs, NJ: Prentice Hall, 1993; Stokstad, Marilyn. *Art History*, revised edition. New York: H.N. Abrams, 1999.

VOLCANO

A volcano is a mountain through which hot, liquid rock (lava) escapes from deep beneath the Earth's surface. The release of lava is known as a *volcanic eruption*, which can vary in character and intensity. In some cases, lava flows out of a volcano in a slow, steady stream, while in other cases, it is expelled with enormous force that can thrust material ranging from house-size rocks to powdery ash high into the air.

Volcanic activity is a localized activity that occurs very commonly in some parts of the planet, less commonly in other parts, and virtually not at all in still other parts. In regions where volcanic activity does occur, it can cause widespread destruction, by igniting forest fires, burning down homes and whole cities, and killing humans and other animals. It is a major source of forest fires in regions where volcanoes and forests are in close proximity, although its effect in this regard on a worldwide basis is relatively minor.

Ancient Views of Volcanoes

It is hardly surprising that ancient people were fascinated and dumbfounded by volcanic eruptions. Such events are among the most powerful of all natural events, matching the destructive potential of hurricanes, earthquakes, and tidal waves. They were viewed in different ways in different parts of the world at different times. In some cultures, volcanoes were thought to be the home of fire gods. Mount Fuji in Japan, for example, was held to be the home of the Ainu goddess of the hearth, Fuji. Mount Kilauea, in Hawaii, has long (and to this day by some people) been thought to be the home of the Polynesian fire goddess Pele.

In such cultures, a volcano may be an amazing demonstration of the power of nature, but not necessarily an object to be feared. In fact, as the home of a revered god or goddess, it may actually be a final destination one hopes to reach at the end of life. In some South Sea Islands, for example, nearby volcanoes were thought to be the homes of local deities and the final resting place for the dead. When a person died, his soul was thought to go into a cave, from which it emerged on the other side. After leaving the other side of the cave, it summoned a boatman by lighting a fire. The boatman then took the soul to the volcano where, according to one legend, "the ghosts dance every night and sleep all day" (Campbell 451).

Structure of a Volcano

Volcanoes may take different forms depending on the way in which magma is released to the surface. In some cases, magma flows quietly and smoothly over long periods of time. In such cases, the volcano usually takes the shape of a gently sloping dome-shaped mountain. This type of volcano is called a *shield volcano*. The largest shield volcano in the world is Mauna Loa in the Hawaiian Islands. Mauna Loa rises to a height of about 4 kilometers (2.5 miles) from its base on the ocean floor to its summit.

By contrast, a *cinder cone volcano* is one formed when magma is ejected with enormous force. In many cases, this force is the result of gas trapped within the magma working outwards and pushing the magma far into the air. As the magma cools, it tends to form larger chunks of rock known as *cinders*. These cinders are too heavy to be carried very far from the point of eruption, so they pile up and form a volcano with steep sides. Paricutin in Mexico is a good example of a cinder cone volcano.

Many volcanoes are formed by some combination of quiet and eruptive lava flow at different times in their history. They may have properties of both shield and cinder cone volcanoes and are known as a *composite* volcano.

Benefits and Drawbacks of Volcanoes

Volcanoes pose terrible threats to communities and ecosystems in which they occur. Once an eruption begins, there is no way that it can be controlled. In the case of shield volcanoes, for example, lava may flow over the countryside for weeks or months, setting fire to everything with which it comes into contact. Fire fighters have no hope of stopping the advance of a volcano, and their best hope is to evacuate humans and other animals, remove valuable property that can be transported, and try to limit forest fires set by the lava.

On the other hand, volcanoes do have some benefits. The rocky material they bring to the surface can produce some of the richest soil found anywhere on the planet. Regions that have been covered by lava flows before now sustain some of the Earth's most productive agricultural regions.

Volcanoes can also be used as a valuable source of heat. After a violent eruption, some volcanoes actually collapse in upon themselves, leaving a huge hole in the middle of the mountain. This hole, called a *caldera*, is an ideal catch-basin for water which, when warmed by the remaining heat of the volcano, can produce hot springs and wells of hot water. In Iceland, for example, the heat captured from volcanoes is adequate to provide the great majority of the energy needs of that nation. *See also* IGNITION.

Further Reading: Crowley, Clinton, "Volcano." In Bridget Travers, ed., *The Gale Encyclopedia of Science*. Detroit: Gale Research, 1996, pp. 3868–3875; Decker, Robert, and Barbara Decker. *Volcanoes*. New York: W.H. Freeman and Company, 1989; Fisher, Richard V., Grant Heiken, and A.K. Morris, eds. *Volcanoes*. Princeton, NJ: Princeton University Press, 1998; Martin, Fred. *Volcano*. Crystal Lake, IL: Rigby Interactive Library, 1996; Miller, Russell and the Editors of Time-Life Books. *Planet Earth: Continents in Collision*. Alexandria, VA: Time-Life Books, 1983; Scarth, Alwyn. *Vulcan's Fury: Man against the Volcano*. New Haven, CT: Yale University Press, 1999; Sigurdsson, Haraldur, ed. *Encyclopedia of Volcanoes*. New York: Academic Press, 1999.

W

WACO (TEXAS) FIRE OF 1993

The Waco fire of 1993 was a conflagration in which seventy-four members of the Branch Davidian religious sect died in a fire that resulted from an attack on their compound by members of the U.S. Federal Bureau of Investigation (FBI). The cause of the fire was still not known for certain a decade after the event, and it has been the source of intense controversy during that time.

The term *Branch Davidians* was used by members of the media to describe a religious group that had originally broken from the Seventh-Day Adventist Church in 1942. The group preferred to call itself the *Students of the Seven Seals*. The group had remained small in size from its beginning, and seldom numbered more than a few hundred. In the spring of 1993, 130 men, women, and children were living at a church compound which they called *Ranch Apocalypse*, near Waco, Texas.

At the time, the group was led by Vernon Howell. Howell had joined the group as a handyman in 1981 and, in 1990, had changed his name to David Koresh. The group believed in a Second Coming that would occur in the very near future, and that it would be brought about by the "lamb" mentioned in Revelations (5:2). The lamb was not to be Jesus, as most Christians believe, but Koresh himself.

On 28 February 1993, members of the U.S. Bureau of Alcohol, Tobacco, and Firearms (ATF) entered Ranch Apocalypse for the purpose of serving a search warrant on Koresh. The warrant was obtained on the basis of claims that the religious group was amassing a large supply of illegal weapons and ammunition. Koresh had also been accused of molesting and abusing children living in the compound.

Shortly after ATF agents arrived, a shot was heard. The source of the shot has never been identified. Both sides claim that the other was responsible for it. The shot initiated an exchange of gunfire in which six members of the religious group and four ATF agents were killed. An additional twenty-four agents and one Davidian were wounded.

Following this event, investigation of the Koresh group was turned over to the FBI, who laid siege to the compound for fifty-one days. On 19 April, government officials decided that they had waited long enough. They decided that the compound had to be stormed and occupied.

The exact events that took place over the next few hours are highly controversial. The facts appear to be that FBI agents used specially equipped tanks to breach the outer walls of the compound. They then injected CS (O-chlorobenzylidene malonotrile) tear gas into the compound. At about the same time, a fire broke out inside some of the buildings. The fire quickly grew out of control and spread throughout the compound. In less than an hour, Koresh and seventy-four of his followers were dead. Death was later attributed to gunshot and stab wounds and asphyxiation from smoke. Eight members of the group escaped from the fire.

For nearly a decade, pro-Koresh and pro-government individuals and groups have argued over the cause of the fire. Less than two hours after the attack had begun, a picture was published showing a tank with a flame thrower attached to its nose approaching the compound. Analysis later proved that photo to be a hoax. Government officials insist that nothing they did during the attack could have led to the fire. They argue that members of the group committed mass suicide by setting the fire themselves. Opponents claim that the CS gas caught fire, bullets from FBI agents set fire to flammable materials inside the compound, or other government actions were responsible for the blaze.

Photographs taken at the time of the attack and later studies of the area suggest that the major fire began as a series of smaller fires throughout the compound. It appears that Koresh may have ordered that stocks of kerosene be placed in strategic locations throughout the building. These stocks may have been set aflame once government agents had breached the outer walls.

In March 2000, the attack on the Branch Davidian compound was reenacted. Airplanes, tanks, and soldiers fired weapons at a structure designed to simulate Ranch Apocalypse. The purpose of the reenactment was to determine whether government forces had, indeed, performed in such a way in 1993 as to have caused the disaster at the compound. The event had been ordered by a U.S. district judge then hearing a civil case brought by supporters of the Branch Davidians victims and survivors.

In September 2000, the judge, Walter S. Smith, Jr., rejected all claims made by Branch Davidian victims and survivors. He said that evidence obtained from the reenactment left no question as to how the fires had originally started. He wrote in his opinion that "the entire tragedy at Mount Carmel can be laid at the feet of this one individual [Mr. Koresh]. . . . The FBI," he continued, "acted with restraint on April 19,1993, despite the deadly gunfire directed at them during the tear gas operation. The FBI did not return fire" (Hancock, p. 2)

Further Reading: Branch Davidians, a.k.a. Students of the Seven Seals," <http://www. religioustolerance.org/dc_branc.htm>; Hancock, Lee, "Judge Blames Sect for Waco Tragedy." *The Dallas Morning News*, 21 September 2000, at <http://www.dallasnews.com/waco/174697_waco_21tex.ART.html>; Kopel, D.B., and P.H. Blackman. *No More Wacos: What's Wrong with Federal Law Enforcement and How to Fix It*. Amherst, NY: Prometheus Books, 1997; Moore, Carol. *Davidian Massacre: Disturbing Questions about Waco which Must Be Answered*. Franklin, TN: Legacy Communications, 1996; Reavis, D.J. *The Ashes of Waco: An Investigation*. Syracuse, NY: Syracuse University Press, 1998; Tabor, J.K., and E.V. Gallager. *Why Waco?: Cults and the Battle for Religious Freedom in America*. Berkeley: University of California Press, 1997; "The Waco Massacre," <http://www.magnet.ch/serendipity/waco.html>; Wright, S.A. *Armageddon in Waco: Critical Perspective on the Branch Davidian Conflict*. Chicago: University of Chicago Press, 1996.

WEEKS ACT OF 1911

The Weeks Act was passed by the U.S. Congress in 1911 authorizing the purchase of private lands for National Forests. Prior to this legislation, the major focus of federal policy had been to transfer public land to private individuals and corporations for their personal use. For many years, there had been little interest in setting aside land for recreation, preservation, or other "nonproductive" uses.

In the early 1900s, however, a number of factors combined to gradually change this philosophy. Congressman John Weeks (R, MA) first showed an interest in expanding the preservation of public lands in 1906 when he joined in support for setting aside forest lands to protect industrially important watersheds. Over the next decade, Weeks became more active in acting to preserve forests not only for industrial reasons, but also for nonproductive purposes.

After a number of tries, Weeks was finally able to draft a bill that received majority support from the House of Representatives and Senate. That bill allowed the U.S. Department of Agriculture to purchase lands to be set aside for public use and to develop cooperative programs of fire control with individual states.

The Weeks bill was important because it was the first step in the development of an integrated national program for management of fire on both national and state lands. It was later to be expanded and improved in the Clarke-McNary Act of 1924 and the McSweeny-McNary Act of 1928. *See also* FIRE MANAGEMENT; U.S. FOREST SERVICE.

Further Reading: Steen, Harold K. *The U.S. Forest Service: A History*. Seattle: University of Washington Press, 1976, pp. 122–131, passim; The Weeks Act became Public Law 61–435, Chapter 186, 36, Statute 961, as amended. See *The Principal Laws Relating to Forest Service Activities*. Washington, DC: Government Printing Office, n.d.

WELDING

Welding is the process by which two pieces of metal are joined to each other. This process often involves the application of heat, although nonthermal welding processes are also available. Probably the oldest type of welding technique uses both heat and pressure. The two metals to be joined to each other are first heated to the highest possible temperature. Then, the two pieces of hot metal are hammered until they are affixed to each other.

In general, two kinds of welds can be made, with or without a filler material. In the latter case, heat and/or pressure is sufficient to cause the exposed surfaces of each metal to liquefy, intermix, and then refreeze. In the final product, the atoms and molecules of each piece of metal have become intimately mixed with each other, much as they would be in a pure metal. This type of weld is called an *autogenous* weld and is possible when the two pieces being welded are made of the same metal or when the size and spacing of atoms and molecules of two different metals are sufficiently similar to permit this kind of intimate intermixing. When this condition is not met, a filler may be needed. In this case, the edges of the two metals and the filler are all melted by heat and/or pressure. When forced together, each piece of metal bonds to the filler, which then acts as a bridge between the two original pieces of metal.

Currently, about two dozen different welding methods are in use. Most of these methods can be classified into one of three major types: (1) gas welding; (2) arc welding; and (3) resistance welding.

Cutting and welding torches use heat to cut or join metal. This torch is being used to cut a half-inch steel plate. *Courtesy of Smith Equipment.*

Gas Welding

Gas welding is a rather straightforward process in which the edges of two metals, with or without a filler, are heated to a temperature high enough to cause them to melt. When forced into contact with each other, the molten materials simply diffuse into each other. As the metals cool, they form a single, homogeneous state that joins the original two pieces to each other.

Probably the most common form of gas welding uses an oxyacetylene torch, a device in which acetylene gas is caused to burn in an atmosphere of pure oxygen. An oxyacetylene torch produces a flame with a temperature of about 3,000°C (5,400°F), sufficiently high to melt iron, steel, and many other metals. The effect produced by an oxyacetylene torch can be modified by using tips of different sizes and shapes and fuel/oxygen mixtures of various ratios.

Arc Welding

Another way to produce the heat needed to melt two metals (with or without a filler) is with an electric arc. An electric arc is an electrical discharge that takes place because of a potential difference between two points in a space gap. In arc welding, a *welding circuit* is created that includes the welding tool and the material to be welded, known as the *work*. When the welding tool is turned on, a large potential difference is created between the tool and the work. This potential difference heats the work, causing it to melt. Temperatures of 15,000°C to 20,000°C (27,000°F to 36,000°F) are not unusual with arc welding.

The welding tool may be fitted with either a *disposable* or *nondisposable* tip. In the former case, the heat produced by the electric arc actually causes the tip to melt and act as a filler between the two metals being welded. In the latter case, the tip simply conducts the electric discharge between the welding tool and the work.

Arc welding is often done in an inert atmosphere to prevent oxidation or other chemical changes in the metals being heated. In some cases, the whole welding process is enclosed in a tent from which air is excluded and replaced by argon or some other inert gas. In other cases, the welding tool releases a stream of inert gas so that only the specific area being welded is covered by the gas.

Many variations of the basic arc-welding process have been developed. These variations may be necessary depending on the kinds of metals being welded and the type of weld desired. For example, in the process known as *submerged-arc welding* a layer of filler is first laid down on one of the pieces to be welded and melted. Then the tip of the welding tool is inserted into the filler. The tool can be moved through the molten filler, melting more material in its wake. Submerged-arc welding has the advantage of penetrating more deeply beneath the surface of the metal being welded than with most other procedures, producing a stronger, more permanent weld.

Carbon dioxide welding is another variation of arc welding in which the gas used to shield the work is carbon dioxide, rather than argon or some other inert gas. The advantage of using carbon dioxide is that it is less expensive than most other gases typically used in arc welding.

Resistance Welding

Resistance welding is accomplished by passing an electric current through the metals to be welded and then applying pressure on those pieces. The flow of an electric current through any metal is always accompanied by the production of heat. The heat is produced as a result of the resistance encountered by electrons in trying to flow through the metal. The amount of resistance differs for each type of material. Copper and silver, for example, have very little resistance to an electric current, while iron and steel have greater resistance.

In all forms of resistance welding, a short pulse of electric current is passed through the area to be welded and is followed by mechanical pressure that forces the two surfaces together. The various forms of resistance welding are named according to the specific method by which the current and pressure are applied. In a *spot weld*, for example, two electrodes are attached opposite each other on the outer surfaces of the materials to be welded. Heating occurs in a specific region on the metals, producing the weld. Other forms of resistance welding include seam, flash, and projection welding.

Other Types of Welding

A number of other methods have been developed for producing the melting and pressure needed to make a weld. For example, laser welding makes use of the highly focused, coherent light beam of a laser to heat the regions on two metals to be welded. Electron-beam welding achieves the same result by firing an intense beam of electrons at the work. In friction welding, the heat needed for the weld is obtained by rubbing two pieces of metal together at high speed. In explosive welding, a small amount of explosive material is coated on one of the materials to be welded. When the explosive is detonated, it is forced against the second material with enough force to bring about a mechanical weld between the two materials.

Further Reading: Finch, Richard. *Welder's Handbook: A Complete Guide to MIG, TIG, ARC, and Oxyacetylene Welding*. Berkeley, CA: Berkeley Publishing Group, 1997; Geary, Don. *The Welder's Bible*. New York: McGraw-Hill Professional Publishing, 1992; Geary, Don. *Welding*. New York: The McGraw-Hill Companies, 1999; Schwartz, Mel M. "Welding and Cutting of Metals." *McGraw-Hill Encyclopedia of Science & Technology*, 8th edition. New York: McGraw-Hill Book Company, 1997, Volume 19, pp. 488–497; "Welding." In *The New Illustrated Science and Invention Encyclopedia*. Westport, CT: H.S. Suttman, 1989, pp. 3131–3135.

WELSBACH LANTERN. *See* Lamps and Lighting

WILDFIRE LANDSCAPING

Wildfire landscaping refers to all of the practices that can reduce the danger of buildings being damaged by wildfires. A wildfire is a fire that occurs in a rural or wilderness area, often spreading rapidly and getting out of control in a short period of time. Wildfires are a serious hazard to many homes and other buildings that have been constructed in un-

Proper landscape planning can reduce the impact of wildland/residential interface fires. *Courtesy of Robert Murgallis.*

developed areas. These buildings may be in the path of wildfires and/or may be too far from fire protection services to receive the aid needed to save the building.

Landscaping for Fire Protection

Fire experts point out that homeowners can take a number of actions to reduce the threat to their property posed by wildfires. Some of these suggestions are as follows:

1. Create a safety zone around buildings. A safety zone is an area that is essentially free of easily combustible materials, such as grass, leaves, trash, and dead bushes and plants. If firewood is used in the building, it should be stored at a safe distance from the building itself.
2. Fire-resistant plants should be favored over combustible plants. Deciduous trees, for example, drop their leaves every fall, creating an annual cleanup problem for the homeowner. By contrast, succulent plants, such as cacti, tend to retain moisture and are more resistant to fires.
3. Trees should be planted at a safe distance from the house. A rule of thumb is that the distance from any tree should be at least as far from the nearest building as the full-grown height of the tree. Thus, any tree expected to grow to a height of thirty feet, should be planted more than thirty feet from a house.
4. A similar rule holds for outbuildings. No outbuilding should be constructed closer to a house than the height of the outbuilding. Thus, a storage shed ten feet high should be placed at least ten feet from the main house.
5. Annual maintenance of roofs, gutters, and eaves is essential. When leaves, needles, and other combustible materials collect in these areas, they provide a ready path for the spread of a wildfire.
6. Trees and bushes should be tended on a regular basis, with dead limbs and branches being removed. Dead branches on tall trees provide an easy pathway by which a fire can "climb" a tree and ignite the crown.

7. Do not use outdoor incinerators for the burning of household trash or debris from the yard. This practice can initiate a fire that spreads both to nearby vegetation and to buildings.

See also INTERMIX FIRES; FOREST FIRES.

Further Reading: Boulton, Joan. "Firescaping: Ways to Keep Your House and Garden from Going Up in Smoke." *Horticulture*, (August 1991); "Hints for Homeowners," <http://www.napsnet.com/garden/39539.html>, 12/22/98>; *Protecting Residences from Wildfires: A Guide for Homeowners, Lawmakers, and Planners.* Technical Report No. 50, U.S. Department of Agriculture, 1981; "Protecting Your Home from Wildfire," <http://www.firewise.org/pubs/protect/>, 12/22/98; "Protecting Your Property from Fire," <http://www.fema.gov/mit/how2001.htm>, 12/22/98.

WILDFIRE MAGAZINE. See INTERNATIONAL ASSOCIATION OF WILDLAND FIRE

WILDLAND FIRE ASSESSMENT SYSTEM

The Wildland Fire Assessment System (WFAS) is a program for producing maps that indicate the risk of fire for the 48 coterminous States and Alaska. The program makes use of the National Fire Danger Rating System (NFDRS) developed by the U.S. Forest Service in the late 1950s and early 1960s. The general purpose of the WFAS is to notify areas of the United States of the relative risk of wildland fire within twenty-four hours.

Input Data

Maps generated by the WFAS are based on data supplied by 1,500 weather stations that make up the Fire Weather Network. These data are sent to fire weather forecasters at the Fire Behavior Research Work Unit of the Intermountain Fire Sciences Laboratory in Missoula, Montana, who use NFDRS mathematical algorithms to produce the WFAS maps.

Data collected at Fire Weather Network fire stations fall into five general categories: next-day weather forecasts, dead and live fuel moisture, drought conditions, lower atmosphere stability conditions, and lightning ignition probability. These data are not independent of each other, but reflect in general the fuel and weather conditions in a region.

Data vary in reliability from region to region depending on the concentration of weather stations. In some locations, weather stations are located fairly close to each other, while in other locations, they are more widely distributed. Forecasters use mathematical models to estimate data in areas between weather stations. The reliability of such estimates depends, of course, on the distance between stations for which they were determined.

The kinds of data obtained from each station for each variable include the following:

- Weather conditions: wind and rain patterns; temperature; relative humidity; dew point for the most recent observation and for the following day's forecast.
- Fuel moisture: time required for ignition of dead and live fuel of various diameters, ranging from less than 1/4 inch to 6 inches.
- Drought: soil moisture based on an average soil capacity of eight inches of water, which depends on the maximum daily temperature, daily precipitation, previous precipitation, and annual precipitation.
- Lower atmosphere stability: temperature gradient and dew point for a given layer of the atmosphere.
- Lightning ignition probability: ignition probability based on fuel moisture for a given area.

Application

WFAS fire danger maps can be accessed on the Internet and provide valuable information for all individuals and agencies who deal with wildland fires. Individuals responsible for fire management at National Parks and National Forests, for example, can find out where the risk of fire is greatest in the regions for which they are responsible and can assign personnel and equipment to those areas of highest risk.

For long-term planning, thirty-day forecasts are also available from the Riverside Fire Laboratory in Riverside, California.

Further Reading: Bradshaw, Larry S. et al. *The 1978 National Fire-Danger Rating System*. General Technical Report INT-169. Ogden, UT: U.S. Department of Agriculture, Forest Service, Intermountain Forest and Range Experiment Station, 1984; Burgan, Robert E. *1988 Revisions to the 1978 National Fire-Danger Rating System*. Research Paper SE-273. Asheville, NC: U.S. Department of Agriculture, Forest Service, Southeastern Forest Experiment Station, 1988; "Wildland Fire Assessment System," <http://www.fs.fed.us/land/wfas/>, 07/26/99.

WITCH BURNING. *See* CHRISTIANITY AND THE BURNING OF HERETICS

WORLD FIRE

World Fire: The Culture of Fire on Earth is one of five books in the "Cycle of Fire" series written by Stephen J. Pyne, professor of history at Arizona State University. The other four books in the series are *Burning Bush: A Fire History of Australia*; *Fire in America: A Cultural History of Wildland and Rural Fire*; *The Ice: A Journey to Antarctica*; and *Vestal Fire: An Environmental History, Told through Fire, of Europe and Europe's Encounter with the World. World Fire* is probably the one book in the series most worth reading partly because it provides an introduction to the other books in the series, and partly because it provides a broad, general overview to the role of fire in human culture.

Pyne's thesis is that the Earth is a "fire planet," a planet covered with carbon-based materials surrounded by an atmosphere of oxygen. Under these conditions, fire is a natural, inescapable part of the Earth's history, in general, and of human civilizations, in particular.

Pyne points out that plant life had adapted to the relative presence or absence of fire in various parts of the planet long before humans appeared. A dynamic equilibrium had been reached in which fire made possible many kinds of plant life and, in turn, the growth of new plants provided the fuel that allowed fire to return over and over again. In this context, fire was neither "good" nor "bad," it was simply a part of the natural order of life on Earth.

Pyne goes on to show that the discovery of fire by humans was certainly an important event in history, but probably much less important than the variety of ways in which various human cultures responded to the presence of natural fire. In some cases, humans adapted to the presence of fire in their environment and built their way of life around existing fire patterns. In other cases, humans chose to become masters over fire, attempting to control and use it as they wished.

As an example of the first philosophy, Pyne describes the way in which Australian aboriginal tribes used the firestick as a tool to shape the environment, much as they used the hoe or any other instrument. To the aborigines, natural fire was responsible for the natural environment in which they lived, and they learned to work with fire instead of fighting against it.

By contrast, Pyne provides example after example of human cultures that never learned—or learned too late—the essential role that fire plays in the environment. In Sweden, for example, he shows that the de-sire to harness all natural resources for the exclusive purpose of bettering human life demanded an all-out, aggressive effort to eliminate totally all natural fires. To whatever extent this effort may have been successful in the short run, it ignored certain unalterable facts about the role of fire in the natural world, with consequences that were eventually disastrous for the Swedes.

The moral of Pyne's book seems to be that too many human cultures have assumed that they could master fire as they have tried to master rivers, animal populations, and other features of the natural environment. Without fail, however, these efforts have been unsuccessful. In the end, it appears, fire will always "have its way," sometimes with disastrous consequences for its human "masters."

Arguably, *World Fire* is the most incisive, most thorough analysis of the way in which humans and fire interact with enormous political, social, economic, philosophical, and environmental consequences. *See also* PYNE, STEPHEN J.

Further Reading: Pyne, Stephen J. *World Fire: The Culture of Life on Earth*, originally published in hardcover by Henry Holt and Company, New York, 1995; republished as part of the "Cycle of Fire" series by the University of Washington Press, Seattle, 1997. Other books in the "Cycle" are also available from the University of Washington Press.

Y

YELLOWSTONE FIRES OF 1988

The Yellowstone fires of 1988 were a series of fires that swept through Yellowstone National Park from June to September, destroying huge amounts of trees, shrubs, brush, and other material in and around the park. The fires were caused by a number of factors, including an extended period of dry weather in which natural fuel accumulated over more than a decade. In addition, the fires spread as they did as the result of a conscious decision by the Yellowstone National Park administration to allow the fires to burn, to a large extent, on their own.

Chronology of the Fires

The first major fire to break out began at Storm Creek in the Custer National Forest, which lies immediately north of Yellowstone. During the months that were to follow, the National Forest Service followed a policy different from that of the Yellowstone National Park administration. It decided to fight fires as soon as they appeared and, as a result, the fires caused relatively modest damage in the areas around the park. Of the 218 fires that occurred on National Forest lands, none covered more than ten acres.

The first large fire to break out within park boundaries started on 23 June, at the south end of the park, in the Shoshone region. Two days later, another large fire began in the Gallatin Range, just inside the park's northwest boundary. Over the next two weeks, The Red Fire broke out in the southern part of the park (1 July), the Mist Fire and Clover Fires started near the eastern boundary of the park (9 and 11 July, respectively), and the Falls Fire began in the southwest corner of the park (12 July).

Up to this point, the park administration had followed a policy of letting the fires burn themselves out naturally. By 15 July, how-

Firefighter in Yellowstone National Park during the 1988 summer fires. *Courtesy of National Parks Service.*

ever, it had begun to have second thoughts and announced that all new fires would be fought. Over the next two months, more than 25,000 firefighters from every part of the United States were employed in trying to keep fire under control in the park. They used 336 fire engines, 57 helicopters, 41 bulldozers, and many retardant bombers. At its peak, firefighting cost the government more than $3 million a day and a total of more than $130 million to extinguish the fires.

These efforts were largely unsuccessful in holding down the spread of the fires. It was not until the first rains of the year began in early September that the blazes began to die out. On 11 September, two inches of snow fell over most of Yellowstone, essentially bringing an end to perhaps the greatest wildfire to have struck the Western United States in recorded history.

Overall, the fires swept across nearly half of the park's 2.2 million acres. Untold numbers of trees were killed, although loss of animal life appeared to be relatively modest. By one count, five bison, about 240 elk, one black bear, two moose, and four deer were killed by the fires. In addition, firefighters were able to save most of the administrative buildings, museums, and other facilities in the park.

The Politics of the Yellowstone Fires

Few fire experts were very surprised that the fires of 1988 occurred. Yellowstone National Park had been a "fire waiting to happen" for many years. In the fifteen-year period between 1972 and 1987, the park had had 235 fires, none covering more than 7,400 acres. Because of earlier fire suppression policies, no large fire had swept through the park in many years. Yet, from an ecological standpoint, a large fire in the park was just what was needed. When the fire did occur, it developed into a natural event that was far beyond the capacity of any firefighting program to control. Who was at fault for this disaster?

Critics appeared on the scene almost as soon as the fires began to appear. A general public that had been educated to hate and fear forest fires were appalled to see blazes sweeping the nation's first and premier national park. Most people could not believe that park officials would stand aside and let the fires burn themselves out. Legislators were also outraged that one of the nation's most treasured assets was being allowed to burn itself up. Representatives from all levels of government stood before television cameras with fires blazing behind them complaining about the policies of park administrators.

Many fire scientists also raised questions about park officials' handling of the fires. They pointed out that officials had allowed their fire management policies to fall behind those of most other agencies, and the park was not prepared to deal with any fire of any size, let alone one of the magnitude that developed in the summer of 1988.

Stephen Pyne, author of the series, "Cycle of Fire," has been one of the most severe critics of the handling of the Yellowstone fires of 1988. He claims in his book *World Fire* that:

> The [Yellowstone] fire committee, which made decisions, became clubbish, haughty over its highly personalized knowledge. Yellowstone did not field a prescribed fire program; it had a let-burn program. Instead of prescriptions, it had a philosophy. (Pyne, 1997, 260)

Outside officials had encouraged park officials to rethink and upgrade their fire management policies, according to Pyne. But they seemed not willing to do so. In 1985, an outside specialist was hired to advise the park on its fire management policies. When the specialist finished his work ten weeks later, he left behind a compromise document that recommended extensive changes in the park's fire policies. According to Pyne,

> [the] proposal [was] quickly discarded. The park never submitted the draft plan for public review. When the fires came it did not even follow the flawed decision charts of that still-unapproved plan. (Pyne, 1997, 261)

Long-term Consequences of the Fire

The long-term effects of the 1988 fires in Yellowstone have been a surprise to virtually none of the experts in fire ecology. The flora and fauna of the park have sprung back as would have been expected following what amounted to a huge disaster for humans, but was nothing other than the normal course of events for nature. Surveys of the park in 1998 found, for example, that some areas of the park contained up to 50,000 seedlings per acre of land. In regions where mature trees had not yet begun to appear, simple plants, such as native grasses, forbs, and other plant life had sprung up. Trees in an early climax community, such as aspens, had also begun to appear. In the 1998 survey, nearly 100 percent of the burned area was covered by some form or another of vegetation.

The patterns of animal life expected after a large-scale fire are also being seen. For example, round-headed woodborers are a type of beetle attracted to recently burned-out areas. As these borers move into a community, they attract woodpeckers, whose drilling into trees provides homes for other types of birds. Eventually, the next stage in the evolution of plant and animal life of Yellowstone will be reached, and the park will move onward once again to its pre-fire condition.

The 1988 Yellowstone National Park fire caused National Park Service officials to rethink their fire management policies. For a period of three years, the let-burn policy was abandoned, and the practice of suppressing all natural fires was reinstated in all national parks. By the early 1990s, however, park officials were forced to recognize that fire is a normal and natural part of the ecosystem, and there is simply no way to prevent the occurrence of large fires on forested lands. As a result, fire managers in the National Park Service have adopted more stringent guidelines as to which fires will be allowed to burn and which are to be suppressed. Such decisions are based on a complex mix of factors, including the proximity of a fire to human habitation, weather conditions, and the availability of firefighters. *See also* FIRE MANAGEMENT; NATIONAL PARK SERVICE.

Further Reading: Chase, Alston. *Playing God at Yellowstone*. New York: Atlantic Monthly Press, 1986; Despain, Don et al. *Wildlife in Transition: Man and Nature on Yellowstone's Northern Range*. New York: Roberts Rinehart, 1986; "Fire Impact on Yellowstone." Special issue of *Bioscience*, (November 1989); "Fires of '88," <http://www.wildrockies.org/fires-of-88.html>, 07/22/99; Pyne, Stephen J. *World Fire*. Seattle: Washington University Press, 1997; Wuerthner, George. *Yellowstone and the Fires of Change*. Salt Lake City: Haggis House, 1988; "Yellowstone Fires of 1988—Jackson Hole Summer Visitors Guide," <http://www.jacksonholenet.com/svg/fires.htm>, 07/22/99.

YULE LOG

The Yule log is a log that had traditionally been burned in a fireplace during the winter solstice, within a few days of 21 December. Burning of the Yule log is a very old tradition that some scholars believe may go back thousands of years before the Christian era. Some scholars believe that the custom is associated with the tradition of ritual fires usually conducted out-of-doors during certain parts of the year, such as the summer solstice and the vernal and autumnal equinox. The burning of the Yule log replaced the outdoor celebrations, according to these authorities, because winter was an unsuitable time at which to conduct an outdoor ritual, like those held at other times of the year.

In many cultures, the winter solstice is a special time of celebration because it marks the end of the old year and the beginning of the new year. The series of seasonal festivals, of which the Yule ceremonies were one part, reflected a belief in the continuity of life throughout the year. The word *yule* itself comes from an old Anglo-Saxon term meaning "wheel," as in the wheel of life.

Many cultures have detailed and complex traditions related to the burning of the Yule log. For example, certain types of trees were preferred for the log, usually ash or birch. Every effort was made to light the log on the first try. Failure to do so was regarded as a sign of bad luck in the coming year. The log was usually allowed to burn for a specific period of time, such as twelve hours, although the fire itself might be tended and maintained throughout the year. In that case, the new Yule log fire might be kindled from the embers of the previous year's fire.

The Yule log, its flames, and its ashes were generally regarded as having special powers and were considered to be sacred. In France and Germany, the ashes of the Yule log were mixed with the feed given to cattle, to ensure their good health and promote calving. In the Baltic states, the Yule ashes were spread on fruit trees to increase the harvest. In Serbia, wheat is scattered on the burning Yule log to improve the health and productivity of cattle.

As with many pagan customs, the burning of the Yule log gradually became incorporated into Christian traditions. This change took place very slowly since Yule and other celebrations of the winter solstice were regarded as heretical and were either banned or discouraged. However, the Christian church eventually discovered that it was easier to incorporate pagan customs and rituals into the new religion than to try eliminating them. For that reason, the Yule log fire and other winter customs eventually became part of the accepted Christian celebrations surrounding the birth of Christ.

Further Reading: Frazer, Sir James George. *The Golden Bough: A Study in Magic and Religion*. New York: The Macmillan Company, 1958, pp. 737–739, et seq.; The Wheel of the Year—Yule," <http://members.aol.com/eponess/yule.html>, 10/02/99; "Winter Solstice," <http://www.picapro.com/picapro/mind.htm>, 10/02/99; "Yule Introduction," <http://www.geocities.com/Athens/5606/yule.html>, 10/02/99.

BIBLIOGRAPHY

Akhavan, J. *The Chemistry of Explosives*. New York: Springer-Verlag, 1998.

Alighieri, Dante. *The Divine Comedy*, translated by Robert Pinsky. New York: Farrar Straus & Giroux, 1995.

Andrews, Tamra. *Legends of the Earth, Sea, and Sky*. Santa Barbara, CA: ABC-CLIO, 1998.

A Quest for Ecological Understanding. Tallahassee, FL: Tall Timbers Research Station, 1977.

Arad, Yitzhak. *Belzec, Sobibor, Treblinka: The Operation Reinghard Death Camps*. Bloomington: Indiana University Press, 1999.

Ash, Michael, and Irene Ash. *The Index of Flame Retardants*.

Bachelard, Gaston. *The Psychoanalysis of Fire*, translated by Alan C.M. Ross. Boston: Beacon Press, 1964.

Barham, P. *The Science of Cooking*. New York: Springer Verlag, 2000.

Baroja, Julio C. *The World of Witches*. Chicago: University of Chicago Press, 1964.

Bartlett, Robert. *Trial by Fire and Water: The Medieval Judicial Ordeal*. Oxford: Clarendon Press, 1986.

Baumgartner, David M. et al., eds. *Prescribed Fire in the Intermountain Region*. Pullman: Washington State University, 1989.

Bealer, Alex. *The Art of Blacksmithing*. New York: Funk and Wagnalls, 1969.

Beaver, Patrick. *The Match Makers*. London: Henry Mellard, 1985.

Berkow, Robert, ed. *The Merck Manual*. Fifteenth Edition. Rahway, NJ: Merck & Co., 1987.

Berkowitz, Norbert. *Fossil Hydrocarbons: Chemistry and Technology*. New York: Academic Press, 1997.

Birks, Tony. *Tony Birks Pottery: A Complete Guide to Pottery-Making Techniques*. Oviedo, FL: Gentle Breeze Publishing Company, 1998.

Biswell, Harold H. *Prescribed Burning in California Wildlands Vegetation Management*. Berkeley: University of California Press, 1989.

Bowditch, Kevin E., Ronald Baird, and Mark A. Bowditch. *Oxyfuel Gas Welding*. Tinley Park, IL: Goodheart Willcox Publisher, 1999.

Boyce, Mary. *Zoroastrians: Their Religious Beliefs and Practices*. London: Routledge & Kegan Paul, 1979.

Boyd, Robert, ed. *Indians, Fire, and the Land in the Pacific Northwest*. Corvallis: Oregon State University Press, 1999.

Bradley, John N. *Flame and Combustion Phenomena*. London: Methuen, 1969.

Brandt, Daniel A. *Metallurgy Fundamentals*. Tinley Park, IL: Goodheart-Wilcox Company, 2000.

Bray, Charles. *Dictionary of Glass: Materials and Techniques*. Philadelphia: University of Pennsylvania Press, 1996.

Bronson, William. *The Earth Shook, the Sky Burned: A Photographic Record of the 1906 San Francisco Earthquake and Fire*. San Francisco: Chronicle Books, 1986.

Brown, Arthur A., and Kenneth P. Davis. *Forest Fire: Control and Use*, 2nd edition. New York: McGraw-Hill, 1973.

Brown, G.I. *Scientist, Soldier, Statesman, Spy: Count Rumford: The Extraordinary Life of a Scientific Genius*. Stroud, Gloucester: Sutton Publishing, 1999.

Brown, George I. *The Big Bang: A History of Explosives*. Dover: Sutton Publishing, 1998.

Bukowski, Richard W. *Smoke Detector Design and Smoke Properties*. Washington, DC: Center for Fire Research, 1978.

Burby, Liza N. *Electrical Storms*. New York: Rosen Publishing Group, 1999.

Burgan, Robert E. *1988 Revisions to the 1978 National Fire-Danger Rating System*. Research Paper SE-273. Asheville, NC: U.S. Department of Agriculture, Forest Service, Southeastern Forest Experiment Station, 1988.

Campbell, Joseph. *The Masks of God: Primitive Mythology*. New York: Viking Press, 1959.

Camporesi, Piero. *The Fear of Hell: Images of Damnation and Salvation in Early Modern Europe*. Translated from Italian. Cambridge: Polity Press, 1990.

Cannon, Donald J., ed. *Heritage of Flame — The Illustrated Encyclopedia of Early American Firefighting*. Garden City, NY: Doubleday and Company, 1977.

Carsten, Lein. *Olympic Battleground: The Power Politics of Timber Preservation*. Seattle, WA: Mountaineers Books, 2000.

Chandler, Harry. *Metallurgy for the Non-Metallurgist*. Materials Park, OH: ASM, International, 1998.

Chase, Alston. *Playing God at Yellowstone*. New York: Atlantic Monthly Press, 1986.

Chavarria, Joaquim. *The Big Book of Ceramics: A Guide to the History, Materials, Equipment, and Techniques of Hand-Building, Molding, Throwing, Kiln-Firing, and Glazing*. New York: Watson-Guptill Publishing, 1994.

Coal Mine Fire and Explosion Prevention. Washington, DC: United States Department of the Interior, Bureau of Mines, Information Circular 8768, 1978.

Collins, Scott L., and Linda L. Wallace. *Fire in North American Tallgrass Prairies*. Norman: University of Oklahoma Press, 1990.

Conant, James B. *The Overthrow of the Phlogiston Theory: The Chemical Revolution of 1775-1789*. Cambridge: Harvard University Press, 1950.

Conkling, John A. *Chemistry of Pyrotechnics: Basic Principles and Theory*. New York: Marcel Dekker, 1985.

Connor, J.A., J.J. Turner, and E.A.V. Ebsworth. *The Chemistry of Oxygen*. New York: Pergamon Press, 1973.

Cook, Allan R, ed. *Burns Sourcebook: Basic Consumer Health Information about Various Types of Burns and Scalds, Including Flame, Heat, Cold, Electrical, Chemical, and so forth*. Detroit: Omnigraphics, Inc., 1999.

Cooke, Lawrence S., ed. *Lighting in America: From Colonial Rushlights to Victorian Chandeliers*. New York: Main Street/Universe Books, 1975.

Cowie, Leonard W. *Plague and Fire: London, 1665–1666*. East Sussex: Wayland Publishers, 1970.

Crutzen, P.G., and J.G. Goldammer, eds. *Fire in the Environment: Its Ecological, Climatic, and Atmospheric Chemical Importance*. New York: John Wiley and Sons, 1993.

Czech, Danuta, and Jadwiga Bezwinska, eds. *Amidst a Nightmare of Crime: Manuscripts of Prisoners in Crematorium Squads Found at Auschwitz*. New York: Howard Fertig, 1992.

DaCosta, Phil. *100 Years of America's Fire Fighting Apparatus*. New York: Bonanza Books, 1964.

Daly, George Anne, and John J. Robrecht. *An Illustrated Handbook of Fire Apparatus*. Philadelphia: INA Corporation Archives Department, 1972.

DeAngelis, Gina. *Triangle Shirtwaist Company Fire of 1911*. Broomall, PA: Chelsea House Publishers, 2000.

Debano, Leonard F., Daniel G. Neary, and Peter F. Folliott. *Fire's Effects on Ecosystems*. New York: John Wiley & Sons, 1998.

Decker, Robert, and Barbara Decker. *Volcanoes*. New York: W. H. Freeman and Company, 1989.

DeKok, David. *Unseen Danger*. Philadelphia: University of Pennsylvania Press, 1986.

Derry, T.K., and Tevor I. Williams. *A Short History of Technology from the Earliest Times to A.D. 1900*. New York: Dover Publications, 1993.

Dickinson, H.W. *James Watt and the Steam Engine*. London: Encore Editions, 1981.

Ditzel, Paul C. *Fire Alarm!*. New Albany, IN: Fire Buff House Publishers, 1994.

Ditzel, Paul. *Fireboats: A Complete History of the Development of Fireboats in America*. New Albany, IN: Fire Buff House, 1989.

Ditzel, Paul. *Fire Engines, Firefighters*. New York: Crown Publishers, 1976.

Donner, John. *A Professional's Guide to Pyrotechnics: Understanding and Making Exploding Fireworks*. Boulder, CO: Paladin Enterprises, 1997.

Donovan, Arthur. *Antoine Lavoisier: Science, Administration, and Revolution*. Cambridge: Cambridge University Press, 1996.

Doremus, Robert H. *Glass Science*, 2nd edition. New York: John Wiley & Sons, 1994.

Dunn, Vincent. *Command and Control of Fires and Emergencies*. Saddle Brook, NJ: Fire Engineering Book Department, 2000.

Ertel, Mike, and Gregory C. Berk. *Firefighting: Basic Skills and Techniques*. Tinley Park, IL: Goodheart-Wilcox, 1997.

Faith, Nicholas. *Blaze: The Forensics of Fire*. New York: St. Martin's Press, 2000.

Faraday, Michael. *The Chemical History of a Candle*, edited by William Crookes. New York: The Viking Press, 1960.

Farber, Eduard. *Oxygen and Oxidation Theories and Techniques in the 19th Century and the First Part of the 20th Century*. Washington, DC: Washington Academy of Sciences, 1967.

Finch, Richard. *Welder's Handbook: A Complete Guide to MIG, TIG, ARC, and Oxyacetylene Welding*. Berkeley, CA: Berkeley Publishing Group, 1997.

Fireclay. St. Petersburg, FL: Artext Publishing, 1996.

Fire Extinguishers, Rating and Fire Test, Ul 711. Northbrook, IL: Underwriters Laboratories, Incorporated, 1995.

Firefighter's Handbook: Essentials of Firefighting and Emergency Response. Albany, NY: Delmar Publishing, 2000.

Fireplaces and Wood Stoves. Alexandria, VA: Time-Life Books, 1997.

The Fire Protection Handbook, 18th edition. Quincy, MA: National Fire Protection Association, 1999.

Fire Protection Systems, 2nd edition. Quincy, MA: National Fire Protection Association, 1992.

Fisher, Richard V., Grant Heiken, and A.K. Morris, eds. *Volcanoes*. Princeton, NJ: Princeton University Press, 1998.

Francis, F.J., ed. *Encyclopedia of Food Science and Technology*, 2nd ed. New York: John Wiley & Sons, 2000.

Frazer, J.G. *Balder the Beautiful: The Fire-Festivals of Europe and the Doctrine of the External Soul*, 2 vols. London: Macmillan and Company, 1913.

Frazer, Sir James G. *Folklore in the Old Testament*. New York: Tudor Publishing Company, 1923, passim.

Frazer, Sir James George. *The Golden Bough: A Study in Magic and Religion*. New York: The Macmillan Company, 1958.

Frazer, James George, Sir. *Myths of the Origin of Fire*. New York: Hacker Art Books, 1974.

Freeman, Paul. *Smoke Jumpers*. Indianapolis: Baskerville Publishers, 1992.

Frome, Michael. *The Forest Service*, 2nd edition. Boulder, CO: Westview Press, 1984.

Fuchs, Hans U. *The Dynamics of Heat*. New York: Springer Verlag, 1996.

Fudge, Edward William, and Robert A. Peterson. *Two Views of Hell: A Biblical and Theological Dialogue*. Downers Grove, IL: Intervarsity Press, 2000.

Fuller, Margaret. *Forest Fires*. New York: John Wiley & Sons, Inc., 1991.

Gaylord, Jessica. *The Psychology of Child Firesetting: Detection and Intervention*. New York: Brunner/Mazel, 1987.

Geary, Don. *The Welder's Bible*. New York: McGraw-Hill Professional Publishing, 1992.

Geary, Don. *Welding*. New York: The McGraw-Hill Companies, 1999.

Glassman, Irvin. *Combustion*. New York: Academic Press, 1996.

Goudsblom, Johan. *Fire and Civilization*. London: Allen Lane, 1992.

Grayson, Martin, ed. *Kirk-Othmer Encyclopedia of Chemical Technology*. New York: John Wiley, 3rd edition, 1978.

Green, H.L., and W.R. Lane. *Particulate Clouds: Dusts, Smokes, and Mists: Their Physics and Physical Chemistry and Industrial and Environmental Aspects*, 2nd edition. London: E. & F. N. Spon, 1964.

Green-Hughes, E. *A History of Firefighting*. Paris: Klincksieck, 1965.

Greenwood, N.N., and A. Earnshaw. *Chemistry of the Elements*. Oxford: Pergamon Press, 1984.

Grimal, Pierre, ed. *Larousse World Mythology*. New York: G.P. Putnam's Sons, 1965.

Guiberson, Brenda Z. *Lighthouses: Watchers at Sea*. New York: Henry Holt, 1995.

Gunzel, Louis. *Retrospects: The Iroquois Theater Fire, La Salle Hotel Fire in Chicago*. Chicago: Theater Historical Society, 1993.

Gutman, Israel, and Michael Berenbaum, eds. *Anatomy of the Auschwitz Death Camp*. Bloomington: Indiana University Press, 1998.

Hague, Michael. *The Book of Dragons*. New York: William Morrow & Company, 1995.

Halberstadt, Hans. *The American Fire Engine*. Osceola, WI: Motorbooks, International, 1993.

Hansen, Gladys, Emmet Condon, and David Fowler Cameron, eds. *Detail of Disaster: The Untold Story and Photographs of the San Francisco Earthquake and Fire of 1906*. San Francisco: Cameron & Company, 1989.

Hasegawa, Harry K., ed. *Characterization and Toxicity of Smoke*. Philadelphia: ASTM, 1990.

Higgins, Kenneth F., Arnold D. Kruse, and James L. Piehl. *Prescribed Burning Guidelines in the Northern Great Plains*. U.S. Fish and Wildlife Service, Cooperative Extension Service, South Dakota State University, U.S. Department of Agriculture EC 760, 1989.

Hilado, Carlos J., ed. *Smoke and Products of Combustion*. Westport, CT: Technomic Publishing Company, 1973.

Hills, Richard. *Power from Steam: A History of the Stationary Steam Engine*. Cambridge, England: Cambridge University Press, 1993.

Hirschfelder, Arlene B. *Encyclopedia of Smoking and Tobacco*. Phoenix, AZ: Oryx Press, 1999.

Hjelm, Norman A., ed. *Out of the Ashes: Burned Churches and the Community of Faith*. Nashville, TN: Thomas Nelson, 1998.

Hobson, A. *Lanterns that Lit Our World: Book 2*. New York: Golden Hill Press, 1997.

Holmes, Frederic L. *Antoine Lavoisier: The Next Crucial Year*. Princeton, NJ: Princeton University Press, 1998.

Homeowner's Guide to Wood Stoves. Menlo Park, CA: Lane Publishing, 1979.

Houdini, Harry. *Miracle Mongers and Their Methods*. New York: E. P. Dutton and Company, 1920.

Huenecke, Klaus. *Jet Engines: Fundamentals of Theory, Design, and Operation*. Stillwater, MN: Motorbooks International, 1998.

International Association of Marine Aids to Navigation and Lighthouse Authorities. *Lighthouses of the World*. Springfield, TN: Globe Pequot Press, 1998.

International Fire Code 2000. Whittier, CA: International Conference of Building Officials.

Kenny, John B. *The Complete Book of Pottery Making*. Iola, WI: Krause Publications, 1976.

Keyes, Edward. *Coconut Grove*. New York: Atheneum, 1984.

Klass, Donald L. *Biomass for Renewable Energy, Fuels, and Chemicals*. New York: Academic Press, 1998.

Kozlowski, T.T., and C.E. Ahlgren, eds. *Fire and Ecosystems*. New York: Academic Press, 1974.

Kresek, Ray. *Lookouts of the Northwest*. Fairfield, WA: Ye Galleon Press, 1984.

Lewis, Nolan D., and Helen Yarnell. *Pathological Firesetting (Pyromania)*. New York: Mental Disease Monographs, 1951.

Lewis, Richard J., Sr., ed. *Hawley's Condensed Chemical Dictionary*, 12th edition. New York: Van Nostrand Reinhold Company, 1993.

Long, James N., ed. *Fire Management: The Challenge of Protection and Use*. Logan: Utah State University, 1985.

Lyle, David. *The Book of Masonry Stove: Rediscovering an Old Way of Warming*. White River Junction, VT: Chelsea Green Publishing Company, 1998.

Lyons, John W. *The Chemistry and Uses of Fire Retardants*. New York: Wiley-Interscience, 1970.

Lyons, John W. *Fire*. New York: Scientific American Books, 1985.

Maclean, John N. *Fire on the Mountain: The True Story of the South Canyon Fire*. New York: William Morrow & Company, 1999.

Mahoney, Gene. *Introduction to Fire Apparatus and Equipment*, 2nd edition. New York: Fire Engineering Book Service, 1985.

McIntyre, R., and C.P. Dillon. *Pyrophoric Behavior and Combustion of Reactive Metals*. Huntington, WV: MTI Publications, 1988.

Meadows, C.A. *Discovering Oil Lamps*. Princes Risborough: Shire Publications, Ltd., 1994.

Mitchell, Donald W. *Mine Fires: Prevention, Detection, Fighting*, 3rd edition. Chicago: Intertec Publishing Inc., 1996.

Morgenstern, Julian. *The Fire upon the Altar*. Chicago: Quadrangle Books, 1963.

Morrison, Ellen Earnhardt. *Guardian of the Forest: A History of the Smokey Bear Program*. New York: Vantage Press, 1976.

Mountcastle, John Wyndham. *Flame On!: U.S. Incendiary Weapons, 1918–1945*. Shippensburg, PA: White Mane Publishers, 1999.

Newman, Jon. *Candles*. San Diego: Thunder Bay Press, 2000.

Newton, David E. *Global Warming: A Reference Handbook*. Santa Barbara, CA: ABC-CLIO, 1993.

Nielsen, Inge. *Thermae et Balnea: The Architecture and Cultural History of Roman Public Baths*. Aarhus, Netherlands: Aarhus University Press, 1990.

Nigg, Joe. *Wonder Beasts: Tales and Lore of the Phoenix, the Griffin, the Unicorn, and the Dragon*. Englewood, CO: Teacher Ideas Press, 1995.

1998 Report of the Halons Technical Options Committee. [Nairobi]: United Nations Environment Programme, 1999.

NFPA 10 Standards for Portable Fire Extinguishers: 1998 Edition. Quincy, MA: National Fire Protection Association, 1998.

Nodvin, Stephen C., and Thomas A. Waldrop, eds. *Fire in the Environment: Ecological and Cul-*

tural Perspectives. Washington, DC: U.S. Forest Service, 1991.

Olsen, Frederick L. *The Kiln Book*. Radnor, PA: California Keramos Bassett Books, 1974.

Orton, Vrest. *Observations on the Forgotten Art of Building a Good Fireplace: The Story of Sir Benjamin Thompson, Count Rumford, an American Genius and His Principle*. Dublin, NH: Yankee Press, 1974.

Partington, J. P. *History of Greek Fire and Gunpowder*. London: W. Heffer and Sons, 1961.

Partington, J.R. *A Short History of Chemistry*. London: Macmillan & Company, 1937.

Partington, J.R., and D. McKie. *Historical Studies on the Phlogiston Theory*. New York: Arno Press, 1981.

Payton, R.J. *A Modern Reader's Guide to Dante's Inferno*. New York: Peter Lang Publishing, 1992.

Peirce, Josephine H. *Fire on the Hearth: The Evolution and Romance of the Heating Stove*. Springfield, MA: Pond-Ekberg Company, 1951.

Pernin, Peter. *The Great Peshtigo Fire: An Eyewitness Account*. Madison: University of Wisconsin Press, 1999.

Peters, William J., and Leon F. Neuenschwander. *Slash and Burn: Farming in the Third World Forest*. Moscow: University of Idaho Press, 1988.

Pomroy, William H., and Annie M. Carigiet. *Analysis of Underground Coal Mine Fire Incidents in the United States from 1978 through 1992*. Washington, DC: U.S. Department of the Interior, Bureau of Mines, 1995.

Postman, Richard A. *Anvils in America*. Berrien Springs, MD: Postman Publishing, 1998.

Prothero, Stephen. *Purified by Fire: A History of Cremation in America*. Berkeley: University of California Press, 2000.

Puchovsky, Milosh T., ed. *Automatic Sprinkler Systems Handbook*. Quincy, MA: National Fire Protection Association, 7th edition, 1996.

Pyne Stephen J. *America's Fires: Management on Wildlands and Forests*. Durham, NC: Forest History Society, 1997.

Pyne, Stephen J. *Vestal Fire: An Environmental History, Told through Fire, of Europe and Europe's Encounter with the World*. Seattle: University of Washington Press, 1997.

Pyne, Stephen J. *World Fire: The Culture of Life on Earth*. Seattle: University of Washington Press, 1997.

Queen, Phillip L. *Fighting Fire in the Wildland/ Urban Interface*. Bellflower, CA: Fire Publications, 1993.

Ramakrishnan, P.S. *Shifting Agriculture and Sustainable Development*. Paris: UNESCO; New York: Parthenon Publishing Group, 1992.

Reavis, D.J. *The Ashes of Waco: An Investigation*. Syracuse, NY: Syracuse University Press, 1998.

Reed, Richard J. *North American Combustion Handbook: A Basic Reference on the Art and Science of Industrial Heating with Gaseous and Liquid Fuels*. Cleveland: North American Manufacturing Company, 1993.

Richardson, I.M. *Prometheus and the Story of Fire*. Mahwah, NJ: Troll Associates, 1983.

Robinston, Glen O. *The Forest Service: A Study in Public Land Management*. Baltimore: Johns Hopkins University Press, 1975.

Robson, Roy R. *Old Believers in Modern Russia*. De Kalb: Northern Illinois University Press, 1996.

Rose, Carol. *Giants, Monsters, and Dragons: An Encyclopedia of Folklore, Legend, and Myth*. Santa Barbara, CA: ABC-CLIO, 2000.

Rossotti, Hazel. *Fire*. Oxford: Oxford University Press, 1993.

Sagan, Carl, and Richard P. Turco. *A Path Where No Man Thought: Nuclear Winter and the End of the Arms Race*. London: Century Press, 1990.

Sawyer, Donald T. *Oxygen Chemistry*. New York: Oxford University Press, 1991.

Scarth, Alwyn. *Vulcan's Fury: Man against the Volcano*. New Haven, CT: Yale University Press, 1999.

Sellers, Richard West. *Preserving Nature in the National Parks: A History*. New Haven, CT: Yale University Press, 1997.

Seton, Nora Janssen. *The Kitchen Congregation: Gatherings at the Hearth*. New York: Picador, 2000.

Shepard, O.C., and W.F. Dietrich. *Fire Assaying*. Boulder, CO: Met-Chem Publishing, 1989.

Shugar, Gershon J. et al. *Chemical Technicians' Ready Reference Handbook*. New York: McGraw-Hill Book Company, 1981.

Sigurdsson, Haraldur, ed. *Encyclopedia of Volcanoes*. New York: Academic Press, 1999.

Smallman, R.E., and R.J. Bishop. *Modern Physical Metallurgy and Materials Engineering: Science, Process, Applications*. Woburn, MA: Butterworth-Heinemann, 2000.

Bibliography

Souter, Gerry, and Janet Souter. *The American Fire Station*. Osceola, WI: Motorbooks, International, 1998.

Spring, Ira, and Byron Fish. *Lookouts: Firewatchers of the Cascades and Olympics*. Seattle: The Mountaineers, 1981.

Staal, Frits. *Agni: The Vedic Ritual of the Fire Altar*, 2 vols. Berkeley: University of California Press, 1983.

Stadolnik, Robert F. *Drawn to the Flame: Assessment and Treatment of Juvenile Firesetting Behavior*. Sarasota, FL: Professional Resource Exchange, 2000.

Stockholm International Peace Research Institute. *Incendiary Weapons*. Cambridge, MA: MIT Press, 1975.

Stone, Richard. *Introduction to Internal Combustion Engines*. Warrendale, PA: Society of Automotive Engineers, 1993.

Sutton, George P. *Rocket Propulsion Elements: An Introduction to Engineering of Rockets*. New York: John Wiley & Sons, 1992.

Swiebocka, Teresa, Jonathan Webber, and Connie Wilsack, eds. *Auschwitz: A History in Photographs*. Bloomington: Indiana University Press, 1993.

Tecton, Mike. *Traditional Fireplaces*. McLean, VA: Mike Tecton Publishing, 1989.

Turns, Stephen R. *An Introduction to Combustion: Concepts and Applications*. New York: McGraw-Hill, 1995.

Tylor, Sir Edward Burnett. *Religion in Primitive Culture*. New York: Harper Torchbooks, 1958.

Uniform Fire Code (year varies). Whittier, CA: International Conference of Building Officials.

United Nations. Group of Consultant Experts on Napalm and Other Incendiary Weapons. *Napalm and Other Incendiary Weapons and All Aspects of Their Possible Use*. New York: United Nations, 1973.

Urban Wildland Interface Code. Whittier, CA: International Fire Code Institute, 1997.

Viemeister, Peter E. *The Lightning Book*. Cambridge, MA: MIT Press, 1972.

Victorin-Vangerud, Nancy M. *The Raging Hearth: Spirit in the Household of God*. St. Louis: Chalice Press, 2000.

Von Baeyer, Hans C. *Warmth Disperses and Time Passes: A History of Heat*. New York: Random House, 1999.

Wade, Wyn Craig. *The Fiery Cross: The Ku Klux Klan in America*. New York: Oxford University Press, 1998.

Walker, John R. *Modern Metalworking*. Tinley Park, IL: Goodheart-Wilcox Company, 2000.

Wallington, Neil. *Images of Fire: 150 Years of Firefighting*. Newton Abbott, England: David & Charles, 1989.

Watkins, George. *The Steam Engine in Industry*. Ashbourne, England: Moorland Publishers, 1978.

Wardrope, Jim, and June A. Edhouse. *The Management of Wounds and Burns*. New York: Oxford University Press, 1999.

Weeks, Mary Elvira, and Henry M. Leicester. *Discovery of the Elements*, 7th edition. Easton, PA: Journal of Chemical Education, 1968.

Wells, Robert W. *Fire at Peshtigo*. Upper Saddle River, NJ: Prentice-Hall, 1968.

Whelan, Robert J. *The Ecology of Fire*. Cambridge, UK: Cambridge University Press, 1995.

White, John Henry. *The History of the Phlogiston Theory*. New York: AMS Press, 1973.

Wright, Lawrence. *Home Fires Burning: The History of Domestic Heating and Cooking*. London: Routledge & K. Paul, 1964.

Wuerthner, George. *Yellowstone and the Fires of Change*. Salt Lake City: Haggis House, 1988.

Yanagida, Hiroaki, Kunihito Kawamoto, and Masaru Miyayama. *The Chemistry of Ceramics*. New York: John Wiley & Sons, 1996.

INDEX

About the Author

DAVID E. NEWTON has published extensively on chemistry and other science subjects. He is the award-winning author of numerous books, articles, and scholarly publications, including *The Chemical Elements*, *Science in the 1920s*, *The Ozone Dilemma*, *Encyclopedia of Cryptology*, *Chemistry of Carbon Compounds*, *Problems in Chemistry*, *Global Warming*, *Encyclopedia of the Chemical Elements*, and *Social Issues in Science and Technology*. Newton received his doctorate in science education from Harvard University.